More praise for *Rough Waters*

"Bravo! A fine-grained and very balanced history of every stage of the struggle against privatization. . . . What could be more delightful for an anthropologist/political ecologist such as myself than a book which is both ethnography and political economy? Mendenhall covers the terrain in an amazingly comprehensive way. She has an insider's view of what it means to be a small-scale fisherperson, since her whole family has fished for generations, in Norway, Southeast Alaska, and the north Bering Sea. But she knows the fish politics of not only these areas, but also of the hot spots of privatization of fisheries: British Columbia, Washington, Oregon, and New England. Her book is a testimony to the havoc wreaked on coastal communities and their fishermen by what is euphemistically called "fleet rationalization" (getting rid of the small boats) and privatization in the form of Individual Transferable Quotas.

But this book is so much more than that. Mendenhall has kept track of the intricacies of policy-making, so she understands at the deepest level how the world of fisheries politics has evolved, and how small-scale fishermen have miraculously survived against all odds by sheer determination. What is most impressive is the maturity and balanced nature of her analysis, which seeks to understand and point ways through the morass, appreciate what has been accomplished, and suggest what needs to happen next. Thus, her work serves as a much-needed bridge between the ever-present critiques, and the big picture of where we need to go and what programs and tools, however imperfect, are helpful along the way. She knows her history and she tells it well, making it accessible and meaningful to a much wider audience than would otherwise have access to this intimate story."

— **Dr. Evelyn Pinkerton,** anthropologist and author of *Fisheries That Work: Sustainability Through Community-based Management*

"Personal and poignant. Mendenhall writes with a depth of intimacy and understanding (of) the roles fishing plays in both the Native and non-Native cultures. The battle (she) chronicles takes place throughout the Alaska and

North Pacific region over the course of a hundred years, but has been most fiercely fought in the last three decades..(Her) meticulous research shows how corporate interests have taken over the pollock and crab fisheries, and how quota management has shrunk the state's halibut and black cod fleets with little or no conservation benefit. . . As in any battle, survival is a matter of luck. But the gist of (her) book is that the surviving small-scale fisheries are the kind that should be intentionally restored and preserved. . . This a war story, and the battle is not over yet."

— *Fishermen's Voice*, **Paul Molyneaux**,
author of *The Doryman's Reflection*

"Covers every angle on the North Pacific small fishermen's battle... Mendenhall documents a way of life with exhaustive research and reporting. This history of the fishing fleet and its struggle is more intimate than what you'll find in the news and official reports and more extensive and accurate than your grandfather's memoirs."

— *National Fisherman*

"Exhaustively researched. . . a definitive review of the current trend toward individual vessel quotas, individual transferable quotas, catch shares, and other versions of privatization."

— *Hakai Magazine*, **Alan Haig-Brown**,
author of *Still Fishin': The BC fishing Industry*

"Fantastic. . . Important. . . I am so glad these kinds of perspectives and voices are having a chance to get in print. I will let people I know in Iceland about it."

— **Margaret Willson**,
author of *Seawomen of Iceland: Survival on the Edge*

"A tour de force. . . a masterful account of the tragic enclosure of the world's fisheries. In the final chapters Mendenhall offers a sobering, but realistic, appraisal of the daunting prospects facing small boat fish harvesters and the communities that depend upon them. While (her) criticism of the drive to economically redesign and privatize common property fisheries is not new in fisheries literature, her first-hand experience with the fishery elevates her voice to an unique and compelling stature. (She) is at her very finest in describing her years as a troller in Southeast Alaska. To anyone who has ever fished commercially her stories will resonant to the deepest possible level. Moreover, her description of her present-day participation in the subsistence fishery, based out of Nome Alaska, is seminal in pointing the way for both humans and wild creatures to sustainably co-exist in perpetuity. Her account of her own sons and their fishing adventures in small vessels in the wild expanse of Bering Sea makes for spellbinding reading."

– Dennis Brown, author of *Salmon Wars: The Battle for the West Coast Salmon Fishery*

"Incorporates vivid first-hand accounts. . . The baneful impact of a modern corporatized economy on small family farms and old-style independent physicians is well known and widely lamented (but the) struggles of small family fishermen are seldom in the forefront of public discourse except in fishing communities. Mendenhall's authoritative and well-written *Rough Waters* adds significant depth to our understanding of these contemporary social and economic issues."

– Dr. Gerald McFarland, author of *A Scattered People: An American Family Moves West*

"I'm a fishery biologist. I care about fish. I also care about fishermen and the communities they support. I have read about the fishermen's problems because of declining abundance of fish, but until I read Mendenhall's book did the dry statistics become real people who work hard to get by on a shrinking resource and have to contend with policies that seem to care little for both the fishermen and the fish."

– Jim Lichatowich, author of *Salmon, People, and Place: A Biologist's Search for Salmon Recovery*

"Heart-warming and heart-breaking. . . Mendenhall comes from a long line of Norwegian commercial fishermen. . . so she writes with sensitive yet accurate profundity, in an intimacy of place that conjures many memories for me. Our story deserves to be told and is told uniquely and honestly in (this) exhaustively researched book up to the present, including insights into fish politics that most have never deigned to imagine. . ."

— **Dave Otness**, Alaska fisherman and activist

"Deeply engrossing. . . Great in terms of suggesting measures we need to win sustainable fisheries."

— **Tim Wheeler**, journalist and author of
News from Rain Shadow Country

"Well-researched. . . offers a unique perspective. . . full of colorful and relevant stories which highlight life as a fisherman (and) relates the struggle of small, family-owned fishing operations against the politically savvy factory fishing fleet. This book brought me back to my time as NMFS fishery observer in the Bering Sea, while it also did a great job of capturing the high-stakes politics that have created the current fisheries landscape in Alaska."

— **Michael Sloan,** fishery biologist

"Mendenhall creates a vivid profile of a classic small business venture with a long history and an uncertain future. Throughout she feeds us first-hand accounts of trolling for salmon, long-lining for halibut and cod, crabbing through unstable winter ice and summer's recent rough seas -- sometimes idyllic, sometimes frightening, always compelling: she puts us there."

— **Dr. William Keep**, fisherman,
English professor and book reviewer

Rough Waters

Rough Waters: Alaska to Oregon

Small Fishermen's Battle

Nancy Danielson Mendenhall

Far Eastern Press, Seattle

Rough Waters: Alaska to Oregon
Small Fishermen's Battle

All rights reserved under International and Pan-American Copy Conventions. For information on permission for reprints and excerpts, contact info@fareasternpress.com.

Quotations and paraphrases from Paul Molyneaux' "The Doryman's Reflection" by permission of Beacon Press.
Cover Photo: Peggy Fagerstrom, Nome, Alaska; showing Robin Thomas, Frank McFarland and Roger Thompson fishing bait herring off Nome, 2004

Copyright © 2021 Nancy Danielson Mendenhall

ISBN: 0-9678842-8-4
ISBN 13: 978-0-9678842-8-8
Library of Congress Control Number: 2015948305

Far Eastern Press, Seattle, WA

Dedicated to the memory of the generations of small fishermen who came before us, to all those now out on the water, and to those who will follow.

Acknowledgements

Special thanks to my father, Torvald Danielson, whose stories made sure I would go to Alaska, and to Jack McFarland, who survived me, our combined children, and cantankerous boats for the years commercial trolling; and to Perry Mendenhall, who created our fish camps and kept them humming over thirty years with my enthusiastic if not so expert assistance, wintertime enduring my abstracted silence as I pondered this book.

A grateful thanks for all the insightful help and moral support for this book: to Far Eastern Press for the tremendous assistance and encouragement that made this book possible; to my editor Lesley Thomas, a world of thanks for editing, formatting, photo layout, research, and constant support. Special thanks to George Morford and "John" for their insights into fishery politics; and to Bill Keep for his encouragement and his critical ear and eye for story. To my uncle Donald Wheeler, who took me back to see the Columbia's Hanford Reach, and to Eric Osborne, Ken Waltz, Barb Amarok and Rich Rusk, who kept checking on me, "How's that book going?"

To all the small fishermen who contributed their interviews, stories, photos, and comments here: Perry, George, Rob, Audrey, Frank, Cherilyn, Dan, Naggu, Lesley, John, Bill, Jack, Esther, Enoch and Mida, Frank K., Linda, Nathan and Virginia, Ken, Johnny, Jonna and Rich, Roger, Sven H., Travis, Scott, Eric, Fred, Conrad, Daniel A., Robert, and Garret. Also to friends so supportive to us in our fishing: Marilyn Jordan-George for letting us buy her beloved *Nohusit*, Frank P. (*Forest*), Al and Mary (*Sudan*), Ron and Cece (*Wanderer*), Norman (*Nordot*), Ben and Louise (*Kitty T.*) and Sven S. (*Comet*), and a thanks to the kids who sailed on the *Nohusit* for their endurance and good humor: Dan, Rob, Lesley, Frank, Jack, Melanie, Hilary and Daniel – they put up with a lot.

And a huge thanks also to those who have contributed at our Cape Nome fish camp: Eli Mendenhall (found the site), Bill Barr (helped build the cabin), Rob Thomas, Audrey and Chris Aningayou, Lesley Thomas, Dan and Marilyn Thomas, Frank McFarland, Cherilyn Kavairlook, Frank and Norma Kavairlook and family, Quincy Iyatunguk, Edward Anasoguk, and Peggy and Percy Outwater and family. And a very special thanks to the future young fishermen who over the years have kept us smiling and even helped haul on the rope: Scott, Sierra, Chris, Melissa, Nicole, Tor, Shannon, Vincent, Colin, Ayla, Braeden, Connor, Liam and Levi.

Thanks, too, to all the scientists, managers, fishery leaders, reporters and politicians that have written, spoken of, or fought so well the threats to artisanal and subsistence fishing, and also to those who took time to explain things to me, including but not limited to (and not in order of importance): Sven Haakanson, David Sohappy, Billy Frank, Jim Lichatowich, Clem Tillion, Harold Sparck, Karen Gillis, Enoch Shiedt, Phil Smith, Paul Molyneaux, Susan Playfair, Charlie Lean, Mike Sloan, William McCloskey, Nancy Lord, Katie John, Leslie Fields, Duncan Fields and others on the North Pacific Council, Linda Behnken, Laine Welch, Richard Gaines, Hal Bernton, Mark Kurlansky, David Montgomery, Alan Haig-Brown, David Arnold, Eric Wickham, Dennis Brown, Roberta Ulrich, Callum Roberts, Tyler Dann, Charlene Allison, Sue-Ellen Jacobs, Mary Porter, Alexandra Morton, Steve Hawley, Daniel Pauli, Evelyn Pinkerton, Andrew Jensen, Einer Eythorsson, Zeke Grader, the teams at these websites especially: Food and Water Watch, Ecotrust, and EcotrustCanada, and so many others.

Table of Contents

2021: An Update to the Introduction

This is a second edition of *Rough Waters: Our North Pacific Small Fishermen's Battle*. Why did I change the title? I decided it was too obscure. Who are the North Pacific fishermen? Why should we care about their battles? The ones this book is concerned with are the small fishermen from Alaska, British Columbia, Washington, and Oregon.

Much has taken place in their fisheries, enough to think an introductory update is in order. But the changes are not so much in management as in larger global changes and how they affect the fisheries. As the reader may have guessed I am talking about global warming. We see these changes every day out our ocean-facing windows. And we hear of them regularly from frustrated fishermen. Most of their problems I wrote about six years ago are connected in some way to these monumental changes. I will list a few below that my community is personally quite familiar with, both in the waters and on land. If any scoffers are reading, here it is from the sea horse's mouth.

But first I will comment about other important changes in the west coast fisheries and their management and politics. The first half of this book is about salmon, and I can't see important change there of the positive kind we all hoped for. Certainly there has been plenty of effort, though not always agreement. The battle between the pro and anti-hatchery supporters continues. The argument that restored salmon runs require a restoring of rivers' natural habitat has gained strength. But how do we carry this out with fisheries management always short-funded? And even if we were to win greatly increased funding, how do we restore rivers that are drying up and shrinking --I'm back to my first topic. Nonetheless, there are serious small efforts going on to recreate accessible spawning grounds. Several outdated dams that are barriers to spawners have been removed, but only when it costs more to upgrade them.

Another movement that to me is less positive: the increasing restrictions on gillnetters, in Washington especially, in favour of the rights of sport fishermen--a far more powerful group politically. It is the cheap solution. As for people who have made their living gillnetting for generations? Find other work. More difficult but more socially positive is the increased attention to controlling the salmon by-catch of the other fleets and especially through electronic monitoring, which on a 40-ft. boat is much easier to manage than an observer on board. But fleets like Nome's small-boat crabbers will often need help with initial costs.

In the non-salmon fisheries covered herein, the movement into IFQ/catch-share hasn't turned around despite ample revelations of how it negatively affects younger fishermen for whom the cost to buy (or even lease, where that is legal) IFQs is impossible. Thus, the cohort of actively engaged skippers continues to age, and their communities are negatively affected as well. In a few fisheries there is an option available for community-owned quota, but the cost for most is prohibitive. It should have been no surprise to fish managers that people with big pockets would before long control the catch share fisheries.

A more positive move is growing among western Alaska's CDQ non-profits that see it can be to their advantage to buy shares now available in industrial fleets' vessels. Ironically these are the very fleets that scoop up the salmon the small coastal communities have depended on. But their revenues can be turned into community benefits like scholarships, stream rebuilding, youth and elder programs, and small harbor development. Norton Sound Economic Development Corp. is one that has gone that direction. Of course, there are mixed feelings about this connection, but if you can't beat 'em, join 'em. The last chapters in this book have more proposals for ways to salvage what we can of the small-boat fleets and their communities.

I will turn back now to the biggest concern of all. I must say I have no idea what fishermen, big or small, or what members of a supporting industry, big or small, can do about global warming acting alone. I certainly don't like the answer some will no doubt hear: "Go find another profession." For the families dependent on subsistence fishing, that isn't even possible without a change of location, or very likely an entire way of life for families, even whole communities. The political and social changes required to slow down global warming require a gigantic world-wide movement, but if it emerges we fishing families must be part of it.

Here are some of the changes we see locally, that are very likely connected to warming waters, and readers probably can add to this list. Our local small crabbers have no crab buyer for the second year. The male crab have crawled off somewhere--or is it overfishing? No clear answer beyond, "It's a cycle." But the crabbers are fishing, just on much cheaper stock. The Pacific Cod that normally hang out in the southern Bering Sea are now up here and so are the pollock. The crabbers just have to tighten their belts--a lot.

At Utqiagvik (Barrow) on the Arctic Ocean they have in the past had a run of pinks; now they see every species of salmon. But here off Nome, although the hardy pinks show up in record numbers, there have been virtually no kings in our set nets for years. This year's regional chum and coho runs were below forecast. At Sitka, a long way south, more change. Halibut boats have converted so they can also chase salmon in separate seasons and are able to fish farther offshore in deeper, rougher waters than traditional trollers can. They catch the coho that used to be along the shores, but now favor deeper waters because that is where their feed is now, in the colder water. I have already mentioned the drought along the Northwest coast and the warming and drying of those rivers, a major impact on salmon spawning. Of course, some people will say, "Just raise them in pens, even land-locked pens."

Along western Alaskan coasts we see changes in other species: a die-off of thousands of diving seabirds, like murres, that have no identifiable disease, but have apparently starved to death--another case of disappearing feed affected by increased acidification as well as temperature. People have witnessed polar bears, walrus, and some seals that depend on presence of sea-ice trying to survive using land. For two summers at Nome we have been beset by the ravages of caterpillars we haven't seen before. They crawl up the sides of cabins and turn hills brown with bare shrubs. Meanwhile, the small fishermen's battles go on, and many of them will stubbornly stick with a way of work and life that I have loved. When you have the opportunity, support them.

Introduction

Our North Pacific small fishermen and the fish they chase are fighting for survival. Their problems are many: a warming ocean, habitat deterioration, earlier overfishing, harvest waste, poorly supported fisheries research, weak management, and powerful political lobbies. Added to that is the government's decision to change the nature of the fleets fishing in the federal waters--to, in effect, privatize the rights to fish. About thirty US fisheries have already taken that step. Anadromous fish like salmon, mainly managed by the states, have different threats, such as deteriorated river habitat, drought, and industrial and agricultural impacts.

While scientists search for more understanding of how fish stocks can be saved, few are funded to investigate how we can save our small commercial and subsistence fishermen and the communities that depend on them. Instead, huge catcher-freezer vessels are promoted in the cause of modern efficiency and competitiveness. Each year more traditional small boats sit at the float, their politically weak owners and crews looking for work. Yet every year new or returning people with fishing in their blood will join the fleets, determined that small fishing will survive as a viable enterprise.

The fleets were almost entirely made up of boats under seventy feet in 1976 when Congress, alarmed that it was falling behind in the race to modernize fisheries, yet at the same time worried that fish stocks could be overstressed, passed the Magnuson Act. It created an Exclusive Economic Zone for the waters within 200 miles of our coast, and the fleets within that but outside the three-mile states' boundaries became federally managed. The Act stressed growth, competitiveness, and fish conservation simultaneously. To help such competitive challenges go smoothly, it created eight regional fishery councils whose job was to recommend to the government regarding planning, policy, and evaluation for the federal fisheries.

At the same time the west coast states, also concerned about "too many boats on too few fish", started to limited new entries to their salmon fleets. Starting in the 1970s you could buy or be gifted a retiring permit, but there would be no new ones issued. That strategy spread to almost all state-managed fleets and soon to most federal fleets on our coasts and Canada's. It would have troubling effects on small coastal communities with their whole economy based on local fishing fleets, such as in Alaska. State fishery managers also began restricting their fleets more in seasons, legal gear, and open areas. However, the federal fisheries boss, the National Oceanic and Atmospheric Administration (NOAA), soon determined that limiting permits and restricting rules didn't go far enough, neither in shrinking the fleets nor modernizing them, nor in protecting the fish stocks.

In the mid-1980s NOAA's fisheries branch, the National Marine Fisheries Service (NMFS) informed the fleets that fishing at a level that was unsustainable for a stock had to stop. Waste through dumping of fish had to be better controlled. NOAA also recommended to the regional fisheries councils that they consider restructuring their federal waters fleets, moving them from the traditional open commons fisheries, or commons with limited permitting, to an individual quota system. That would give each qualified boat owner the right to harvest a certain percent of the fleet quota and eliminate the commons completely. It was a system already used in a few fisheries outside the US. Economists rightly called it a "commoditized" system, as it would create a market in valuable individual fishing quota shares that they believed would consolidate the fleet to the most efficient operators. By 1995 a very different system of federal fishery management began, with the federal waters off Alaska the region at first most impacted. It awarded larger amounts of the quota to the larger vessels with most production, which worked against the smaller boats.

The salmon fleets escaped this restructuring, as they fished in state waters and stayed with the system of limited permits and traditional management in a commons. But the salmon fleets had endured a sea of their own troubles ever since Euro-American agriculture and the incredibly profitable Northwest canning industry had replaced Native American subsistence fisheries along the rivers. Over the next decades, the growth of urbanization, federal dams, agricultural and lumbering pollution, and in some regions drought, in some weak hatchery policy and overfishing, have been adversaries for all species of salmon and for the fishermen chasing them. In the last twenty years climate change, increasing competition from recreational fishing, and foreign farmed fish have joined the list. Many salmon stocks are now listed as endangered or threatened, while salmon fishermen, except in some regions of Alaska, are an endangered species themselves. Neither system of management, federal or state, promised a good future for small fishermen.

As I researched for this book I discovered that although our federal government floods us with messages promoting healthy, sustainable fisheries, its management is full of contradictions for the small fishermen, the majority, which are carried over to the regional councils' work. But outside of coastal communities where a fish processing plant is an obvious part of the economy, the general public is oblivious to the economic and cultural threats to the coastal communities that depend on small fishermen. Even residents of coastal towns are probably not aware of all the forces that have combined to favor industrialized fishing and to drive the small fishermen toward the rocks. It is not the simple version of that history: "They caught too many fish."

The argument that small-scale fishing should be salvaged, that the boats tied at the docks growing seaweed should be out fishing, is best told in the words of the fishermen that I know that I interviewed for this book. But though commercial fishermen have written interesting memoirs of their working lives out on the water, few, until the advent of fishermen's blogs, have been willing to tackle in writing the aggravating, even enraging political element of fisheries politics and management. Maybe it is too complex and depressing, or they feel they won't win the arguments, and they are beat down by decades of scapegoating.

My purpose here is to present the small fisherman's view of that side of their working life that very few take up in their memoirs. A good place to start is a brief trip back to the decades when salmon fishing was still an entry into the economic engine of the west coast for the poor and the young--the 1950s-1960s. (Other entries were halibut and Dungeness crab.) It was not that long ago, and the fishermen of those good times all along our west coast and British Columbia's are still alive to talk about them--and the changes that have been sinking so many. Alaska, the Northwest, and British Columbia waters have been the source of livelihood for thousands of medium and small boats, many of them from communities based entirely on fishing and sometimes fish processing.

Individual Fishing Quota, and similar forms of privatized fishing rights, now known by the friendlier term, "catch share", by its structure will always favor industrial fleets. Unfortunately the gear used by most of these fleets, the trawlers, can be the most destructive to fish stocks and their ecosystems. But it is the most efficient way to make low-value fish such as flounder profitable. The conflict between traditional small-boat fleets and industrial fleets can be most easily seen in Alaska, the home of so many small-boat fleets based in small ports. But the problem exists on all of our coasts and in larger fishing ports like Kodiak, Alaska, Newport, Oregon and New Bedford, Massachusetts.

The fishing industry has always had its political and social conflicts, but the conservation issues grow with each generation. Fishermen and environmental groups have recently sued the NMFS, claiming that not only is it not doing updated, accurate assessments of fish stocks, but that poor policy has allowed "essential fish habitat" to be destroyed. Such cases are frequent and reflect the general frustration over fishery management. NMFS responds that it doesn't receive adequate funding to do the needed assessment, and that is also true. A recent example of what the fleets label mismanagement is Atlantic cod, once the money fish of the North Atlantic. And catch share, the system in use off Alaska since 1995 for halibut, and since 2005 for king crab, had not cured the problems of those stocks. In the last two chapters, I include suggestions I've gleaned from many sources as to how catch share can be, if not entirely fixed, at least humanized.

For fish stocks like salmon under Northwest states' management, the problems are quite different, yet also have been destructive for the fishermen, especially since the 1990s, and difficult for fish managers to overcome. Oregon fishermen describe how climate change and drought have compounded the changes in the ocean and put salmon fishermen in conflict with politically powerful big irrigators. For four or more years trollers have been left with almost no harvest. Cuts to both federal and state support threaten corrections needed for hatcheries and habitat restoration. Yet in Alaska, where droughts, dams, and agribusiness are not an issue, Chinook salmon runs are nonetheless in serious decline. The news gets worse: many economists believe our traditional "catch fisheries", big or small, are soon to be obsolete, and that the world's future protein depends on fish farming.

A great deal of information about our fisheries is available, far more than one can assimilate, from the coastal news media, government and scientific reports, and conservation groups. But with few exceptions, for the perspective of the small fishermen you have to go to essays in a few trade magazines, websites, and a growing number of fishermen's blogs. I am indebted much to them. I write from the fishermen's perspective, calling a lot on the experiences of my extended family and friends that chase salmon, halibut, crab, herring roe, and bait fish in Alaska and Oregon, and from my own times during the 1960s-1970s as a commercial salmon fisherman in Southeast Alaska, and as a subsistence gillnetter for thirty more years in western Alaska. For a comparison with other regional fisheries, I looked to New England, British Columbia, and Iceland. I don't give much time to the perspectives of NOAA or that of industrial fleet owners, as they are readily available on the web. I know, however, that their crews have the same concerns as my family: a livelihood they need to protect and families to raise. I also talked to and read many commentaries by fishery scientists and managers who struggle with their charge to save and rebuild threatened fish stocks.

The news for our small fishermen is not all bad. The last re-enactment of the MSA in 2006 was a turning point for them, as it was again stronger in its mandated protections for fish but also for small-boat fleets and their communities. And though they are still the minority, a growing number of boards, councils, consortiums, and environmental groups insist that both federal and state management do more to preserve small-scale fishing. And there are also other groups that would be happy to see these efforts fail.

To explore how we got to such a crisis for our North Pacific small-boat fleets, I divided this book into two sections. The first looks at salmon fishing, including subsistence fishing, starting with fishermen's stories from the 1950-1960s before the large changes took place, and moving on into the present. The second section describes the changes taking place in our North Pacific federal fisheries, again in part through fishermen's own stories, and with comparisons from other regions' fisheries. I have included my family's experience with start-up fisheries in my own region, Norton Sound, that are modern versions of what commercial fishing was along the whole coast sixty years ago.

Today's boats look very different, but fishing still embodies classic themes honored in America of independence, challenge and risk, hard work, cooperation, and respect for our natural resources. Behind the voices in these stories are families and their small coastal communities, or the ghosts of them. My hope is that twenty years from now global mega-corporations will not rule the fishing grounds, nor that the fish in our markets will not all be labeled "Quality Farmed Fish", raised somewhere else. I hope the fish will still be swimming and that many traditional fishing people with their small boats and gear will still have their own cultures and communities intact.

Part I

The Salmon Feast

"Today there is no morning bite at Bamfield. The old village was all about salmon and they are gone. There are no small fishing boats, no big seiners, no commercial fishing left at all... [it's] a holiday place. It has websites devoted to it praising it for its scenery, peace--and yes, fishing (find them if you can)...If we can see it clearly at Bamfield, the same applies to those hundreds of villages along the coast...."

ERIK WICKHAM, RETIRED BRITISH COLUMBIA TROLLER /BLACKCODDER, FROM HIS *DEAD FISH AND FAT CATS: A TRIP THROUGH OUR DYSFUNCTIONAL SALMON INDUSTRY*

Pacific Chinook salmon is the glamour fish of the North Pacific, a living icon for the western conservationist public. For eons, salmon of all species was the economic base of coastal and river-dependent Native American people and often figured in their spiritual life as well. Then, following the European-American fur trade era, it was the renewable resource for the first large industry of the Northwest and Alaska--salmon canneries. Commercial fishing and processing is still number three in the Alaska economy, and the subsistence fisheries continue to be essential in rural Alaska. Yet, outside Alaska and other west coastal regions the general public knows almost nothing about the salmon fishermen themselves or the communities dependent on them. There is no hero for children to romanticize, no Davy Crockett, no Hollywood cowboys, no glittery social history of overnight gold rushes.

The Northwest salmon stocks have battled one adversity after another, almost to their extinction on many rivers, one more story of riches gained through resource rape. The struggles of the "mighty Chinook" to get past the dams on the Columbia to their upriver spawning grounds is the one story the general public may actually know. They may even know that there are fish hatcheries, but nothing of the politics swirling around them. Some may know that the cheaper salmon fillets in the grocery display case are now from foreign farmed fish that compete with our native-born "wild" stocks. But that's probably the extent of their knowledge. It's a different story in Alaska, where a recent poll found the 60% of people still think that salmon is very important to the image of the state. But even in Alaska the commercial fishermen themselves can remain obscure, out in the fog somewhere.

As the salmon stocks began a serious decline along the west coast in the 1970s, so did the salmon fleets and so, in many cases, did their coastal towns and villages. Yet salmon fishing represents the best of small-scale business, though more in touch with nature, more physical than most. I won't romanticize it; it can also be backbreaking, family-busting, boring, dangerous, and vulnerable to exploiters. And it can be uniformed, as I once was. A stubborn nineteenth century belief lingers among a few fishermen that overfished stocks can always renew themselves given time. Yet salmon still could be a very renewable resource. A coho salmon can hatch, grow, be harvested, and sold in just three years. That contributes to the species' trouble, as today several countries already reap big profits from coho salmon farming. But fish farms are just one reason the commercial fishing life gets more complicated with each decade.

In Chapter One I describe a protest fish-in by a small group of Alaska Native salmon fishermen from a remote western village. They are in almost all ways typical of small-scale fishermen: members of extended families that depend on local fisheries, using small boats and simple gear, methods that rarely threaten fish stocks, and dependent on a single fish buyer. They are also subsistence fishermen, each summer drying or smoking much of their catch for their winter sustenance. They have access to a system of regional advisory fishery councils, but it is hard for them to see their influence on state management. The decline of the Chinook salmon they chase has gone on for over six years, in some districts longer. At many locations the residents are now forbidden to take a single salmon commercially, at others they can't even catch one for dinner. Yet offshore industrial fleets inadvertently scoop up thousands. Those villages' small fishermen could be us.

The problems of our west coast salmon fleets have lessons for all of us in the value of natural resources, work of dignity, and healthy communities that we shouldn't allow to be lost. Will these small fishermen still be out on the water ten years from now, and does it matter? Today scientists are pressured from all sides to find out where the missing Yukon kings, Columbia coho, and other salmon went. There is not nearly so much public interest in where our missing salmon fishermen went.

Chapter 1

TARGETING SMALL FISHERMEN

In the summer of 2009 Nick Andrews and five other men from the western Alaska Native village of Marshall conducted an illegal fish-in for Chinook salmon they had legally fished for each summer for decades. They netted one hundred--enough, they thought, to make their point. Their six gillnet skiffs had been tied up all summer for a second year due to a record shortage of Chinook headed up the Yukon and other rivers along the west coast. The fishermen hoped to get public attention for their economic crisis they blamed mainly on a fleet of corporate pollock trawlers fishing off the coast in the Bering Sea. With their great trawl nets these 100-300 foot floating factories caught tons of pollock, but also inadvertently scooped up and dumped over the side prohibited unsellable salmon as "bycatch".

The Marshall men wanted the attention of the pollock trawl industry and the federal North Pacific Fisheries Management Council (the Council) that develops management plans for the trawlers and other federal fleets. The Council sends its plans on to NOAA for approval. The Dept. of Commerce and Congress then get their chance to award pass or fail, and Congress appropriates needed administrative funds. It's a system hard to fight, so the men wanted the ear of the Alaska governor, whom they hoped would support their complaint to NOAA and on up the bureaucratic ladder. They thought the Council's targets for reduced salmon bycatch were far too easy on the trawlers.

The Council, which shares duties with NOAA Fisheries (NMFS) over all the fish stocks in the federal waters off Alaska, has struggled with the pollock fleet's bycatch problem for years. It is no easy task to keep unwanted species out of a trawl net, and salmon bycatch is one of the most troubling. Halibut bycatch by bottom trawlers is another. Now NOAA had given an ultimatum to all of the regional councils to create plans that assured major cuts in bycatch by 2014. The bycatch rate of the pollock fleet was actually one of the lowest, but they were such a large productive fleet that the number of salmon they wasted was shocking and was getting increasingly bad press.

Control over bycatch was forced not just by the Magnuson-Stevens Act (MSA) but because intermingled with the trawlers' salmon bycatch were wild Chinook stocks from the Columbia River that required special protection under the Endangered Species Act. The US had also been shorting the Canadians on their Yukon River Chinook quota that we shared with them through treaty. The Marshall fishermen had a strong push for bycatch action on their side, but it hadn't been enough to overcome the well-endowed trawler interests. I was in sympathy with the Marshall protest because I myself have been a subsistence salmon fisherman for thirty years in western Alaska, and many years earlier I fished Chinook and coho commercially in Southeast. But for Marshall and all of the western Alaska Native communities along the coast and on up the rivers the issue was an immediate one: would there be food for the winter or not? A Council environmental impact assessment in 2009 had found that from the Kuskokwim River north to Norton Sound, the value of the commercial Chinook harvest had dropped to between one and two percent of the 1980 value. And not only had the commercial Chinook season been closed or greatly reduced, in some years much of the subsistence fishery was closed or reduced as well. The western Alaska families did not look forward to another winter of requesting federal disaster relief.

By 2014 the balance of power would change, though, as the worry over missing Chinook spread and morphed to anger. The Alaska Federation of Natives and west coast tribal groups submitted formal protests to the federal government. Tribes along the rivers under State Fish and Game management demanded more input into salmon management, as the Inuit already had with sea mammal management. Now the collapse involved the whole west coast of Alaska, while fishermen's associations, both commercial and recreational, were screaming over similar losses in Cook Inlet runs. The Council felt forced to repeat regional meetings where fishermen would give most the blame for missing salmon to industrial pollock trawlers' salmon bycatch.

Scientists reminded protesters that normal cycles, warming waters, and other ocean conditions had to take some of the blame, maybe most of the blame. And why was it that the small commercial chum salmon fishermen at Kotzebue, the country's northernmost commercial salmon port, had their best season ever, both in numbers and price, in 2014? Did those fish take a different route through the Bering Sea and bypass the trawlers? But the Yukon and Kuskokwim River fishermen knew there wasn't much they could do about those issues, whereas the Council had the power to reduce the pollock fleet's bycatch quotas.

The Marshall fishermen's protest, and those to follow, were small but spreading acts in a much larger drama. The battle of a small-boat fleet's interests versus that of a huge industrial fishery has been played out for decades along all of our coasts. Pacific salmon have been under long-term fishing pressure since the western canning industry began in the late 1880s. Pollock fleets weren't the problem then. After a few years of intense fishing by traps, wheels, and seines, by the 1920s the commercial salmon pack was seriously down. The government recognized the signs from similar overfishing in the East and started a western hatchery program. From then on each decade brought more puzzles, more impacts, more experiments, and more frustration with efforts to save and rebuild salmon runs.

At certain points the management of state stocks like salmon conflicts with management of federal stocks like pollock. Bycatch is one of the best examples. The Council had held public meetings all over western and even Interior Alaska to get reactions to its proposed alternative plans for reduced Chinook bycatch before it set its "preferred alternative" at 60,000 bycatch hard-cap quota for the Bering Sea pollock fleet. The Marshall men probably knew western residents had a slim chance of getting their overall hard-cap request of about 30,000 through, but they wanted to be on record that the larger figure was unethical. The way the Marshall fishermen interpreted it, the pollock fleet could legally toss 59,999 unsellable salmon over the side for the crabs, or in some cases donate part to a food bank, while the west coast salmon fleet could not catch and sell a single one, and in some local fisheries could not even take one to eat. So they carried out their small protest, gave the illegal fish they had netted to their elders, and hoped to stir things up across the state while waiting for government response. It never came. When the Council's quota came through in 2011, it was at 60,000, and there it stayed.

Up the coast in northern Norton Sound where I live, we didn't hear more of the fish-in incident other than one half-hearted ticketing by the state troopers, later dropped. Though several of my extended family are subsistence salmon setnetters, we don't catch Chinook; we fish local coho, chums, sockeye, and pink runs. But our chum salmon runs had recently gone through a long spell of weak returns. We were anxiously waiting for a pollock fleet's chum bycatch plan to come up next on the Council agenda, but suspected that the pollock companies were encouraging sympathizers on the Council to stall however they could, as the more time they spent avoiding bycatch meant the fewer revenues at risk. The trawlers were hoping for a generous 60,000 hard cap on chum, too, or no cap at all. They had the numbers on their side. The Council voting seats are eleven, six of them from Alaska. The trawlers' home base was Seattle and other southern ports. All that the pollock group had to do was get one Alaskan to vote with them. The trawler associations regularly collected the needed votes for their causes. Chum bycatch quotas had been stalled fifteen years.

Fish "interception"--one group catching fish that another one claims as rightfully its own--is not a new problem. It has surely been taking place since humans started fishing. Today it causes many of the battles that fishery managers must deal with, and always with plenty of emotion expended. A fishing season, after all, means a year's sustenance. This battle between fleets is a particularly exasperating one when the inequity of the fleets' power is so obvious: skiffs versus giant factory freezer ships.

The salmon bycatch battle in the Bering Sea was even more complicated than usual. Because of an unusual twist in regional fishing history, residents of the same west coast village, even from the same family, could take different sides. In the early 1990s a group of big US trawlers wanted more control of the valuable Bering Sea pollock. (More about this later.) They had earlier won an unusual deal at the Council whereby the economically struggling west coast Alaska Native communities on the Bering Sea would get ownership of seven percent of the fleet quota, while the Bering Sea pollock fleet would soon be getting more control of the pollock.

The royalties from the seven percent pollock quota were now being used by the west coast communities for small fisheries development and more general community projects. These were managed for the communities through regional non-profits called the Community Development Quota (CDQ). Thus, Norton Sound and five other coastal regions had local families that could be protesting salmon bycatch by the pollock fleet, but at the same time be fishing salmon or crab from a boat purchased through a loan from their CDQ, that loan made possible from pollock royalties passed on to the CDQ. I knew families with this very situation.

People, to one extent or another, recognized the dilemma, but no one wanted to turn down a CDQ's help to their region. The royalties created a chance for development on an economically stagnant coast. But at the same time their salmon fisheries were economically and culturally vital for the communities and of long tradition. The North Pacific Council had played a major role in creating this unique arrangement, and now, twenty years later, it was charged by the MSA with cleaning up a long list of problems--bycatch waste just one--to save our fish stocks. The Council's meetings were long and each year more tense, with bycatch a topic at every one.

Commercial fisheries can often become adversarial when so much bounty is there to be scooped up. Of the North Pacific fisheries, salmon has been number one in value for Alaska, while pollock has grown to be number one in volume for the entire country, and may be reaching number one in value for the state, surpassing salmon. Halibut, crab, cod sablefish, and many kinds of "flatfish" and rockfish have their own rich histories. Not just salmon but all of them, especially the federally-managed fisheries (in waters out beyond three miles), have undergone major, stressful changes in management in the last few decades, and the process continues.

A Small-Fishing Perspective

My interests lie with salmon, not pollock, for personal and cultural reasons. The salmon fisheries off Alaska and the Pacific Northwest started many thousands of years ago with the indigenous people, my husband's ancestors being among those harvesters. My own land-poor family joined a huge migration in1906-1907 when three of my paternal uncles and my oldest aunt arrived from northern Norway. They went directly to Astoria, Oregon where they could find a way into the fishing industry, the only cash occupation they knew. They sent money home, and by 1910 the whole family of eight was on the west coast, all of them involved in commercial fishing in some way. The Pacific Northwest and Alaska drew thousands of Scandinavians for the same reasons. It was as beautiful as their homeland, but with a chance for a better living--just what they had come to America for.

My oldest uncle found the gillnetting at Astoria too crowded and chose a new fishery-- salmon trolling off Neah Bay, Washington, while two brothers went north for cod and halibut. Another became a seaman and worked his way up to captain for Alaska Steamship Company, freighting salmon. My aunts married other Norwegian fishermen. My father, the youngest, was a storyteller in the old tradition, and though he didn't stay with fishing his nostalgic tales and oil paintings of the north country stuck with me as a child, and I knew I would eventually head for Alaska.

The tradition on my mother's side was farming, not fishing, but her family's orchards were right on the Columbia River, so rowing boats was a favorite pastime. From that side I have a cousin who has spent his whole working life fishing salmon. He is another storyteller, though his stories are hard realism, no romance. He is my source for a picture of commercial salmon fishing beginning in the 1950s when it was a livelihood that a young person unafraid of challenge and with common sense could find a way in with a few hundred dollars. His father certainly was an influence, for he built his sons two small skiffs and at an early age they were out rowing in the Tacoma Narrows, I along with them.

I turn now to my cousin George's history for a close-up picture of commercial salmon fishing before salmon problems were so overwhelming. George saw his first view of Southeast Alaska when he arrived from Washington with a larger skiff in 1955. Stretched out before him were what appeared to be endless, rich salmon trolling grounds. Not obvious to him, Alaska was struggling with a simpler problem than today's: an outside-controlled salmon industry and its army of fish traps. The salmon runs, the base of the territory's economy, were in major decline, in part due to the super-efficient traps. To George, though, the "Southeast" fishing looked fine enough; he knew nothing of the politics. He was eighteen, and it was his second summer commercial fishing. His father had renovated the 22-foot skiff with an inboard motor and a semi-cabin. It had already proven it was seaworthy fishing in the ocean off La Push, Washington before George turned his bow up the Inside Passage.

The salmon fishing commons was open to all who could scrape together enough for a hand-troll rigged skiff like his. They could promise all their future fish to a broker on the dock for credit toward for the license, fuel, grub, bait, and a few lures to try their luck. Hand-trolling was a special life, special even among fisheries, the most accessible financially, and with almost no restrictions. It required consistent energy, observation, agility, a little mechanical know-how, and a basic level of seamanship. It was perfect for George. He survived the season and even made some money.

About 1995, forty years later, I received a rare Christmas letter from him. He was living near Newport, Oregon, still salmon trolling. I was living at Nome and decided that looking him up would be a fine excuse for a brief vacation outside Alaska. In July I flew south with one granddaughter, picked up my daughter and another granddaughter, and we rented a car and drove down the coast. I hadn't seen my cousin in almost twenty years. By luck, as we arrived at the port George had just sold his fish, and from the parking area I spotted him at the float, cleaning his deck. It wasn't difficult to find him, as the *Helen McColl* was unusual for a troller: a salty, sixty-four ft. 1912 vintage converted herring tender he'd brought around from Maine. I gave a shout, and we each had to assess how the other had weathered as he came striding up the float. Enough fish were coming aboard to keep him trim, his hair no grayer than mine, his face more sun-wrinkled.

We stood on the float looking over the fleet, and I was full of questions, trying to catch up with all the changes in the commercial salmon fishery since my own stint in the 1960s. I took photos of the *Helen McColl*, and I noticed the boat had not gotten the prescribed annual coat of paint, which surprised me, but I soon found out why. I thought the *McColl* would be a great photo subject for tourists, but George objected. "I've got some much nicer photos--when I was able to keep her up, not like this."

I soon found out George had a new project that took all his spare cash. He and his wife Pat had embarked on a cheese-house venture with prize alpine goats. Farming was Pat's passion, not fishing. She was in charge of the goats and everything connected with them while he provided start-up funds from fishing. We headed on out to see the farm, admire the young goats, and barbecue a king salmon. The talk of the new farm was optimistic, not so George's comments on fishing. The previous season had been an especially poor one in a run of weak ones, with each year seeming to bring another new low in harvest numbers, a new high in fleet restrictions. He was glad to have a caring audience.

"Fishing's not like it was when I started out. That trip I just brought in barely covered the fuel. There are so many restrictions now we can hardly make it. You know I used to sometimes catch 100 to 150 salmon in a single day? Now it's like twenty to thirty."

That was a shock to me, coming from George, the highliner. He shrugged, "The seasons were cut back, then the gear, then the areas we could fish, the days. They don't want us to survive."

I didn't understand what he meant. Salmon used to be so important to the Northwest economy--why wouldn't it still be? I had really lost track of the fisheries news.

"Well, salmon is still important, but you wouldn't think so. They've closed the Oregon coast coho entirely; we're fishing kings only. And the price is terrible."

"But you keep fishing anyway."

"I'm fishing for this farm."

George was disgusted with both the state and federal government policies, which seemed in their different ways to be aimed at eliminating the fleets. I would find out later it was a common view among salmon fishermen. He gave me quite a list of facts and theories about what was wrecking commercial salmon fishing. There wasn't a single insight I could offer; I was too remote from the economics or the politics. I hadn't commercial fished for many years, and his description made me glad I was just a subsistence fisherman now. I thought that with so many years of trolling he ought to write about it, but he shook his head, "No, it's too painful. But I can talk to you, and you can write it."

I stayed in contact with my cousin, and took notes. I had my own questions as to why our western Alaska chums were going downhill, and perhaps he had some clues. I talked with him often by phone, and made it down to Newport a few more times. I gradually wove together pieces of his story of years of salmon trolling and how he saw the changes, both to him and the fishery. It was obvious that he had spent some time pondering just how he had ended up, after over forty years of trolling, now trying to fish a boat he couldn't afford to keep up. I knew many fishermen had started out about as he did. At Newport and other ports I saw too many salmon boats now tied, long-term, to the dock, their automatic bilge pumps keeping them afloat. What went wrong?

Alaska is the last place in the country where there are enough salmon left in the streams to provide healthy subsistence or commercial fisheries. But even their survival is tenuous. Projects like the proposed huge Pebble Mine tempt many in the state. But Pebble would dump waste into the headwaters of the river system flowing into Bristol Bay, the home of world's most bountiful sockeye run. In the Northwest the rivers are controlled by federal dams and huge irrigation projects, with industrial and agricultural pollution and urban water demands all threatening salmon survival. I had been away from it all in remote Nome and was to find out that George's observations were just scraping the top. He later volunteered a comment that was interesting coming from a commercial fisherman. He said that he believed the subsistence fisheries were the most important of all, and what above all we had to save. He reminded me that though subsistence fishing is past for most of the US, around the rest of the world those fisheries still are vital for vast populations, most of them poor people. He'd had a lot of time to think beyond the day's catch, out there on the ocean.

George's comment brings me to good place to pass on another personal history, this one from a traditional Alaskan subsistence fisherman. On a summer morning in the late 1950s, as George was probably heading out for a long, foggy day of trolling, my husband Perry was a youthful part of a team on the beach near Nome pulling in a seine net, learning subsistence fishing from his Inupiaq grandmother. Here is his view of fishing in those days. Though today's fishing is skimpier, the work crew and the technology we use are much the same.

Chapter 2

An Inupiaq Subsistence Camp: 1950s-1960s

At five years old Perry Mendenhall washed fish, at eight he hauled firewood and water, at ten he was part of a chum salmon beach-seining crew, a helper to his grandmother every summer when she went to her fish camp at Fort Davis near Nome. He remembers it all as he helps create the scene again at our own camp today. Their 1950s fish camp was typical for Eskimo subsistence fishing of that time. It wasn't much different from forty years before that. Our fish camp is on the same ocean beach about ten miles east of Nome where we fish for salmon and pick wild greens and berries--all we can squeeze into northwest Alaska's short summers.

During Perry's childhood, some of the families were commercial gill-netting, but all were concentrating on getting their essential food for the year. When he was still very young, his grandmother chose him to be her companion at camp, and through her he learned the lore that he is able to use and pass on today. In Nome winters' long nights I coax personal history from him that gives a picture of subsistence activities fifty-sixty years ago. I'm sure one reason I'm interested in the fishing he did as a child was that I, too, had been a early, modest version of a subsistence fisherman. As kids, my cousins and I hauled our catch of rock cod and sole home for dinner and were always well praised. His story is far more serious version.

Grace (Allelaaq) Mendenhall, a widow, made her living as a skin-sewer of parkas, mukluks, and caps all winter and watched her grandson Perry while his mother worked in restaurants. But as soon as the snow started to melt, the grandmother headed out for the tundra and hills behind Nome for her real life, her camping life. She packed Perry along on her seasonal round of subsistence activities, and in a few years he would hike along on the tundra on his own. This was a common arrangement for grandparents who didn't have cash jobs--to take over care of grandchildren all summer, most of them at the Fort Davis and Nuuk encampments. The parents with jobs tried to join them for the long evenings and weekends.

Perry remembers: "I was a lucky kid because of my grandmother. She saw to it that I learned the subsistence life. She started early in spring with squirrel trapping, living in a white-wall tent when there was still snow on the ground in patches, in the hills near Penny River. Then she was picking greens, mainly *surra* [young willow leaves], then fishing, and kept right on, moving around to different camps until fall, finishing up with the berries. I was her only companion when we camped way out in the hills, but during salmon season we were at Fort Davis in cabins with many families from villages who had come into Nome to live. Quite a few families still camp at Fort Davis, and it's still popular for grandparents to take the kids out with them.

"In those days a family that had a camp with a cabin, a stove, a fish rack, and storage place was considered wealthy. But they were working camps, of course. The first work at Fort Davis was the sea mammal hunt that started as soon as the sea ice broke up, probably mid-May. This provided fresh and dried meat and the blubber to render into seal oil. Most women were also drying and tanning the hides for use in skin sewing later.

"There wasn't much work for very young kids at the early hunting camp as it took too much skill and strength, but later, when the salmon started running, a kid about five could start carrying water and wood, picking a net, carrying fish. Older boys helped the men with the net, hauled fish to the tables, fixed the poles on the rack, and took the cut fish from the women to hang on the rack to dry. Later we might start a smudge fire to keep off the flies. Meanwhile the women just cut fish as fast as they could go. We were a busy crew to keep up with the flying *ulus* of those women, in a rush to get all the fish cut before they got soft from lying around. The women complained to the men if they had a dull *ulu* or had to cut soft fish.

Rough Waters

"The salmon season started with the chums for us, in mid-June. We almost always seined our salmon in the Nome River, up from the mouth half a mile, above the bridge on that gravel bar. With a seine you get a slew of fish all at once, so it got very busy with people cutting and hanging fish into the middle of the night, then sleeping late the next morning. That was the fun part, the seining and cutting. Once the fish were on the rack, if you were lucky and had a good breeze and clear days, they would pretty much take care of themselves. But you couldn't just go away and leave them. Sometimes in a good run there were racks on top of racks. Someone had to be watching those fish, and it was often a grandma.

"If the wind stopped and the fish didn't have a good dry coat on them yet, here came the flies, several kinds. Then it was important work to keep the fish clear of fly eggs. This was mainly the women's work--I myself never had to pick fly eggs, don't know why I was so lucky, but I noticed some kids who were tall enough to reach the racks or could climb up to the second tier were assigned to fly patrol. And some people were a lot more ambitious than others with that job. But in those days we always got plenty of fish, and if some spoiled they weren't thrown away. Most people had at least one dog that needed feeding, and others had several.

"I never did cut fish either, except once I remember, someone said that they better see if the boys knew how, just in case they needed to someday. And of course we had seen it done a thousand times so we knew the cuts. We made ragged attempts, just to get out of it, but enough to show we understood. 'Okay, you know how.' And we boys were released back to our other work. The girls stayed on to learn more. In those days all the girls mastered it before long and took pride in sharp, even cuts that dried best.

"While those fish were drying, we were seining more, because you were never sure about the weather and how many would successfully dry. A chum salmon is a big fish, and even in good windy weather, it can take a couple weeks to dry well, so several days of rain or mist, or very calm overcast weather could wreck fish that weren't well along. A little start of white mold didn't hurt anything, but too many days of this and people got nervous, knowing the fish would get the gray-green mold on them no one would eat.

A couple years we got caught with poor weather and had to wash off and eat moldy fish all winter--the white mold, not the green mold. But there is a limit to that.

"One year in the 1950s I remember that the entire catch of drying salmon was ruined beyond salvaging by rain and fog. Many of us families trucked up to Salmon Lake about forty miles into the hills and set up camps along the west shore. The red salmon come up the Pilgrim River to spawn in the lake. Just because of the barrier of the hills, the weather could be dry and hot up there when it was totally fogged in or stormy down on the coast for weeks. Some families had the custom to stay up there all summer. We seined those almost-spawning red salmon, and that's what we had for dry fish that year, and they were good.

"Salmon Lake was a nice change for the kids. Just like at Fort Davis, we had our work to do, but plenty of time later to roam around the hills, warned to keep an eye out for bears. We would also be encouraged to pick some greens, mainly willow tips, *surra*, that are great preserved in seal oil. Salmon Lake is beautiful with the high green hills and rocky peaks circling it, and years later I would spend a lot of time up there hunting and berry-picking.

"Thinking about bad weather wrecking the fish reminds me of something my grandmother said, 'You always put away enough food for two years'. One time there was no summer for two years.' That could have been in 1908 when a meteor crashed in Siberia. She would have been a teenager then. The dust made a haze over the Arctic, and the ground and water would have stayed cold. No summer berries. Or it could have been two summers with hardly any salmon for some reason. She'd known famine times, too, and most of her age group had. Some had eaten ptarmigan droppings then, she told me. She said she had no yen for the old times, going hungry. There are books of oral histories that affirm some of the events that she remembered. But she also said that according to our legends the climate was going to get warm again, like during the period when mastodons and horses lived here, as we know from the bones we find.

"The Inupiat had their own beliefs about conservation. Here are other things I remember from my grandmother. 'You don't waste the natural resources, you use them. Use every bit of what you take, or store it carefully. If you don't make use of what's there, it could go away.' And, 'You feed everyone that comes to you, because you know that one day you might have to go to them.' And another, 'You never act or even talk disrespectfully toward nature, not even plants, or they could leave you.' And 'When you see a place out in the country where people can get drinking water, you clean it up.' A lot to think about today."

The old traditional practices worked for the Ft. Davis campers as they have worked around the world until western commercialization interferes. With the increased human population around Nome the fish needed modern management. But modern management itself is changing from a single species to an ecosystem-based management. The impression I have is that indigenous people in general had, and still have, an ecosystem-based understanding of nature, that everything is connected and should be managed that way. In Alaska and probably elsewhere scientists today give more attention to what elders have observed about the changes in the weather and in the ocean, how they affect all of nature, and how we must manage for them.

The fishing regulations in Norton Sound were quite liberal when I arrived in the 1970s. Any Alaska resident could get a free subsistence permit and seine or gillnet during openers in certain stretches of the rivers or estuaries as well as in the ocean. Rod and reeling was also popular when the coho arrived. Perry remembers that earlier at Fort Davis there were unwritten understandings about how to share the shoreline.

"The families took turns at the spots in the river where they could beach seine. As I got older, I was the man at the oars, taking the net out to make the set. And sometimes if the fish were slow coming into the river, and we got anxious, we would set a gillnet in the ocean. I got the hang of launching the skiff in the surf. If it's not bad offshore, you can make it out through quite a bit of surf. But people preferred the fast, efficient seining in the river. Also, fish in the river a few days have lost some of their oil and will keep better over the winter dried without getting rancid, so most people preferred them. We didn't have freezers then, but tied the dried fish in bunches and hung them in cool sheds in gunnysacks.

"The chum fishing happened in the best drying weather in June, so they were an important fish for us. There was always a good run, and you could get enough dried, if all went well. August is supposed to be the rainy month, but some years the mist, fog and drizzle could start in mid-July. By then, the chums could be dry. A half-dry chum is good too, and we would always boil a few to eat that way, for the change in flavor. But dried fish was our staple, all year round. Farther down the coast and up the rivers they also smoke the salmon, due to more rainy weather, and they counted on kings, which are so oily and harder to dry. Smoking is more work, and we only resorted to it when we had to. I don't remember any real hard-smoking, just some smudge fires. If we wanted hard-smoked king or silver strips, we traded from Unalakleet. We really counted on those early chums and the good drying weather in June, and they looked beautiful hanging on the racks.

"As the chum run was winding down, here came the pink salmon into the Nome River, and we seined for them too. Fourth of July was the magic date. They came in quantity every even year, much fewer on the odd year. Pinks also make good dried fish and dry fast, so they can save you when you have poor weather. They give less trouble with flies and don't get rancid so fast when put away in bundles. But they only run strong in the even years, and they run later so you will often have to fight the rain. Then there were fall chums too, and a good run of silvers [cohos] on the Nome River, but by then many people had their fish and were busy doing other things.

"I don't remember that my family ever netted silvers--the cohos--though there were openings. But by then it was too late to get big silvers dry, and we had no way to deep freeze fish, so dry fish was the important staple. My grandmother had all the fish she needed by then anyway and was off to her berry camp across the head of the Nuuk lagoon fifteen miles east, taking me with her. The families had to get berries too, not just fish.

"She had her son-in-law haul her stuff down to Nuuk by truck. Then she packed a tent and everything else by foot around the head of the lagoon to a spot near a creek on the other side, about a mile or so. We set up our tent and stayed out there a couple weeks while she picked *aakpik* [cloud berries] on the flats. And meanwhile she had a short whitefish gillnet she put out with a long pole. She cut and dried them there till she had a gunnysack full. For supper we'd have boiled fish or maybe a duck I had caught, greens in oil, pilot bread, and berries. Even if it rained and blew she picked. I was never much of a picker, but I had to carry many gallons of berries in a backpack. She did take a break on Sundays. Then I could do whatever I wanted as long as we had enough water and firewood. By September each year we were headed back to town so I could go to school. By then, I was glad for a little town life.

"All of the subsistence activities were important. Ice-fishing for tomcod and crab were big in fall and winter, and according to stories had kept people from starving during some years. People would go down to the lagoon at Nuuk--it freezes up early--but you could also jig tomcod right in the river mouth in Nome. Tomcodding and crabbing were available to even the poorest family, no license or permit needed, or truck, or boat, and no limit, just a line and a lure or bait. Even if people didn't have a tool to make an ice hole, they could wait until someone else was done with their hole and hop right over. It's a good family activity in calm weather because if kids are dressed warm, from age three and up they can jig for tomcod. Usually people like to do it in groups and socialize while they are hauling up their fish. Sometimes they get close to a hundred, sometimes a handful.

"I jigged tomcod to use for crab bait. I would go out on Friday nights and find the crab holes that had been abandoned. I could be all alone, or maybe I'd see only one other group sometimes, but it didn't bother me. I'd try to find four open holes, drop my baited lines until they reached the bottom, then raise the bait up a bit, wait, keep checking the lines, and when I felt the weight of a crab I'd haul it up nice and smooth and usually the crab would hold on. When I got back into town a couple hours later, I'd go bang on someone's door that I knew had money and hold up a couple big, still wiggling crab and I'd get five dollars! That was big money then, like twenty now for a kid. We'd also get some dollies or grayling in a net or ice fishing, or jigged in the rivers for freshwater ling cod [burbots], and we scooped up cigar fish [eulachon] out of the surf when they were spawning. For smelt and sheefish, we traded from other areas. All these were important for feeding families, but salmon was the most and still is.

"I was part of this life until the fall I was seventeen, when I left Nome. I had told my grandmother that I wanted to be an Eskimo and not go out for training, that I just needed to learn how to make a compound bow. But she rejected that idea totally. 'Go out and learn all the white people can teach you and come back and help your people,' was her strong advice and I followed it. It was over ten years later that I returned, and all these subsistence activities started coming back to me, and then I really appreciated those times with my grandmother. There were some poorer kids in town that had no camp and hardly ever got out in the country to learn all those things. You don't forget things, like when I had to mend a net, I can't tell someone how to do it, but my fingers remember. Everything from my grandmother comes back to me. I remember how she even made salt one time--we must have been out. It is not just the activity itself but the life that goes with it that's so valuable."

Commercial and Subsistence Combination

Subsistence fishing in Alaska was for decades mixed with semi-commercial fishing. Old-timers from the Nome area recall when families earned cash by selling dried fish for winter dog- team freighting until small tractors replaced the dog teams. Perry remembers people bartering dry fish to the "Native Stores" to pay for their other staples like tea, flour, and sugar. The state looked the other way on this bartering for the sake of the rural economy until about 1974. After that you had to have a commercial fishing license.

The sale of salted barrels of salmon for export began in Teller, up the coast from Nome, by the 1920s. Perhaps the runs there were overfished, or the buyers stopped, because that came to an end, and the next effort was in Norton Sound in the early 1960s when a freezer ship came up the coast buying. The fishermen were restricted to setnets, no drifting allowed. A private buyer began buying and air-freighting salmon from Nome in the 1970s, but attracting a buyer to come as far north as Norton Sound was always at risk. Japanese freezer ships filled the need in the Lower Sound between 1984 and 1988. When the Magnuson Act closed out foreign buyers except for emergencies declared by the governor, there were again years when the only way to sell Norton Sound fish was to fly them out. Local coops started up and gave up, unable to make an adequate profit. After limited permitting began, the Nome district commercial fleet averaged only sixteen active permits. The price was never great; the average season's catch for boats selling at Nome in 1983 was $1300.

In 1990s the new non-profit "CDQ" program (told about more in the second section) was able to create stable local buying ventures for lower Norton Sound fleets totaling from 100 to 200 boats. Then salmon fishing was again a more guaranteed source of cash for many families, but subsistence was essential on the entire west coast. In some districts like Unalakleet and Kotzebue, commercial fishing was a main way to get cash that helped buy subsistence gear like gas, bullets, nets, and so on. The other way to manage for frugal subsistence fishing is to keep costs down. That is what we have done by building our own structures, hanging our own nets, using a rowboat, and using local wood for smoking. But our frugality was overpowered by a serious bear one year, and we were forced to a modern innovation--an electric fence around our fish rack. Once the bear family learns where the meals are, they will be back.

Chapter 3

A Salmon Troller's Start

My cousin George's start at commercial fishing would be most unusual today. It wouldn't be possible financially; it wouldn't even be legal without a hand-trolling Limited Entry Permit, and then only if you could find one for sale. But in the 1950s hand-trolling was a start for many young coastal people--for him very young.

"I was about eight when I got bewitched by the salmon fleet during a family trip to La Push, down at the ocean. I watched those little double-ended trollers slipping into the dock to tie up and unload their fish, and from then on I knew exactly what I wanted to do. You remember the little skiffs my dad built? I felt I'd made a big step forward when I was nine, and I could hoist one end of my skiff up over one shoulder and my head and drag it down to the water all by myself. And I remember one day my mother rowed me down the [Tacoma] Narrows a mile to where those old retired fishermen were living and introduced me to them. Remember Mike? He was the one who liked kids and showed me all kinds of things about fishing. After my mother, he was the one who really got me going."

The summer George was sixteen he went out to La Push and got a job washing dishes at the café where fishermen hung out. He kept his eye on the fleet of trollers, both the hand-trollers using rods or hand-operated cranks called gurdies, and the bigger boats with modern power gear. He watched them in the morning as they headed out into the swells and fog, and later when they came back to unload their fish and count their cash. On the stormiest or foggiest days he listened to them as they drank coffee at the cafe and told their fish stories, and he filed it all away. By the next summer he was out on the ocean in his own hand-troller. Though he may have been unusual in his early start and lack of family mentors, from then on his story could have been told by many west coast youth.

"In the beginning, you know, it was the romance of fishing I was after. Making money hardly counted at first--I just wanted that fishing life. In those days I prided myself on being tough, needing very little. I could sleep anywhere, eat out of a can. I thought I could live on fish, rice, and applesauce, but found out that didn't work when I got scurvy. But I'd saved enough dishwashing to buy a skiff and motor. My dad put an addition on the boat to make it 22 feet, with a little half-cabin, a well for the motor, and an insulated fish box. My buddy Ron helped me rig it with two poles and hand-gurdies. The next summer I fished with Ron--he'd dropped out of school and had his own little boat. We fished out of both LaPush and Neah Bay.

"I didn't want to go home at the end of the season, so I spent the winter at the Quileute Indian village. Steve Penn gave me a cabin to stay in and took me under his wing for the fall gillnet season. He'd lived on the Quileute River his whole life. He had a beautiful dugout, I bought the net, and that worked out as our partnership. I was interested in the way the Quileute people fished--they fished until they had what they thought was enough and then quit. When I asked them why they were quitting, they just shrugged and said they didn't need any more. That impressed me.

"The next summer I was out at Neah Bay again--I don't remember now where Ron was. I heard some trollers talking about leaving for Alaska to fish. On impulse I decided to follow them. Of course I lost them halfway across the Strait [Juan de Fuca], I was way too slow, but I was on my way, no turning back. I hadn't even stopped to get water, or charts or anything, just my beagle dog."

The first time George shared his entry into Alaska story with me, I was amazed. Having traveled the Inside Passage myself about ten times in a forty-three foot boat, I found it hard to imagine that trip in a skiff. But I found out that in the early years it was common enough.

"Somehow I made it as far as Nanaimo, on Vancouver Island. The Canadians thought I was a deserter from the military! But someone decided to help out and loaned me a dog-eared copy of *Capt. Lilly's Guidebook* for the Inside Passage. I made my way on up the passage as far as Port Hardy. But then, leaving there, heading toward Queen Charlotte Sound, I was having a little trouble with the engine, and out of nowhere here came a big Canadian troller. I was smart enough to accept his gruff offer to tow me all the way, clear across Dixon Entrance. He dropped me off at Egg Island, and I made it on my own into Ketchikan."

George arrived in Southeast Alaska when salmon were still plentiful enough-- that a lucky hand-troller could actually make a small living. You found a protected drag and followed someone around until you caught onto the routine and the unwritten rules of the fleet. George's start was like others I talked to later who had come to Alaska with nothing but a basic boat and found out that old-timers didn't mind talking about their fishing strategies.

"After I got gas and fixed up my engine, I had one dollar left, so I ate breakfast. Someone pointed me in the direction of Meyers Chuck, the closest little fishing village from Ketchikan. You've probably been into the Chuck? A little cove carved in the rocks with spruce-covered hills. It had a float and a fish buyer in those days that carried gas, bait, water, and a few supplies for small day-boats and hand-trollers like me. I caught a king salmon on a herring fillet that day, what a gift! When I sold it that night I was solvent again. I knew I had it made.

"We were fishing kings, but I heard that more sure money was to be made in the summer out on the coast with coho, so later I went on out to Hole in the Wall. It's another little buying station hidden in the rocks, not even a village, just off the ocean by Noyes Island. You've been there.

I think we were getting .16/lb, nothing to get rich on, but I was doing all right. Everywhere I went, people were helpful, I was a novelty for them, I guess."

I imagine George in that setting and how he fit in. Small fishermen felt generous, as he would have had a friendly smile, but not pushy, respectful in his questions, modest about his small successes. Elders like to help out young people like that. I experienced the same hospitality myself a few years later, just as ignorant, starting out in Icy Strait.

The first time I traveled the Inside Passage, ten years after my cousin's start, I was going southeast, not northwest. I viewed history I couldn't comprehend at the time. Between Ketchikan and the clear-cut areas on Vancouver Island I saw what appeared to be almost pure wilderness. We passed long days along channels bordered by nothing but steep rocky shores and craggy-topped hills of spruce. I remember seeing only three living settlements, at Klemtu, Bella Bella, and Alert Bay--all active Native fishing communities. The only other signs of human occupation were a few log raft anchorages and falling-down cannery remains at Butedale and Namu. Later I learned there were many more of those deserted places up in the inlets. But that vision of almost-wilderness was an illusion. The long inside passage between Seattle and the beginning of the Alaska panhandle, and along the west coast of Vancouver Island, and farther out on the Queen Charlottes had once been home to many Indian settlements. Epidemics partly did them in, then the move to cannery-built villages as part of the transfer to a cash economy.

Soon tiny settlements of Euro-Americans and mixed communities were also scattered in settlements along the spruce-lined shores to subsist much as they probably had in the "old country." Canneries that needed workers had to provide housing, so villages or even little towns gradually developed wherever there was canning going on. For two years my father, just turning 18, was part of that work force in 1917-1918 at an Icy Strait fish trap. Still later, the most remote cannery settlements disappeared too, as fish processing moved to centralized towns and boats had to travel there to sell. Thus, at least three societies based on salmon had flourished in those waters and then faded. The larger towns like Ketchikan that survived still rely on salmon as a major part of their economies.

One can still see vestiges of the same social ebb and flow on Puget Sound decades earlier, starting with Native subsistence fishermen, then a mix of Native dugouts and Euro-American skiffs combining subsistence with commercial fishing and selling to canneries, with the cannery towns later to close down or metamorphose to the cities of the industrialized Sound. The lonely Inside Passage I saw in the 1960s had passed through all those changes except the last. Each change in the salmon industry has caused a major shift for the families, small businesses, and communities involved. Each decade brings stricter regulations for the fishermen. Yet salmon fishing from small boats, despite history full of socio/economic change, has survived fairly well in Alaska, far better than elsewhere. And each year new people will invest in the fishing life, no matter the expense. The explanation for salmon fishing's deep attraction, despite its troubles, is something I hope comes through in these pages.

During the summer that George pulled cohos off Noyes Island, the area was still a territory with the fish traps still operating. But another change was coming fast with statehood just three years away and with that the outlawing of the traps, and the first serious efforts at salmon conservation. But such affairs were far from his 18-year-old mind.

"I caught enough cohos out at Hole in the Wall to think I had a real living. That summer I found a huge Japanese glass float in the tide-rip. It was a real find--romantic for me, an artifact from beyond the horizon. But I had no room for the float in my skiff, so I left it with the buyer's scow. When the scow went back to Petersburg for the fall, I had to go up there and retrieve my glass float. Petersburg was a neat little fishing town, but right then, I wasn't sure what I would do next. I could stay in Petersburg for the winter, live at the dock, and winter-fish kings right there in Wrangell Narrows, or I could go home. It turned out there was a barge in port willing to put my skiff aboard and take it down the Inside Passage to Puget Sound. It was too perfect, so that's what I did; I went home. But my dog ran off at the last minute when the barge was leaving and became an Alaskan."

George was sure he would be back but other events interfered. Instead he became a Northwest troller, and I was the one who went to Alaska to stay.

My Own Start

My seduction by commercial fishing I can blame a good bit on my father's Norwegian side, far back in the foggy history of the North Sea. But my earliest boating memory is from age five, being cordelled up the Columbia River's gravel shore in a big rowboat pulled by many women with my grandfather steering. They were looking for a place to net whitefish. It was slow going, and the sun was hot, but the river and bluffs were beautiful, and the boating adventure stuck with me.

Actual fishing time came a couple years later as my father rowed me around on mountain lakes on the Olympic Peninsula, trolling for trout with hand-lines. I don't remember that we caught anything, but it didn't seem to matter to him. He told me it was much like the deep fjords and lakes of his childhood Norway, and he was in obvious bliss as he rowed, his very long legs cramped into the oarsman's place, a big smile creasing his face. He wanted me to row until the wind came up; then he took over and was ecstatic as we headed back, bouncing along in the chop, taking spray, laughing, "I love this!" So of course I was ecstatic too.

An early subsistence fishing memory is my cousins and I wading the sandy tide flats of Bainbridge Island, Washington on the incoming tide, feeling in the sand with our toes for the crabs, and when we felt movement, we would quickly reach down and grab the shell. If it was small or female we let it go--we knew the rules. The big ones went home for dinner, along with the clams we'd dug. I think now what a good start I had with all my subsistence fishing--cod, sole, crab, and about four kinds of clams.

Although George went to Alaska on impulse, my move was through conscious planning. My father's tales of a magical North were not of the dangerous, dirty work his brothers endured as they baited thousands of halibut hooks and rolled around in the slop in traditional dories. Longlining has changed a little since then--no more dories, for some boats maybe even an automatic baiter. My father also fished at least one summer on his oldest brother's troller out at Neah Bay, then worked as a deckhand for a few years on the salmon-freighting sail/steam schooner, the *St. Paul.* And for one lonely winter he babysat that Icy Strait fish trap, no place for an 18-year-old, and he didn't return. Long months out fishing with a bunch of growly, bossy older brothers wasn't for him either, and he became a skilled craftsman ashore. He grew up to be a glamorous man who like to sing, dance, and later painted scenic pictures of the North. But a part of him regretted losing Alaska, and I heard all about it. He did his best to assure I would make it there, soon to become a fisherman.

I knew what sent me north, but George? I asked him to recall more of his early influences.

"Part of it was those little skiffs my dad made--you remember we rowed all over the Sound in them, just had to wait for the tide changes. But my mother was the one who climbed down the long bank with us and showed us how to catch rockfish and sole. How did she know this stuff? She never grew up on Puget Sound. I remember we caught lots and hauled them back up the bank to her, and I'm sure we ate fish for dinner more often than we wanted. But it taught us to be providers, didn't it?

"I think maybe because of their life on the river, my folks saw nothing wrong with letting us take the boats out in the Narrows, once they were sure we'd learned all about weather and tides and safe boat behavior. We never got in any trouble out there, and I suppose it was partly because they were such good little boats and partly that even though we could swim, we knew it was damn cold in that Narrows current. To the grownups it was probably no faster or colder than the Columbia. I remember being told my mother made a sensation as a teenager when she swam all the way across the river, the first one to do that in anyone's recall. But a rowboat went with her. They weren't crazy."

I'm sure he was right, that a lot of river influence came through--salt added to the water, different fish. But though I became a permanent Alaskan, it was a long time before I had more than a narrow picture of what was taking place in fisheries management. When I went fishing commercially for a few years, I thought only about the fish right out there, the ones we wanted to catch. A boat payment was waiting. As for my cousin George, I will get back to him. After his trip to Alaska he would put in over fifty more years of salmon trolling, a player in many of the changes coming for small fishermen in all fleets and on all coasts. His take on it is valuable for his decades of direct observation but also his insights into how the salmon industry is a player in the larger economy and the government's role in it. But first, more background on how Alaska looms large in the story.

Chapter 4

CANNING ALASKA

Though all regions have their dramatic stories of rich and hungry times, I have discovered that the stories of families and communities dependent on Pacific salmon fishing are much the same everywhere. After the 1920s, however, though all salmon stocks were over-exploited, boats fishing in Alaska had the advantage, as those stocks moving through held onto their numbers longer. Yet, by the 1950s they were at serious risk there too. For decades the Guggenheim Trust, Alaska Packers Association, and other outside corporations treated Alaska as a colony with the focus on "the silver hordes". Even before the canneries seized the inlets, salteries at such lonely places as Forrester and Coronation Islands were in the path of incredible runs of Chinook on their way to big rivers to the south. Companies brought their saltery crews to such remote spots to camp, fish, and salt the fish in barrels all summer. Those fabled Chinook runs off the islands are today a shadow of what they were, a possible warning of what we see now in the Alaska rivers' runs.

Alaska's cannery epoch began in the 1870s as companies moved north from rivers like the Columbia and soon replaced the salteries. By 1880 there were already about 40 canneries operating in Southeast. They processed mainly sockeye, a species that had the fat and flavor to can very well. The crews were at first mainly local Native people that found fishing and cannery jobs the way into the cash economy, but soon immigrants from Asia became the main work force. As David Arnold explains it, local people would leave for home when they felt they had the cash they needed, while imported Asian-Americans had to stick out the season. Eventually Alaska was home to hundreds of canneries from Metlakatla to Emmonak, with cannery-owned boats fishing for them.

In Southeast Alaska territorial days the canners' super-efficient traps were the dominant way to catch salmon, and with feeble management, trouble was coming for the fish and the fishermen. The Part I photo # 2 shows my father at age 18, 1918, knee-deep in salmon as he took care of that Icy Strait trap, an ordinary day for him. David Arnold (p. 101) reports that by 1927 there were as many as 800 such traps in Alaska waters, three-fourths of them in Southeast. The Indians worried that they were becoming a serious threat. Arnold describes (p.92) how leaders from the Alaska Native Brotherhood saw how equity for Natives in the Territory's economic life had to include, in attorney William Paul's words, " '…protection of small Indian fishermen from economic and environmental exploitation of large salmon canning companies.' " But the big canners dominated the political as well as economic life of Southeast.

Then, in the early 1880s another group more quietly began to arrive. The Tlingits seining and gillnetting in the spring at the mouths of rivers spied a curious sight of unfamiliar dories moving toward them. They turned out to be from far away Puget Sound. People who had survived the Russians and then the American corporate takeover were witnessing yet another invasion: independent small fishermen who had rowed and sailed all the way north to join the salmon feast. It was a four to six-week trip for the oarsmen, depending on where they dropped off to start their season. How did they find their way through all those endless forest-lined inlets? Perhaps they had first come as crew on a cannery ship, as my father did.

The major story of the West was almost always about Big beating out Small, but this oar/sail fleet of independent hook-and-line fishermen was one of the exceptions made up of idiosyncratics who figured out how Small could stay small and yet survive. There were enough of them that soon there were boatyards that specialized in trolling skiffs for BC and Alaska waters, leaving it to the fisherman to row them to remote BC or Alaska. Hardy folk like my oldest uncle, Conrad, who could build a troller like the *Frida* and then fish it, they were mainly transplanted Norwegians. Like the Tlingits, they had no fear of ocean waters. Some traveled back and forth each season, but others stayed north.

Before long there were small fleets of independent hand-trollers and gillnetters fishing salmon along with the Natives, selling to the canneries, carrying on the life they had known in the old country, but with the best fishing in the world. By 1910 they were joined by bigger boats using gas engines. One Slovenian seine fleet later came from my hometown Gig Harbor, Washington, where they were the town's elite. By the time my cousin George arrived in Southeast the salmon fleets had gas-powered engines, some even had diesel, and many had a two-way radio. A technology surge fueled by WWII surplus was well on its way.

The halibut and cod fleets, out of Seattle and Tacoma--bigger boats, tougher fishery, mainly Nordic crews like my uncles and cousins--would also join the bonanza off Alaska. The halibuters were on course to deplete their fishery as had already happened on the Atlantic when the International Pacific Halibut Commission was formed in the 1920s, the first attempt at serious fish conservation on the coast. The herring reduction plants of Southeast, all owned by outside interests, fished out those stocks for their oil and for fertilizer manufacture until, by the time I arrived in Southeast, the herring plants we saw in the inlets were all falling down. The cod stocks somehow stayed healthy, perhaps because the American market favored Atlantic cod.

Throughout Alaska, though Native families often took cash jobs for the canneries, they continued to make subsistence the focus of their economy. The new immigrants joined in the subsistence activities, squatting where land and harbor were available, and in some cases pushing out Native families. Gradually the newcomers added other small-scale resource development: logging, fur farming, and gold mining. Where they could, they grew small gardens. When I saw Bergen, Norway, I understood why those immigrant families felt so at home in Alaska and BC.

Arnold (p. 134) estimates the numbers of non-Native fishermen in Alaska by 1935 at 1800, and by statehood at about 4000. Today's fishermen should enjoy his report that a complete hand-powered troller outfit, ready to fish, could be bought then for around $100 and an equipped gillnetter for $400 or so. Before long, every Euro-American group with a fishing culture had its fleet of salmon gillnetters, trollers, or seiners in the North, each with its network of support and its code of conduct. Coming from their various heritages of small fishing or farming, they adapted easily to the Alaskan waters, working directly for the canneries, or staying independent to compete with each other as well as the fish traps. Even with all the spacious Alaskan water, arguments over who controlled what fishing grounds developed. I'm told by old-timers that a few shotgun blasts overhead were usually enough to drive off competition. Fishermen had their ethnic clans, just as they had earlier at Astoria, just as the Tlingits had their fishing networks based on families and clans, and though today we might think the fleets' methods overly aggressive, it no doubt helped them squeeze out their corner of the fishing grounds.

For decades the big canners like Alaska Packers Association virtually ran Southeast Alaska, and much of the west coast. I caught the tail end of the canners' glory days on my first job in the 1950s as a newly married college dropout at nineteen. I was lucky to experience the best of it at the unionized CRPA salmon cannery at Bellingham, Washington, sliming and stuffing sockeye into cans moving down the can-line. It was clean, modern, and at $1.50 an hour plus time-and-a-half for overtime I felt I had joined the working wealthy. I lost that status when the cannery closed in 1956. I am still in touch with one of my cannery pals from that era and we marvel at the innocent kids we were, with no knowledge of the boom economy salmon canneries had created along the west coast and most of coastal Alaska. We were oblivious of how the building of cannery towns helped encourage a lumber industry, then gold mining and fur farming, to open up Southeast Alaska to more waves of adventurous Euro-Americans. It was the "Western story" we all studied in school, just transferred north, but we cannery workers had never read a thing about it in our high school classes.

Competition between canners and fishermen in Alaska grew steamier as each spring more and more independent boats geared up. The independents,

European and Indian, had good reason to hate the canners' traps. Arnold (p. 114-115) describes how the Indians tried repeatedly in court to get exclusive fishing rights in certain traditional locations but failed. The runs were already dwindling, and a well-situated trap could capture what a whole fleet of net boats could produce. Arnold (p. 97) comments that the battles were "a contest of the fundamental rights of fishermen in an 'open access' fishery." The independent fishermen complained that a fishing commons was traditionally and legally open to all, and that the canners were closing them off with their traps. But the canners argued that the independents were trying to close off the canners' rights in the commons and were able several times to win legislation in the companies' favor. The battle continues. The value of the fishing commons and who has a right to them resurfaces every few decades and is still a fervent topic today somewhere in the country and the world.

Fishery battles today take place in boardrooms and in court, but then people tended to take matters into their own hands. A class of rebellious people who saw the canners as robbers became themselves "fish pirates". The fish traps therefore needed full-time caretakers, which is why my father was hired for the winter near the Dundas Bay cannery and village in Icy Strait. Not far from them the east shore of Excursion Inlet was home to 21 traps that must have filled the entire shoreline. He was never attacked by fish pirates, or I'm sure I would have heard the story. But he described an odd form of synchronicity that evolved between the independent boats and the traps. A fishing boat would pull up to a trap at dark, and its skipper would bribe a lonely fish trap tender like my father to unload salmon from the trap onto the boat. The fisherman would then steam down the Strait to the cannery that owned the trap and sell them. No one, not even local judges, had much sympathy for the millionaire canners.

Unlike the independent trollers, many gillnetters and seiners stayed cannery employees, fishing company boats. But the largest Alaskan gillnet fishery was not in Southeast but far out on the western coast at Bristol Bay.

The huge shallow bay had strong currents and tides, was--and still is-- no placid scene, always crowded with boats, but so productive. Just as on the Columbia, the Alaskan gillnetters at first drifted with two-man row-sail dories, or used anchored setnets, and waited for a cannery tender to pick up their fish. The advantage for the fishermen was the patronage of the canners with direct employment or with boats they offered for lease and seasonal start-up loans. The disadvantage was that there could be no illusion of freedom with the cannery as boss, so there was always the dream of buying a boat and going independent. The canners discouraged this and by 1900, had maneuvered passage of a territorial law to limit the Bristol Bay boats not only to 32 feet but to sail power only. They claimed it was for the sake of fish conservation, but critics say it was more likely for control over the fishermen. You wouldn't want to stray too far with just sail-power in that typically rough bay. The sail-power restriction was finally undone in the 1950s, and the Bay fishermen embraced independence and technology as fast as they could afford them. Though the boats are still limited to 32 feet, fishermen get around the rule by building them ever beamier.

Sail power and a 32-feet limit may have helped preserve salmon, as the Bay runs do keep their robust health while many others are struggling, but more important, according to biologists, is the Bay's great stock diversity due to the many rivers running into it.

By 1930 the Southeast salmon runs, especially, were showing harvest pressure but the territory was still prime grounds for small-scale fishing as well as big industry and continued to draw people. I have talked to people whose parents came north during the Depression, nearly broke, and still were able to outfit a hand-troller with a gas engine, get credit for startup supplies, and, like George, head up the channel for Meyers Chuck. But overharvesting is always a threat wherever humans encounter impressive fish runs and was bound to become a problem in Alaska eventually. Fish traps already had done much damage on the lower Columbia and through interception of British Columbia's Frazer River runs. They were simply too efficient. Efforts to correct overfishing never got far until traps and other efficient gear had done their damage.

In 1935 gillnetters won a stormy political battle when the Washington legislature finally outlawed the traps, and a few years later Oregon outlawed traps, fish wheels, and horse seines. The Canadians at last began to harvest a better share of their Frazer fish. But the corporations weren't likely to give up the super-efficient traps. Their solution was to move more to where there was no state legislature: Alaska Territory. Alaska Packers and its cohorts would gain twenty more years of magnificent profits.

Fishermen's hatred of the traps drew an ever more sympathetic public, and when Alaska became a state in 1959, the public speedily by popular vote outlawed the roughly 240 traps still active. The new state's constitution declared the state's residents' priority right to natural resources such as fish, and using the principle of "sustained yield". Many of the state's salmon grounds were closed or severely restricted for a few years to rebuild the runs. But by the time I arrived in Juneau in 1962, and started walking the docks and gazing at boats, almost all of the grounds had been reopened, at least to trolling. Meanwhile the canners had discovered they didn't need the traps, as a new immigration of people to fish in Alaska had begun. The number of fishing boat licenses (all gears) in Alaska, all of them needing buyers to sell to, would surge from 6,000 in 1945 to 22,500 by 1970. Most of them were salmon boats. How did the stocks endure this? The Southeast salmon hordes by then were mainly travelers from the Columbia River, and they again were showing the pressure.

Ice Revolution

The arrival of ice technology a short time later changed the salmon drama once more. It made fishing and processing both even more profitable--one more time technology plays a major role in our fisheries drama. The new larger trollers could take on loads of crushed ice, keep their catch for days, and deliver direct to cold storages in towns for a better price than to a packer or local scow. Trollers produced the highest priced, best quality fresh or lightly cured salmon on the market, to be flown everywhere. They were the elite, but soon even the smallest troller would carry ice. Other commercial fisheries--halibut, herring, sablefish, and cod--all went through a similar ice revolution, and bigger or beamier boats for carrying more iced fish would became the norm. Gradually the packer boats were fewer as even seiners and then gillnetters iced. Most of the remaining remote, smaller canneries of BC and Alaska closed down, as canners moved their operations to centralized towns and eliminated the overhead of crew transport, housing, and meals.

In the *Alaska Atlas and Gazetteer* one can see from the number of place names marked "abandoned" just how much the state changed as the remote canneries closed. Still, several canneries operate in Bristol Bay today. The Dundas Bay cannery village in Icy Strait near where my father guarded the trap completely disappeared, but not far away Excursion Inlet cannery surprisingly re-opened recently. A few smaller places like Meyers Chuck and Pt. Baker are villages that survive only because a handful of people like their ambience and choose to endure the long, isolated winter. Most villages that depended on a packer boat coming by regularly to pick up fish were forced to close down when the packer stopped its service.

Fishing and cannery towns like Ketchikan, Cordova, Kodiak, Petersburg, and Prince Rupert survived the industry changes, but 2010 census figures tell us that rural flight is not yet over in Alaska. Today overfishing is far less likely, but foreign competition in the market, such as from salmon farms, ranks high. So do fishery management problems, as described in chapters further on. And there will always be the lure of bigger towns for youth. I am chagrined to read of another village school closing because that usually means the end of a place. I hope that there will be enough inflow of new families to keep the little places alive.

British Columbia's Evolution.

B.C.'s salmon industry history is similar to Southeast Alaska's, but even fewer of northern BC's tiny settlements survived the remote cannery closures. Today its northern region has one real city, Prince Rupert, and only a handful of smaller settlements. Eric Wickham, a life-long BC fisherman who grew up at Bamfield on the west coast of Vancouver Island, population then about 300, writes of how technology changed the life there. Between 1910 and 1950s Bamfield was a settlement similar to many along Vancouver

Island and the Inside Passage. Wickham's photos show a community--he says about half white, half Native--on a cleared-off hill carved from the spruce, scattered with cabins, gardens and chicken yards, a store, a post office, a school, and a weather station. The little harbor had a dock and floats full of small workboats, skiffs, and a fish-buying scow. In the days before ice, the fish dropped off at the scow were picked up by a packer boat every day and delivered to the cannery at Ucluelet across the inlet.

Wickham remembers Bamfield as a simple but semi-thriving community. Probably almost everyone was lower-income, and most young adults left for high school and the city lights. They would often return later, seeing it as a place to raise a family, far from urban problems. But still, they had to have some way to sustain themselves. Fishing communities' fate followed that of their fleets. When the packer boat stopped running, the options for survival were few. Sawmills were a solution for a time for bigger settlements. But for many little places the downslide of not just fishing but mining, fur farming, and finally timber markets left few commercial fishermen; others turned to hosting sport-fishing visitors for a living.

As fishing crew I made the trip up or down the Inside Passage many times. We would tie up at tiny places just to stretch our legs or wait out a storm, every one of them charming in the summer sun. Even in the summer rain I would think, what a great place to live, especially for raising young kids. To Canadian economists, the same as our own, such beautiful little hidey-holes are of no importance at all, but I would think social scientists would see their value, as their fading is part of the larger disappearance of American villages and small towns and the kind of living that is possible only in them. All along the coast, as in Alaska, the only B.C. locations still alive are those that were able to keep processing salmon for a profit, or turned to other northern industries like lumbering, mining, or recreational fishing.

The big canners and fleets have a good share of the blame for the earlier salmon runs' decimation all along the Pacific, but logging companies' destruction of the banks of salmon streams and through sawdust dumping gets a large share too. Cruising the Passage to and from Alaska we passed through miles of clear-cutting. What we didn't realize was how heavily it affected the salmon we chased, all headed for those streams to spawn. In the Northwest and California more destruction yet came from the federal government's western river dam systems and the introduction of thirsty irrigation projects, as described further on. The fishery managers' strategies for dealing with all this were mostly a failure, but we can't blame that entirely on the managers. Salmon were important, but for our industrial leaders and government the industrialization of the West was more so. How much power did state fish managers have to challenge that?

There is one salvaged piece to this tale of a disappeared society. Despite all of the changes in salmon processing, the salmon fleets themselves have kept their small-boat character. Salmon, because of their anadromous nature, don't adapt to industrialized factory fleets. No big trawler is likely to steam into an inlet or river mouth, even if it were legal, to sweep up a catch. High Seas drifting for salmon is internationally illegal, or at least everywhere strongly discouraged. A 58- foot seiner is quite change from its ancestors, but still, it is a third the size of a medium industrial trawler.

Chapter 5

Ekuk: A Cannery Village

My friend Esther is one who experienced life in a cannery town, Ekuk, on Bristol Bay, during the period when canneries were no longer at their peak but might still be the center of a lively community. Ekuk was a Southern Yupik village turned cannery town. Esther's family has used their commercial setnet site there for four generations. She was there beginning in the 1960s, but probably the experience wasn't much different from that in the 1930s. I imagine that the Ekuk cannery was typical of many in providing all that fishing families needed in order to hold a stable fleet and workers for the cannery. The family camp she describes is much like our own today, though with many, many more fish hauled in. Her mother and the children camped in a cabin near the beach all summer and carried out the fishing while her father ran a tender for the cannery. Here is how she recalls "cannery town" life in its heyday.

"When we were kids in the 1960s at Ekuk there were about twenty families fishing there on a stretch of beach about six to eight miles long. The main part of Ekuk where the cannery was located was like a little city to me then since I came from a small isolated village (150 population), Aleknagik. The cannery had two docks where 32-foot power-boats were launched and where scows brought in fish and unloaded to the cannery. Bunkhouses housed cannery workers and some of the fishermen, and there were four or six houses where important cannery people lived. A couple warehouses, a mechanics' building, a woodshop, and a huge laundry house with a store and an office completed the cannery buildings.

"As I recall, in the traditional village itself there were about a half-dozen year-round houses and a school. The school closed in 1974 because of a lack of families living there year-round. There were also cabins that people used during the short summer fishing season, and a tent village of about a half-dozen, with most of those families fishing and the women working in the cannery.

"After the school closed, with only two year-round families left, Ekuk became a cannery village just busy in summer. The women and children ran setnets from the shore while the men fished from boats. But my father owned a scow he used as a tender. The sites were traditional sites that were understood to belong to the families using them. They had winter homes up at Lake Aleknagik like my family, and others were from villages on the Nushagak River and from Dillingham, with a few coming up from the Lower-48. My mother remembered fishing that same setnet site as a child with her late grandfather that raised her, so we were the fourth generation to be fishing at Ekuk.

"We were catching and selling all species of salmon, beginning early June through the early part of August in this order: kings, reds, chums, pinks and silvers. The fishing was, and is, probably like nothing you've ever seen! You know that we have really big tides down there. At the beginning of the season, on the low, low tide of minus 3.4 in late May or early June, we would walk way out to attach our running lines to a couple of screw anchors placed in the gravel bottom. If they had survived the winter tides we'd just reuse them, otherwise we would have to replace them. As the currents were strong we had to use two screw anchors to keep our running lines secured for the whole season. We were required to put a buoy at the end of the running lines securing it to a screw anchor with a line about twenty feet long, The buoy had a pulley tied to it, and the 100-fathom running line ran through the pulley, then to the shore where it was tied to stakes above the high-water mark.

"On the shore we had a gasoline-operated winch tied to three or four stakes, and before we had 4x4 trucks, we used the winch to pull the net, didn't have to use a boat. We'd crank in the net onto the beach, pick the fish, pile them on tarps for the cannery truck to come down the beach and load them up, and then run the net back out on the pulley. The cannery provided all this equipment to the setnetters.

"A very early memory I have of fishing at Ekuk was one day when we went down to check the net and discovered it so full of fish it was white. Thank God for the rain that day that kept the fish in great condition, as it took the four of us about twelve hours to pick the net! We began as the tide started to recede and finished when it reached the high tide line again. What an education for us kids to work so hard so long, with just little breaks to re-energize! We always helped pick the net, but nothing like that day.

"But really Ekuk was a paradise for children, and it still is, it's not just work. As a child I would dream of going down there. The yearning would come in the spring and then I could not wait until May when we moved down. My children also had this same yearning and couldn't wait to get down to Ekuk. I believe it was because of the natural freedom--to play and enjoy the environment, and also to participate in the family commercial and subsistence fishing. There were lots of kids there. When we weren't working we swam in the muddy waters of Nushagak Bay, slid down the mud banks, got lost in the tall grass fields, and walked the long beaches searching for agates we could sell for a dollar a butter can to a jeweler that lived at Ekuk.

"In those days the cannery provided almost everything needed in the way of supplies and equipment to keep us setnetters happy and working. There was a store we could charge at, later a free laundromat, and the cannery brought drinking water to us. When it was a bad year they even took our big seasonal order on credit for the next year. My mother remembered a time earlier when they even provided the nets, and sometimes there were so many fish, and they needed to get them to the cannery but couldn't pick them fast enough. The cannery told them to just cut the web and go get new nets for the next set. Meanwhile, all the families were also doing the important work of putting up their own fish for the winter. We kids helped with this--we cut, brined, dried, and smoked and even canned them, using mainly cottonwood driftwood, and we salted the king heads.

"The cannery offered so many services to keep us fishing. There were three coffee breaks a day for the crew there, and we were welcome to walk over and join in: home-baked goodies and anything left over from breakfast, and at 3 PM--we just helped ourselves. The 9 PM break was the one not to miss with all of these things plus sandwich bread, meats, and cheese, plus anything left from dinner. These were fine times for socializing too. One year the cannery began to invite us to purchase meals at the mess hall. About once a week or so our family would go down and buy a meal ticket, about $9.00 for our whole family. Most people would wait until steak night. The meal was always topped off with ice cream--what a treat, since we had no refrigeration, and at home all we faced were meals of fish, fish, and more fish!

"The cannery also showed free weekly movies--like Walt Disney--in the upstairs of a warehouse. They also sponsored a big Fourth of July celebration. They would load up the jeep with bags of goodies--pop, candy, gum, an apple---and drive through the village tooting the horn and waving the American flag, hollering 'Happy Fourth of July!' handing out bags of goodies to the kids. Eventually this changed to just allowing the kids to pick up pop and ice cream and books from the store.

"As I said, the store had everything we could need. There was a mail center and a net loft at the store where we were able to purchase all our gear for fishing on the beach, and could order the nets we would need for the next season. The stockroom held every nut and bolt that you could use, and buoys, floating night lights, blocks, come-alongs, anchors, rock salt, and all types of oils to keep your machines running. There was also a space where we could store our gear over the winter. This went on until the cannery was sold to Ward Cove about twenty years ago.

"This tradition of the cannery taking care of everything went back in time. My mother remembered when setnetters could go down and have their complete dinner, free, after the cannery workers had eaten. The ownership of the cannery changed hands several times that I remember, but all the services stayed the same until the 1970s. Really, it's quite amazing to me now what a good arrangement it was for us. Did other canneries have similar services? I don't know; I only know about Ekuk."

Listening to Esther tell about this, I know that as a kid I would have loved a fish camp like hers, loved being part of an important work team with plenty of time-off. But in the late 1970s Ekuk changed just as things were changing throughout the fisheries. In Alaska it included a federal land claims settlement with Alaska Natives. The state received land as well. In 1975 came the salmon limited entry permits. The canneries also began to change their policies, simplifying their own operations, cutting back on services, and making life more complicated for setnet families. The cost and logistics of maintaining a camp made it into a life many families couldn't afford. During the 1970s the salmon were also in a cyclical downturn even at never-fail Bristol Bay. Esther's family operation was actually in the red a few years. She was in college and in order to cover her costs her dad told her she needed to get a job at the cannery.

"The hours were long, working in rotten, cold conditions, not like setnetting, but it kept me in college out in Colorado.

"Meanwhile we became aware that the state was now the owner of the beach land, and they said we had to lease the sites, and eventually we had to have them surveyed and pay for the surveys. The sites were about 450 ft apart. Right now a lease is between $300-400 a year. People want those valuable sites, but no one can come in and jump your site as long as you keep up your lease payments to the state.

"The miles of sites are all filled up now. The cabins we built for ourselves were originally on cannery land, but now we are supposed to pay a lease for that land too. We received our limited entry permit for the site in the late 1970s, and we also have to pay an annual fee for that. The permit price keeps going up until it's worth a lot now. But it's no use buying a permit worth many thousands if you don't have a site for it. All this adds up to a financial obligation people never had before. [Due to inflation in permits, by 2015 the state valued a setnet site at $41,000.]

"Gradually all the great services from the cannery that we were so fond of went away, until today it provides only ice for our fish, drinking water, and a laundromat. There's no store, no credit, no warehouse of supplies. There's no truck to come get the fish, or deliver ice. You have to provide everything else. But I still continue to fish, so every year I order my nets, and whenever I am in Anchorage I load up on all the supplies I'll need for the fishing season----everything, even drinking water--and bring it back on the jet. Then I barge my 4-wheel drive truck out to Ekuk, with all the supplies packed into it--that's $1000 round trip. I still don't have a generator for electricity because I like it that way.

"So Ekuk is a different place now. From twenty families, there are now about fifty families with about eighty sites. Today you need to have the resources up front for your season. The community is mixed now, with more white families. Professional people can afford to buy these sites and fish for a working vacation. It's much harder for poor families to get outfitted, and some have given it up. Many families do continue to setnet, but you can't live on that alone; you have to have another job during the winter.

"My partners now are the younger ones of our extended family. But no matter whether I make money or not, I think it's worth it to keep fishing, for me, but for my partners too. It's a family tradition and has become part of our culture. Next time I'm in Anchorage I'm sure I'll be stocking up again for the fishing season. Even if I don't fish, someone in my family will."

The cannery towns' glory days are over but canneries will exist as long as there is a market for canned fish. The commercial setnetting or beach seining life with a camp on a nearby beach such as Esther describes can still be a family-oriented activity. They are both physically demanding within the short, hectic openers, but the problems now, as she describes it, are the shortage of sites, land leasing costs, and the price of limited permits.

As for drift gillnetting, with its shallow rough waters and strong tides, and crowded fleets, everyone stressed to make the most of the short openers - that does not sound like a fishery for children aboard. Boats can drift into each other, and when the fish hit, boats can get overloaded and swamp, or be stalled with a simple mechanical problem, the tide turns, and they are grounded on the mud with a storm rising. Setnetting from a boat instead of the shore is also a different scene from Esther's; like drifting, it can be a hectic, even dangerous fishery. Leslie Fields describes in her starkly honest memoir of family setnetting on Kodiak, *Surviving the Island of Grace* how it would be hard to decide for the older kids what they should be doing. They were able to help, but should they even go out on the many days with rough seas? Yet I believe there are drift and setnet fisheries in more protected areas of Alaska that could be fine for children as long as they had a place to go ashore for the periods between openers.

Of all the marvelous salmon fisheries in Alaska, Bristol Bay is one that has produced steadily, survived the downturns whatever the cause, and is still a place for fishermen and processors to do well. The Bay has important lessons for us as to what makes successful fisheries: diversity in its wild fish stocks; no impacts from drought, or dams or agriculture, or industry; no hatcheries; a historically strongly regulated small-boat fleet; and State support to the fleets when they needed to retool to keep competitive in the market. All fishery managers are aware of these Bristol Bay qualities, but that doesn't mean they are easy to reproduce. Yet there is never a shortage of boats at the Bay with the going price for a limited drift gillnet permit $140,000. There is one place, Prince William Sound, where the price asked for a drift permit is higher--$212,000, with a seine permit similar. The Sound is home to huge runs of hatchery pinks.

I wish I'd had the chance to try setnetting when my children were small. That was my dream, a fishery for the whole family to be happy in. I found out they did exist, for those that could find the narrowing way.

Chapter 6

FISHING FAMILIES: KOTZEBUE SOUND, NORTON SOUND, KACHEMAK BAY

Personal stories from small fishermen in traditional settings that still exist is the best way to get a true feeling for the fishing life, why it continues to draw people, and what we lose if it disappears. When I was salmon trolling in the 1960-1970s I was sure that was the only way to fish. Now I've been setnetting for thirty years in a fishery that has been around for thousands of years, and this even better in many ways. But in the meantime I was out winter crabbing on the ice many times a few years back, and have been an observer of summer crabbing and skiff-fishing for halibut where observing was all I was good for. I even was a volunteer cook on a family-operated seiner for a week just to get a feeling for that fishery. These experiences have given me more insight into the satisfactions but also the adversities of small-scale fishing today.

Every summer I'm up on a bank above Norton Sound at a table cutting subsistence salmon for drying, and down on the beach below me a pack of little boys and girls is racing up and down the sand. Down the shoreline the same thing is happening at a dozen setnet camps, often run by grandparents, tending children while they catch and process subsistence salmon, just as Perry's grandmother did. I want to believe it's also possible for commercial fishing to survive, involving the whole family as subsistence fishing does. I have no plan to be commercial again; I just want to know that it's still possible.

Kotzebue Sound Setnetting

Perry's cousin Mida and her husband Enoch, retired commercial setnetters, have a story much like my cousin George's of starting from scratch, but with confidence that they could succeed. They began fishing as newlyweds and continued for forty years, only recently turning over the permit to their sons, who began fishing with them when they were about ten or twelve. Kotzebue Sound has two claims to fishery fame: it is the northernmost US commercial fishing port, and its chums are the best quality. For us that fishery is also enviable in that while our own chum runs fluctuate, Kotzebue's stay strong. That fleet does have one problem that remote fishermen often suffer--a poor price. Many years recently when the price hovered around .25/lb for their fat bright ocean fish they couldn't attract a buyer. When one finally appeared about 2008, not many fishermen wished to fight Arctic seas and burn $7/gallon outboard gas for such a price. Enoch and Mida fished just enough to keep their permit active. Then the recession hit and .25 looked better. Their sons filled the gas tanks and hauled out the nets again. Today the chum market is healthy, the fish buyers are back, and the price--tripled--would attract any fisherman.

When Enoch and Mida started gillnetting together in 1965, the fishery out of Kotzebue provided most of the cash seasonal income for the mainly Inupiaq residents. They are another example of how starting from scratch you could in those days become successful fishermen. Enoch says they would do it again if his health permitted and the price was fair. Instead they are actively subsistence hunting and fishing. Mida has retired from her city job. Enoch is recognized as an expert on many aspects of Northwest Alaska subsistence, and earns a winter salary in that field, plans to retire every year, then changes his mind. He recently co-authored a book on the Inupiaq whitefish harvest of the region.

Enoch, like many of his neighbors, recalls starting out young as a commercial fisherman:

"I started fishing as crew on Snyder's boat when I was about seventeen. Then, the year I was nineteen Mida and I got married. I built a small house for us the next year, and then I built a nineteen-foot skiff and got an eighty horsepower outboard for us to go gillnetting. It was an open skiff, no cabin, like all the boats up here. They double as hunting boats. Mida and I started out together as partners and that's the way we continued. Some guys said to me that I ought to get a man to fish with me, and I would be able to handle more, but I thought no. We are too good a team, always know exactly what the other one wants done.

"We aren't allowed to drift up here, so we usually anchored both ends out, in fact sometimes the current was so strong we had to use three or four anchors, but we used light ones in case we had to pull them when a storm was coming. Sometimes we did tie off to the beach; it depends on where you think the fish are going to be. I always take time to try to think like a salmon. Mida and I were high boat eleven times and with my sons with me we were high boat two more times. Once when we didn't feel like fishing three shackles [nine hundred feet] we fished only two and we were second highest boat that day.

"When we started having children---we have six boys and an adopted girl---either my mother or my adopted mother would babysit, so Mida could keep on fishing. We only fished that first boat for a year because we discovered it was too small. You're allowed 300 feet of net, three shackles, and with that boat we could manage the fish from only one or two shackles. There was money to be made out there, a busy season from July 10 through August. So I bought a twenty-one foot boat and we fished that one four years.

"This is a chum salmon fishery. Our chums are superior to the ones farther south, because ours have more oil and more flavor, so they can up nicely, or nowadays, freeze. Quality fish. The old timers remember a commercial king salmon fishery in the Sound, but it was open seven days a week, twenty-four hours, and they fished them out. Now we get just a few kings, but we hope they may be building back up. We do get some coho, too, but our commercial fish is chum. At its peak Kotzebue Sound had over two hundred thirty active licenses, all fishing out of open skiffs like ours. Then for a while with the prices so bad, only ten to twenty boats were fishing seriously, and the others fishing a little to keep up their permits and hoping for a better market.

"The prices made sense when we started out in the 1970s and for years. At peak we got 1.85/ lb. (in the round), but we also got .50/lb. for the roe. Chum roe is considered high quality. The people from the villages were often out of money by spring, and they couldn't always wait for us to negotiate that price, but it could go that high. Back then the fish all went to Japan. We delivered to a processor that was anchored out in the Sound. But when we chased out the foreign processors in the late 1980s, we had to rely on an American buyer coming in. Then for years the price was nothing: .22-.25/ lb. No one was interested. Then we went some years with no buyer. Now it's great again [2014]--three buyers last summer.

"After that first four years Mida and I decided we needed a much bigger boat. We could carry more fish, but also get out in rougher weather, which you have to do on the Sound if you want to make money. We designed it ourselves, thirty-two feet, thirty-eight inch high side, with a higher bow. I used 3/8ths plywood, with two layers of fiberglass on the inside and three on the outside. We powered it with two ninety-horse outboards. You sure don't want to be out there with only one engine. The boat turned out just right. We could fish rough water and could carry 1700-1900 fish.

"I remember one time especially, we were going out to pick our net and it was rougher than usual. It was so rough, really bad, that Mida didn't want me to take my hand off the steering at all---she wanted that bow pointed into the seas, and she said she would just haul the net herself--we have never had a reel. There was no way we wanted to stay out there to pick the net, so she just pulled in that whole net, fish and all into the boat, while I kept us steady into the seas. Then we took it in to calmer water and let the net back out, picking as we went. She had hauled in 730 fish in that net, big chums, by herself. I won't forget that day and how she hauled those fish in--she's not a big woman. Another time our three shackles were so plugged we knew we couldn't pick them all before the first fish would get soft, so we just got seven boats to help us and told them to keep their fish they picked. We just wanted to get our net cleared. That's how it could be in those days.

"One reason we needed that bigger boat was that some guys got the idea through to push the line farther out into the Sound and have an earlier interception. That worked out for the bigger boats like ours that could take the rougher seas out there. But when the line was at Cape Blossom, it was better for the smaller boats. The fish moving farther into the Sound were by then fanned out, going to their different streams, so the boats were spread out more in calmer water. Everyone could get some. It was a bad idea for the smaller boats to move the line out, and some of them gave up. And it was more competitive out on the new line, as well as rougher water. You had to anchor your net 100 feet from the next net, everyone trying to get onto the school as it passed by. I remember there was one guy that thought he owned his spot, though of course he didn't. He got into it with me one time when I beat him out there and set there. He threatened me with murder, but I gave it right back to him, told him they'd never find his body. He backed down. He was the only one that bad."

"That good boat of ours got ruined one year. The fleet was out picking their nets in rough seas and some of them started to swamp. Mida and me were not in trouble, just picking our net, when a boat near us all of a sudden headed right for us and rammed us, cracking our boat. They were sinking and maybe panicked, figured that was the way they could be sure to get aboard ours before they went down. So they were okay, but the crack in our boat was too big, and we thought it would weaken the structure of the boat too much. Pretty soon we scrapped it, and bought another one.

"The bottom dropped out of the chum market later because of the farmed fish. In 2004 chums were down to .15/lb. The value of the Kotzebue Sound harvest went from $2.1 million in 1985 to $65,000 in 2004. No buyer came for years then, so even if people wanted to put out a net, their fish weren't going far. But finally by 2010 we've had buyers again. They pack the fish and ship them by air to Anchorage, and from there I guess they go to Japan. The boats seriously fishing are getting bigger numbers than we ever did. They even had to have short openers to handle all the fish. A retired buyer earlier told me that with our quality of chum, so nice and fat, we should be getting a far better price, and now finally we are."

I asked Enoch how much their kids were involved with fishing. He said fishing had worked out well for them too.

"Our six boys started fishing with us when they were about nine or ten, and by twelve they were the best help. Were they fast! They're grown up now and some of them fish our boat. Other captains try to get our boys as crew because they know all about fishing.

"All the years I've also worked at other jobs in the winter. For twenty-five years I did maintenance for the schools, and that worked out perfect since I had my summers free. If the schools needed something fixed right away, I could do it during a fishing closure. Then for ten years I was the subsistence coordinator for Maniilaq, our regional Native non-profit. That's where my heart is--protecting subsistence. Two of my sons especially like commercial fishing and they've used my permit. But Mida and I fished together for thirty years, and she knows as much as I do about fishing.

"For years it was hard for our fleet. What could we do with the farmed fish competition? An inferior fish, but cheap to produce. When the virus was at its peak and fewer farmed fish on the market, the price for us went back up. The most important thing the state of Alaska can do for its fishermen is to promote every way it can the superior Alaska wild salmon. And I know it's doing this, but it needs to do even more. When the virus fades, I know our price will drop again. But in 2014 we had a fantastic run of big chum salmon, so good I went fishing again. Three buyers showed up--competition! And that drove the price right up--almost four times what it was in the worst years. Everyone was back out fishing again, including me. Our only problem then is sometimes the buyers can't handle all the fish and they have to tell us to stop. In 2014 the chums flooded the Kobuk River so heavy they ran out of oxygen the scientists think, and hundreds died in the shallows before they could spawn. Why such a big run up this way? We don't know."

More Gillnetters Speak

Recently I found books on the setnetting life written from the point of view of women that reminded me of Mida, a small woman, pulling in over 700 fish. The scenes from *Fish Camp* by Nancy Lord on Cook Inlet, and the books by Leslie Fields ring true. However, reading of Field's many years setnetting, first the years without children and eventually with six, my view of commercial setnetting got more realistic. How could anyone fishing as hard as she and her whole extended family have energy left for games with kids? Then, with the advent of farmed salmon, her family was forced to fish each year harder. They had to put more boats and nets in the water, work faster to pick before the closures, and take more chances with weather. Being part of the shore-side crew was no vacation either, with huge meals to prepare every few hours, miles of net to mend, tubs of fishy clothes to scrub and hang. Wintertime far from camp was a complete and welcome break. But Fields, and Lord too, chose to keep fishing each summer even though they didn't have to. I understand.

Frank Kavairlook, another setnetter, my son Frank's father-in-law, is now a crab crewman out of Nome, in the summer on the *Mithril* and in the winter on the ice. He spent most of his fishing life, starting at 14, gillnetting salmon from his home community of Koyuk. A couple years ago he had to have a knee replacement, but the next winter was out crabbing again, and went on to crab in the summer for the third year, as he knows he can't make a living commercial salmon fishing anymore. Frank is an example of how for decades people of western Alaska combined commercial with subsistence fishing as a livelihood. Now over sixty, he says:

"I'm glad to be crabbing, it can be good money, but I miss salmon fishing. What I really miss is going down to the Iglutalik River and smoking king strips, but I can't work it in right now. I used to take my wife and kids down there camping every summer to catch subsistence kings. I took my daughter Cherilyn one year when a bunch of us went up to Cape Darby and camped out to fish commercially for herring roe. Right now one of my sons is using my commercial salmon permit, and one is commercial crabbing, but mainly my kids have done subsistence fishing. Last summer my wife was able to go back to our Iglutalik camp for a couple days and she sure enjoyed that.

"I've always been involved with fishing. Right out of high school I started working for Fish and Game on some research they were doing on salmon, and a couple times I worked on processors in the region. Salmon fishing really got going strong in Norton Sound in the 1980s, with Moses Point area the big producer. Earlier, starting in the 1950s, people were fishing with 18-foot boats, and a buyer, Bill Bodie, brought in a processor and some 24-foot wooden boats with 40-horse motors. People could pay for them out of their fish sales. Commercial gillnetting really picked up then. People from Elim, Koyuk, Unalakleet, and Shaktoolik all were fishing. There was always a buyer of some kind. Usually there was a floating processor, either American or Japanese, and tenders to pick up the fish. If no floating processor showed up to pick up the fish, someone would buy them and air them out. The price was about 25 cents, in that range, for silvers.

"There was a processing [freezing] plant at Moses Pt. in the 1980s and there is one operating there again now. NSEDC [the regional Community Development Quota non-profit] has tenders that take the fish there or to another NSEDC plant at Unalakleet. But this year Moses Point wasn't the place with the best coho harvest. Norton Bay, where I'm from, with only thirteen permits, beat it--I don't know why.

"Later a man named Tweeto and some others began building aluminum skiffs at Unalakleet, and people came in from other villages and learned to build them. They use them now for both salmon and herring. But lately our salmon runs have had problems. Cohos used to always be strong in Norton Sound. And a few rivers were good for kings too, but not lately. Now there has been no commercial season for kings for years, though we can still fish them for subsistence. But the kings are also much smaller. You don't see the great big ones as high as my shoulder like we used to. I don't know what's causing them to get smaller. The scientists and the boards and councils should spend more time talking to the people with traditional knowledge. I don't know if it's for sure, but the trawlers' salmon bycatch is probably part of it. It also seems like the Council [North Pacific Council] and the Fish Board don't listen to subsistence issues. I don't like to see the salmon harvest in trouble from whatever cause. I am now on the aquaculture planning group for Norton Sound."

Like many people in the region, Frank Kavairlook is most concerned about the youth, the experiences they are missing about how to survive in the future.

"One thing I mentioned at the State Food Policy meeting last week was that we need to have technical courses at the high schools that include subsistence education regularly, like reindeer herding and food-growing. The kids need to know how to prepare subsistence dry fish from start to finish--the whole process. I believe they have camps like that up at Kotzebue. And the greenhouses out at the high school need to be used more for teaching food-growing. The kids learn from hands-on things like that and they are proud of what they produce. The youth are losing that knowledge they need to have, and there are people with that knowledge that needs to be passed on."

Kachemak Bay Seining With Kids

Seining, like setnetting, is a type of salmon fishing that can work well for a family with children as soon as they can help with the work. A seine boat expects to carry a crew so there is more space for kids than on a gillnetter. Many Native families, in particular, probably are using the boat for subsistence fishing as well, and may have the whole family aboard part or all of the time. I have never seined, but about 1992 Perry and I and our youngest son Eli were aboard the seiner belonging to Perry's sister Virginia Wise and her family, making briefly a lively crew of nine. They fished for hatchery-raised sockeye in Kachemak Bay on the Kenai Peninsula. They'd hired our son Eli, sixteen, as their skiff man, while Perry and I came for one week just for the experience. The Wises were fishing in small inlets and bays off the main bay, perfect for a family operation. The *Silver Streak*, which they had built the year before, was forty-three feet, with nine bunks in the fo'c'sle, planned deliberately for a family. Nate, Virginia's husband, and his brother had the fiberglass hull built locally and then finished the rest themselves.

The brothers had grown up in a family fishing operation. Their dad bought their first boat, the *Icelander*, a seiner less than 30 feet, back in 1963. The plan was that Nate's folks and the older kids would fish while the grandmother would do child-care for the younger ones at a rented cabin at Tukta Bay. The crew would come in at night. Nate tells what really happened.

"Our grandmother wanted to be on the boat herself, so many days she got up early enough, and had the toddlers ready to go, so that the fishing crew could hardly refuse to take everyone along. That meant eight people aboard, five bunks. We kids slept on the floor sometimes and got wet as the boat was always leaking a little. They fished the calm bays then, and made expenses. Later in the summer, without the toddlers, they would go out to the more challenging drags: Port Dick, Rocky Bay, Windy Bay, and could make enough to carry them through the winter.

"As children we thought the whole experience was great, and when our dad bought the *Kachemak Lady* in 1970, a boat of 36 feet, later lengthened to 40, we had the perfect boat for a family, with a big cabin, and a hull built for the hand-seining that was still the rule in Kachemak. Eventually my brother John and I were the ones most often fishing, later with wives, and often our youngest sister. In those years it was still possible for us two couples to make most of a year's living seining salmon, though some years I also went as a deckhand in the spring on a herring boat."

In 1975 the state had introduced enhanced red salmon into the bay from a government sponsored hatchery near Seward. The juveniles were dropped from an airplane into lakes above China Poot Bay and came down the creek. Years later they returned to the seine fleet waiting in the bay. Another hatchery produced pinks. The future looked good for salmon fishing for hatchery fish. When Nate and Virginia had baby Jessica, they assumed she would be going along, that the family fishing adventure would continue for another generation. But then came the *Exxon Valdez* oil spill disaster. The *Kachemak Lady*, hired out to do boom watch, hit something while running down the bay that punched a big hole in the hull. Nate's brother, skippering alone, made it to the beach, but the boat was a total loss.

"We were determined to keep fishing and managed to borrow the money to build and outfit the *Silver Streak*. Then came years when there was money in seining. Sometimes we used the boat for crab fishing too, but mainly it was a continuation of the family enterprise we had grown up with. Later in the season, as we did before, John and I would take the boat and the older kids out to Fort Dick to make serious money. But for Virginia and me it was more important to keep the family together than to make huge money. Like my own parents did, I chose places to fish where it wouldn't be disagreeable or dangerous for the kids. I wanted them to have fun, not like some families I knew that took their families out to places more lucrative to fish, but not really where you'd want your kids to be. Some of those kids, as soon as they had a chance to get off their parents' boats, they were seen no more. My idea was, I'd rather work more in the winter so we could have a better time fishing as a family, and it worked out that way."

The week that Perry and I were on board the *Silver Streak* crew consisted of Nate and Virginia, his brother, three children, and Eli. Perry and I took over the remaining two bunks. I helped with the meals, spelling the other women who would be at it all season. The Wises' style of seining was different from what I had seen in Southeast. The family anchored the *Silver Streak* in the chosen inlet to use as a home base. They used a jitney, a small fast boat with a spotter platform like a crow's nest. The jitney driver perched up there searching for a school and a good place for a set. When he thought he had a fair school located, he shot out to make the circle and the waiting skiff man circled the other way to make their purse. When they brought the net over to the big boat, the crew drew the net in by hand over a block. Nate prefers this to a power seine.

"It's the old way of doing it, more labor but more flexible, and we can get into small shallow places. I'm a hold-out. It used to be we kept the big boats out of the bay with a regulation for hand-seines only. Now that's not the case, and I am one of the few that still do it. I just like it."

I thought hand-seining could never work in the rough seas such as longliners, gillnetters, and trollers often work in, yet that was how seining started, and made a lot of sense as they used it. They certainly burned a lot less fuel with the mother boat being anchored most of the time. Younger kids aboard the big boat could be out of the way during the last stage. The weather was warm so the Wises kept the fish fresh by icing. Even so, it was quantity they were after, and the hatchery run was poor that year. We didn't get a true impression of the rigorous, long hours that seining can be. My trolling kids would have been happy indeed to fish those calm waters, and to be able to go off in the skiff every night to play along the shore. The gnats were the worst suffering those seining kids had to endure. The Wises admit they could have an occasional rough trip going across Kachemak Bay, but for them the misery of seasick kids was soon over.

The Kachemak runs got poorer but the price did as well. The Wises couldn't afford to fish every year, much as they wanted to. To take the *Silver Streak* away from the dock meant thousands up front just for the insurance and the fuel costs. Virginia and Nate regret that their youngest child missed that family fishing.

Now with small grandchildren, the Wises decided a few years ago to lease their boat and permit to their oldest son Jake, who by now had been a crewman for others for years. I am always glad to hear of a child that grew up fishing and liked it enough to stay with it. I asked Virginia for her view, in retrospect, of family seining. Her answer comes quickly: "We always enjoyed being on the boat with the kids along. They learn they are productive and valuable. And I loved fishing! Except for the financial side, and that shouldn't be everything, it was positive in every way."

In 2014 Jake was still skippering the *Silver Streak* and had a good year, seining salmon from June through August with his younger brother and a cousin part of the crew. In the winter he crews on a groundfish boat. But many years due to the farmed fish competition, seining, overall, hasn't done well financially. The years when salmon prices dropped so low were hard on seiners with their large expenses. Alaskan fleets voted several times for voluntary buybacks, and hundreds of licenses left the fleet permanently. That saddled the remaining boats with big annual loan pay-offs to the government. What has saved them is the far better marketing that both the state and regional fishing associations, like "Copper River Salmon", have invested in, such as the frozen pan-ready salmon fillets, a good outcome from what was a bad development. The fish farms won't go away, but neither will the more sophisticated marketing, so seining in Alaska still has a future.

Chapter 7

Northwest Trolling: Everyman's Grounds

I need to go back to the 1950s and pick up my cousin George's history. The stories of salmon fishing so far have told of hard times and good times, but George goes more into how fishing affected his family life and vice-versa, and even more into how larger changes in the economy influenced changes for the salmon fleets.

I left him headed down the Inside Passage for home after his Alaskan summer, thinking he would be back the next spring with the rest of the northbound fleet. But it was not to be, and from here on his story is one mainly of hard work and stubborn determination that he would succeed as a fisherman. He had no idea how success would ebb and flow, not just with fish runs and shifting management, but with shifting personal life too.

"I never got back up to Alaska. I trolled off Washington, and later off Oregon from then on. I was married soon after I got back from Alaska, and Farrell and I had a baby to take care of. That changed everything. I needed to stay close to home and had to look for something to do in the winter, since I had no other skills but fishing. But I knew I needed a bigger boat with power gurdies if I wanted to make a real living. I was able to sell the skiff and bought the *Pup,* a 24-footer with a four-cycle Atlas engine, and got it rigged with power gurdies, and I also enrolled in college, thinking I'd have better chances with two plans. The next summer I was back at La Push, with my wife and baby camped in a little trailer in the sand dunes just outside the harbor."

Though fishing as a livelihood was new for his extended family, the rest of George's young adulthood was typical for our times. We left home early, got married early, had children early and found some kind of work early. At eighteen--I was just a year older than George--I too was married, working in the cannery, waitressing in the winter, starting a family, and trying to take college courses at the same time. For George the romance of fishing faded fast to the same hard facts.

"I had a tougher time than I expected learning how to power-troll on my own. I couldn't afford a crew, and the *Pup* was too small for another person anyway. And that boat turned out to be a submarine; it couldn't ride the waves. And there was still no way I could make a living for us just by fishing, so I dropped out of college. I had a wife and three kids to support. For three years I fished summers and pumped gas and set chokers [logged] in the winter. Setting chokers was more dangerous than fishing, I think. There was a big turnover--many guys got hurt, but I was lucky. And I saved enough to trade in the *Pup* on a larger troller yet, a thirty-two footer, the *Grampus*.

"I was working my way up, and finally I got back to where I could enroll in school in the winters again. I stayed with that to finally earn a degree in English literature. [He laughs at this.] I never did use that degree for a job, but it did help me learn to love to read, and that turned out to be a good thing! Trollers have too damn much time on their hands on stormy days or dragging up and down between bites. I can recommend a lot of good books to you, even though I never did give you the most helpful advice on fishing--which probably should have been to just stay out of it. But for quite a while in the sixties salmon fishing was good.

"Then, when I'd almost reached that mythical success I'd dreamed of, I found out Farrell was tired of being married to a fisherman. We had drifted apart. It was give up fishing or give up her. But I couldn't give up fishing--how could I? I had no idea how to make a living from an English degree. I had to fish I thought. So, she said, divorce.

"You know, there is this romance about fishing you start out with, the idea of being close to the elements and the good honest life. You dive into it, put all your time and resources and energy into it, and then come the hard realities. But for a time, fishing seems quite possible. You get trapped in it. A lot of marriages have broken up over it."

A problem all too familiar to commercial fishermen had caught up with him. Fishermen have many critics, often people working at respectable 40-hour week jobs they may hate but feel they can't leave. Often their critics are the partners left on shore, who in retaliation abandon the fishermen. Some women adjust to having a husband gone such long periods, but George's young wife, with her three little kids to care for, could not. He did take his children fishing when they were older, and the eight or ten years that one or another trolled with him he says were a pleasure for him, but none of them chose to stay with fishing. For his family, that livelihood began and ended with him. For many people, however, the children do want to follow the tradition. They can't imagine a better occupation, a better life. And no doubt for many it is as George found it--a way of life he both loved and hated, and felt he had no better choices.

Modernized Fleets

By 1957 there were 866 active Washington troll licenses, up from about 90 in 1950, a good many of them fishing Alaskan waters. The surge in modernization included larger boats with most of the comforts including living areas and galleys that were modern kitchens, soon even with refrigerators. Men could invite their wives and kids to come along and believe they would stick out more than one trip. Southeast Alaska's protected waters, especially, were ideal for a family operation. By the mid-1960s the majority of the troll fleet were "ice boats". The new electronics boosted production greatly. With a Loran the toughest of trollers could fish the newly discovered Fairweather Grounds, 40 to 60 miles out in the Gulf of Alaska, load up their boats with the Chinook passing through toward the Northwest, and live well all winter. It had its price. If a gale came in, the only anchorage to run into was Lituya Bay, a spooky place I remember well, where you could enter only at a short slack water no matter how wild the storm outside. Those macho Fairweather highliners became the culture heroes of the fleet.

For a time, though, even shore trollers were doing well. The markets in New York and Europe offered .55 to .60/lb for red Chinook, a good price then. Much of that catch was lightly brined for the gourmet treat, lox. Troll coho brought a poorer price in comparison, from .17 to .26 lb., but made up for it in quantity. Meanwhile the same expansion went on with the gillnetters and seiners, and for other fleets too. The number of commercial licenses, including all fleets along the coast, kept climbing.

Today, looking at the restrictions trollers face, it is hard to believe the freedom they had then: open access, and in Alaska where I ended up fishing no fleet quotas, very few closed areas, and a long season from mid-April to mid-September. A minimum size on Chinook was the only real restriction for us, and we could keep and sell our incidental halibut and groundfish. Compared to other fisheries trolling also required fewer resources to start and attracted people from many backgrounds. Compared to other fisheries it was safe. In the ten years I fished starting in 1963, there was no one lost in the troll fleet working Alaska waters, not even on the Fairweather grounds.

Other work was disappearing in the Northwest, and on the dock you would meet former loggers and sawmill workers, along with small farmers, construction workers, retired truck drivers, teachers on summer leave, and small businessmen who loved sport fishing and thought they could combine two interests. Many older trollers were on social security, some of them with disabilities that would have kept them from regular employment, but who were able to handle trolling's independent hours. I knew an older trolling/shrimping couple out of Petersburg, the husband in a wheelchair but able to carry his weight fishing.

By the late 1960s, women who wanted non-traditional work previously denied them came aboard, along with college students needing summer earnings. This openness was socially positive, but with such a diverse cultural mix it was hard for trollers to organize themselves against adversaries. And adversaries would appear. Though I didn't know any gillnetters or seiners until later, from reading their own accounts it's clear those fleets were expanding in much the same way, for the same reasons, but they often had close-knit traditional communities and were better able to organize into associations to defend their profession.

George himself was still fishing a very modest boat, so upgrading again was in order.

"When I was 24, 1961, I bought the *Rocket*, an Alaskan troller. It was 37 feet, and I installed a Volvo Penta diesel, the engine of choice then. The *Rocket* was a great sea boat. It had too small a hold for the Fairweather, though it had fished out there, and it had a tiny cabin, but it was plenty for what I needed. I barely invested in all the new electronic gear, just the basics. I never did bother with radar. I believe some people have special senses, like I could sense another boat coming in the fog and would turn aside without knowing why. Other boats told me they could watch me on radar doing that. And I could sense where I had caught some fish and hunt them down again. With the *Rocket* I was finally making a whole living off fishing, and at the time I thought of myself as a self-made success story. I realize now just how much help I'd had along the way, starting with the little rowboat my dad made me, and then the commercial skiff he rigged up for me. And Steve Penn, at La Push, didn't have to let me be his gillnetting partner on the Quileute River after my first disappointing season. He didn't have to give me a cabin to stay in. I didn't even realize until later that all those hatchery fish we caught were a project of the government to help fishermen and build coastal economy. We'd have starved without them, but then I took hatchery salmon for granted.

"Helping me with my success, too, was that I found a woman to be my partner on the *Rocket*, Karlene. She was a fine boat-puller, probably the best crew I ever had. Women have all it takes to be excellent trollers. Later I learned what a frustration it could be when you couldn't get a good boat-puller. But that was later. During the 1960s the beaches were crowded with young people wanting to crew on boats. It was part of the natural life style everyone wanted then, and some bought small trollers. A few of them are still fishing--I talk with them every now and then."

I commented that I had often heard younger fishermen who were just starting out speak of him with admiration, that to them he was a highliner. George shrugged it off.

"Oh, they were friends, young people who had gotten into fishing after I did, so I enjoyed talking to them about boats and myths about fishing.

They hadn't seen the real highliners--they never came into LaPush. But all the little shore boats were making some kind of living for a time. Well, it's true, I did well with the *Rocket* and was top boat at La Push sometimes, but that was mainly because I put out a lot of effort. I never thought of myself as a highliner. For every boat that you think is a highliner, he thinks there is someone better that is the real highliner that he has to somehow catch up to. They eat themselves up with this. The longer I fished, the more I saw this in fishermen. I had friends that couldn't do the scratch-scratch I did. They had to have the big trips or they suffered. Sometimes they wouldn't even go out if they thought the fishing was no good that week."

He saw me smiling at this and protested.

"Nancy, it was nothing mysterious, believe me. It was no secret gear or anything like that, though the *Rocket* did truly fish well. I just fished longer hours and rougher weather, so I was a high boat. If it got too rough to fish the westerlies some afternoons, I would pull the gear and drift and sleep for a while, which I could do on the *Rocket* with comfort, and then put the gear out again when it calmed down. That way, I was out there when others were still at the dock waiting. And until the mid-seventies I thought that there would always be plenty of salmon. We had the strong hatchery programs to guarantee it. You know, there would be good seasons and bad ones, but if they fished hard, even the smaller boats could make at least a good share of a living. I don't think any of us realized how good it was till we lost it.

"I remember it was in the 1970s that things began to go downhill for salmon trolling. I didn't see it then, but now I realize that there was a conscious effort, from the federal government, from NOAA, not just the states, to reduce the numbers of fishermen, and not just the salmon fishermen, eventually all gears, and so of course it affected the smaller boats, the smaller fishermen the most. When they closed off new permits in the 1970s it really hurt the younger people, the deckhands hoping to someday have their own small boats. Now no more, unless you were personally wealthy enough to buy an existing permit, or got one handed down."

George is right that everything changed for American fishermen in the mid-1970s. Though it was the states that instituted limited permitting, even bigger changes would come from the federal level. The Magnuson Act (MSA) finally passed in 1976, but was on the planning table years before, affecting every fishery and fisherman. The bureaucracy that managed the fisheries became more complicated. NOAA's new National Marine Fisheries Service (NMFS) needed to work cooperatively with the U.S. Fish and Wildlife and state fishery managers. It had to coordinate its interests with other federal agencies: the Coast Guard, the Bonneville Power Administration, and even the Bureau of Land Management, due to river issues for salmon. With the enactment of the MSA, fishery politics became a snarl of agendas hard to untangle, much less come to agreement on. Small fishermen had strong currents to maneuver that they'd never had before, and it took time to learn them.

The Northwest had a particularly complex situation. The federal dam system kept growing, with its effect on salmon runs one area of dispute. Hydropower users that depended on the rivers--the aluminum industry, the irrigators, and the growing urban centers--saw commercial fishing as an adversary as it protested the effects of dams on migrating salmon. During the same period Native American "treaty tribes" in Washington won a victory in federal court that awarded them 50% of the commercial fishery harvests--more on this is in Chapter 11. And as described in Chapter 16, competitive sport fishing associations lobbied to reduce the size of commercial salmon fleets with some success. Canada had earlier determined to cut its British Columbia salmon fleets to half, and had introduced a vessel buy-back program focused on eliminating the smaller boats. That set a precedent for the US to take the same tack in the Northwest.

During this time the environmental movement shifted its focus to the Northwest's old growth forests and to the wild salmon runs that appeared to be shrinking. They weren't separate issues, as the logging methods in use had a heavy impact on salmon spawning areas. The environmental non-profits (ENGOs) took on the wild Chinook salmon as a symbol for Nature at its best that must be protected. Hatchery programs that the commercial fishermen depended on came under much more scrutiny, both for their costs and their effect on wild salmon stocks' health. Alaska became an easier place to fish salmon. Under pressure it had halted its plan for a big dam, it had no hatcheries as yet, and oil development in the Arctic was just starting to be an environmental issue. The salmon cycles were not alarming. But it was only a matter of time until Alaska would feel its own pressure on the stocks.

It is easy to understand why George says that the 1970s salmon fishermen started to feel under attack. The MSA recognized the need for fish conservation, but in the first years after its enactment little effort was made to address the negative effects of the growing industrialized fleets and other industry on salmon habitat or even the more obvious loss through unsellable salmon bycatch dumping by mid-level trawlers. After those issues were brought to government attention more strongly, little still was done in corrections. The regional fisheries councils struggled hard with their problems, especially those of declining stocks, but the industrial fleets had political power on their side. It was no simple challenge for the government to take on huge hydropower, agricultural, timber, and industrial fishing or recreational interests, and at the same time deal with the growing environmental interests.

It was especially easier to go after salmon fishermen and possible overfishing than to fight powerful federal agencies like the Army Corps of Engineers (the Corps) that operated the federal dams, and inevitably the weakly organized salmon fleets became the scapegoats for shrinking salmon runs. For years even many scientists were willing to blame run failures almost completely on overfishing. But the salmon fleets were important to the West's coastal economies, which meant that state politicians did not always see eye-to-eye with federal or even state fishery managers' positions. George recalls that then, in the middle of so much economic and political stress, a piece of even more disturbing news came to him and his partner.

"Karlene saw a newspaper story when we were unloading a trip at Garibaldi that some large US corporations had gotten franchises to start ocean salmon farming. She said, 'That's going to be the end of salmon fishing!' Well, she was right, though it turned out the end for us didn't come as soon as we thought. But you can see how farmed salmon affects our market now."

Yet Northwest fishermen didn't put up much of a struggle as their restrictions increased, and then increased more, and more. They didn't have the political connections that the salmon themselves had, or that dam operators, or BC fish farmers had. With the exception of the Indian Treaty tribes that fished, they had no federal law on their side to help them to survive what was coming, no movement demanding that both fish and fishermen be cared for. George's take on the stormy events of the 1970s, that everything started downhill for them during that time, probably reflects what many commercial fishermen thought in retrospect. But he was not a meeting goer by nature; few of the trollers were. The fishery planning coalitions and formal consortiums that formed seemed for years to include everyone except the small commercial fishermen. Whether or not that was their choice, they weren't at the conference tables often.

Obviously it is hard for fishermen to attend meetings during their working season, but my own observation was that they had to get angry to get organized, and the changes coming in salmon fishing restrictions and hatchery policies came so gradually through the 1970s and 1980s that it would have been difficult to get a fishermen's protest movement launched. Karlene was prescient, more than anyone knew, but salmon fishermen generally just kept fishing.

Chapter 8

MORE OF FISHING WITH FAMILIES

When I arrived in Juneau in 1962 the Southeast salmon fleets were in transition. I had three young children in tow and no longer a husband, but I had always loved boating so I ended up spending time in Juneau boat harbor, hauling the children along. I saw trollers, gillnetters, seiners, in every state of repair and modernization. The beam-trawlers and halibut boats were still big old wooden boats operated by extended families. About half of the salmon fleet and many of the halibuters traveled up from the Northwest to fish each summer. I had landed in Juneau in a roundabout way, but my father's urging to head to Alaska had a lot to do with it, that and finding work.

The job I'd landed in Juneau was a humdrum one in a library, and of course before long my goal was to buy a boat and be out on the water. But I soon saw the only way to afford serious boating was to go commercial fishing. How did you manage that with three little kids? Yet a family venture was no new idea; farming families have been working as units since prehistory. In the early 1960s there were thousands of counter-culture people experimenting with this idea, including some in commercial fishing. But for the majority of US fleets the women with small children stayed ashore, the men went off fishing, and families endured the problems it brought.

My new partner, Jack, agreed we should both flee the shore life. We saved hard all winter, he got a bank loan, we pooled our savings, and we bought a Juneau troller the next spring. We made a good choice in the 43 ft. diesel-powered, well-built *Deanna Marie* for the money we pooled and the fish we thought were out there waiting. Alas, though we had no way to know it, we went fishing just as one of the natural salmon cycles was peaking and starting down. It would have made sense to talk to George before offering that down payment, but I had no idea how to reach him, and would I have listened anyway? I wanted a boat, a real boat, not a rowboat, and trolling was not a crazy idea apparently, as the Juneau harbor was full of active trollers.

Jack and I spent several years and on two different boats trying to work out the combination of profitable fishing with kids aboard, not just for my three but his four. The family aboard, the kids taking turns, worked out somewhat, at least there was little complaining, but we rarely made much more than a bare winter's expenses. Boat maintenance needed any extra money we had. We survived the winters, actually quite well the kids would say, on old farms whose families had fled for the cities.

Forty-five years later I would read of the failure of the fall Chinook run on the Sacramento River, one of the biggest runs that California and Oregon trollers count on. The Klamath River run had crashed a short time before. I try to imagine how I would have felt if I'd heard such shocking news when I was about to jump into trolling. Would we have understood the implications for the whole coast? Even if we had, I doubt it would have caused us to back out of our plan. We were too energized by the trollers we visited with on the floats as they hustled up and down, readying everything for the spring Chinook season, full of optimism.

We were innocents when it came to the fisheries, and even after the years of fishing we embarked on, I was still bewildered by salmon and their management. Our radio didn't reach a land station with "news", and we had no way to get a real newspaper during the season. We went five or so months each year depending just on news at the dock, and that was usually news about where the fish were or were not, and what they were hitting on. George says it was almost the same with him. Today, if they wish to, fishermen can learn hourly all the world news, fisheries included.

My entry to the fleet was eight years after George's trip to Alaska, and trolling was still the poor man's entry into commercial fishing. Probably there were only two cheaper ways: hand-trolling and hand-jigging for halibut and bottom fish. A trolling start-up is certainly not cheap now, yet for every Alaska power troller for sale there is someone who will find the money for it and the required permit, and will work all winter if necessary for the privilege of short months on the water. As there are fewer fish and often weaker prices now--considering inflation--the new skipper will just fish harder.

Today the commercial salmon skipper could very well be a woman, for though trolling hours are long and the routine demanding, there is little of the brute physical effort that is part of winter crabbing or halibuting. There were only a few tasks on the boat that I couldn't perform, and I am no muscle-woman: I couldn't horse the forty-foot poles up and down, and I couldn't completely lift the heavy stabilizers that we dragged in the water when fishing in rough seas. I found out later that on smaller trollers with lighter gear, or with hydraulic help, small women could do these tasks too. I still know nothing about mechanics, except to ask basic troubleshooting questions, but today there are women fishing who can fix anything on a boat. Trolling also doesn't have to be a dangerous fishery that you wouldn't introduce children to.

I created conflict aboard in that I believed I had equal authority, as even though Jack was listed as boat owner, I had contributed equally in cost and soon, labor. I refused many days to recognize the old truism that there can be only one captain on a boat--and that it had to be Jack. So when the fishing was great we were too tired and happy to argue, but when it was poor, there was a lot of steam in the cabin. Do we fish over here or over there? Do we quit or keep going?

My extended family of elders couldn't understand why I wanted to go fishing or take such financial and physical risk, a woman with a degree, a mother of young children. And thought I was totally beyond comprehension later when we took kids aboard with us. The elders were polite, though, probably because my mother always defended my choices whether she agreed with them or not. But how did my mother really feel about me, her only child, and later her grandchildren, out on the ocean? She, and their father, took care of them during the fishing season and I know they enjoyed them. But she could have said no any time and she never did. I hope she be-lieved me, that shore trolling was a safe fishery and that she didn't just keep quiet because she believed in my rights to adventure.

It was years later that I read fisherman Leslie Field's book, *The Entangling Net*, based on interviews of women fishing Alaska waters, and found out the motives of those women were no different than mine and no different than men's. They, like me, loved the independence, the challenge, the sense of being close to nature, and the gamble that you could make good money. But some also talked of the chance to prove they could do something that had been off-bounds to women--I seriously doubt any female Norwegian ancestor of mine went commercial fishing. From personal experience as a waitress and schoolteacher, I knew that much of the work women do is as debilitating as fishing, just in other ways, and the stresses and insults of fish-ing were easier to recover from. However, I wasn't pulling fish all day, Jack was, leaving me to clean and often ice them, and because he wouldn't allow me to pull fish and take the chance of losing a single one, he soon suffered from a common tendinitis, "troller's elbow".

I had time to think, steering all day, and realized the quarters of the *Deanna Marie* were too small for more than two people, and I missed my children too much. By year three Jack and I were more than just fishing partners but domestic partners, too, with a new baby, making eight children between us. Yet I had no intention of staying ashore. We agreed we had to have a different boat so the kids could go, a few at a time. We sold the *Deanna Marie* and for months we were boatless and depressed, searching for a "family" boat, thinking of the next summer. That spring up in Petersburg Jack found a boat with quarters large enough for a crew of six. The *Nohusit* was a 45 ft. 1926 vintage boat that needed a lot of fixing-up but it was fish-able, and a family just as large as ours had fished it many years. We bought it in May, worked on it frantically, and off we went in June of 1966 for a protected drag that we thought would be a good one for four children to get their sea legs. In the next years we met or heard of several trollers that had small kids along.

Rough Waters

Though we did fish fairly hard, we didn't just fish. Thinking back on those times, and comparing it to other salmon fishing, I understand why trollers were willing to work all winter in a bad season in order to keep fishing in Alaska where, in those days, you could move freely from area to area of world-class scenery, fishing where you pleased. The pleasure of Alaskan boating and exploring was an important part of trolling's lure for many people. For the sake of the children we tried to stay in inside waters, and if fishing slowed down we went exploring--probably along every channel between Petersburg and Cape Decision, from Roller Bay to Lituya Bay. We dropped our lines in most of the popular troller drags, and tied up at least once at every tiny fishing village. When we came into a harbor, if there was no float we tried to drum up energy to get the raft down, and the kids would race off to the beach and woods and leave us in peace, or often, to make the usual repairs. I realize now what a lot of exploring we did when fuel was cheap and daylight long. But I have hardly any photos; it's all just filed memories. Was I too busy fishing, or could we not afford film?

A bigger question: had I reached my goal, to find the perfect fishery for families? No. Though we tried to stay in the calmer waters, we soon realized that to make a season we must go back to the rougher outside capes where all those Washington hatchery fish passed by. The capes had a big drawback for the children. They never complained, but it was too physically confining for them, they couldn't always go ashore in the evening, and since Jack and I were never seasick, we had not factored that in enough for them. We would send one set of children home and bring aboard another for the next season. But we never overcame this conflict of wanting kids aboard versus the tough fishing required to make a living.

The previous owners of the *Nohusit* had fished almost always in the inside waters, and that's where we should have stayed, but the catch kept getting smaller, no matter how much we learned, so we would end up out on the capes each year, trolling up and down for fourteen hours in sloppy water, circling with irritated boats, not making expenses. We each year got more got bored and grumpy over the disappearing salmon. We should have gone clam-digging, done anything but fight a bad season fretting over the salmon mysteries.

Jack would get regularly furious over the limitations of our boat, "Why can't I have a decent boat? I could make it at Fairweather." Yet we never could afford a Fairweather boat. The *Nohusit* was definitely a shore boat, and just right for us, but not for making a whole living. So we worked part or fulltime winters. Probably most trollers did--certainly today's survivors do--or they take on much tougher winter trolling. One year we tried out fishing the Washington coast out of La Push with two of our now teenaged kids aboard, following George and Karlene on the *Rocket*. We found it a poor alternative for what we took for granted in Alaska. The nights were longer, harbors few, the fog so constant we might not see the shore or hardly anything else for the entire trip, and I was glad for our radar purchase. The unpleasant afternoon westerly chop started early and ended late. I thought the La Push fleet (the hippy fleet, some called them) more than earned every fish they got. That experience taught us how great the Southeast fishing life was, even in the worst years. Many people had learned that same truth.

In 1972 we gave up commercial fishing. We weren't patient enough, I know now. We'd learned to fish kings as well as coho, but still couldn't make a whole living. That last year was the bottom of the coho cycle, and the harvest would slowly start back up again, but we weren't there. Yet before long the coho runs to the Columbia started down again and would stay down for a long, long time.

Family fishing--what I was so dedicated to--is still possible. People write about it frequently now. But in too many regions the fisheries are industrializing, and are no longer something families can do together. In many cases the crews aboard the factory ships outside us are from around the world, very far from the families they support. But in the 1960s in the Southeast inside waters I saw families with children, not only trolling but seining, gillnetting, jigging and even longlining for halibut and sablefish and running crab or shrimp pots on a small scale. The fishing families with children that I talked to more recently for this book also figured out a way that children could join in as respected, productive members of a team. If the parents were smart they tried to mix fishing with some fun time. If they succeeded, many of the youth will set their sights on being boat owners or skippers or setnet operators one day. We shouldn't let that promise fade in the mist.

I missed commercial fishing a lot. But that was my view. I wanted to know what it had really been like for my stoic children who never complained. When I asked my oldest sons, Dan and Rob, forty years later, what they recalled of trolling, they said similar things. Fishing could be fun, especially if they were allowed to do the macho stuff--run the gurdies and pull in fish. Dan said his biggest thrill at fourteen was landing a large king by himself. Even cleaning them was "sort of fun" they admitted. Exploring the little anchorages and fishing villages, all agreed, was great. But fishing off the capes, to wake each morning to a Dramamine pill forced down their throats before we pulled anchor--that was ghastly.

I could write one more fisherman's memoir about the trolling life and all the details of fishing and boating, but many people have written captivating personal accounts of both the working day and the natural world they encounter. Outside Alaska, the wooden boats are about gone and the restrictions have grown, but trolling itself doesn't change that much. If you get into a good run of fish you are going to wake up in the morning with hands like frozen claws you must soak in hot water to get moving, and in that it is like any fishery for those working the deck. You will feel the old thrill when you see the tattle-tale lines shake. You'll have big pride in a successful trip, and if it's a poor one, well, the next one will probably be better. That's part of the draw, the gamble. I don't need to add to that memoir collection. What's missing from those accounts is a child's view. I asked Lesley, though just six to eight years when she was aboard, but a writer blessed with a good memory, to give this troller kid's recall of the fishing life:

"I knew we had to make money and that everything else must be subsumed to the pursuit of fish, the cause of much pain and angst in the folks… I felt the thrill when we hit a good run and were pulling in coho on every line. I was interested in the fish--thought them beautiful and mysterious. But I was not allowed to clean them with the sharp knife, only scrape out the kidney and wash them. I was a perpetually seasick landlubber, often in my bunk long mornings, but if not sick I learned to stay busy and out of the way. I read everything that our mom ordered in boxes from the Sitka library. That was a joy, to open the new box and dive in. But I couldn't read when I was seasick, so I had to go into a stoic trance and wait it out. I learned it was much better to be out on deck washing fish and puking over the side. Ashore in Sitka, I couldn't wait for the unloading business to be over, so we could run uptown with our earnings, about 25 cents apiece, enough for a candy bar, and then explore the totem park.

"Traveling was fine as then I wasn't sick. I liked to sit up by the front window and feel the speed, feel like we were getting somewhere. Traveling often meant shore-leave in a village like Elfin Cove, racing on boardwalks, cooking in communal steam baths. When we went into an anchorage, I remember the longed-for sound of the anchor chain rattling down, and the sudden wondrous quiet after a day of the roaring engine. Now I heard birds, and the lap of waves, and we kids could take the raft ashore and cut loose into the most beautiful wilderness and awesome playground that could be found on earth...eagles, otters, wildflowers, butterflies, baby frogs by the millions, devils club taller than us. We just had to watch out for bears, keep toddler Frank safe, and make it back to the boat in time for sleep. The dead-tired parents seemed to be already asleep, unworried about us."

It sounds as if that experience wasn't all endurance and Dramamine pills!

The trolling opportunity for newcomers is pretty much gone in the Northwest. But Alaskan salmon trolling can still be a good small-boat fishery for people who can afford to enter it, and if they can accept the likelihood that they will never hit that huge run of salmon and have a big, fast season. For that, you must go seining or gillnetting. As for the other benefits of commercial salmon fishing, except for some views of clear-cutting marring the hill slopes, the natural surroundings are still there: the awesome capes and cliffs, seabirds drifting the ocean swells, magical green passages, tiny anchorages with trollers swinging on their anchors--all that is still there for the fishermen.

Rough Waters

A few years after their youthful trolling days ended, two of my sons and Jack's own two boys were back long-term commercial fishing again, and my oldest son was out boating and subsistence fishing with his wife every summer weekend, so it seems our venture helped mark them for it. When Lesley comes north in the summer she is always ready to join in with setnetting at our camp. As for the next generation, so far only two of my grandchildren put a hook or net in the water, so perhaps the magic lure didn't pass on. Subsistence setnetting in front of our fish camp is just right for me now. The salmon price is irrelevant. The biggest repair is a hole in the net. Since we use a thirty-year-old rowboat there's no worry about gas price, just oar replacement.

As I have admitted, fishery politics mainly went over my head during my trolling years, and I expect the same is true today for many. I didn't see the significance when the skippers from Washington told us that the cohos we caught on the ocean capes were from Columbia River hatcheries, and that it was too late to save the wild salmon, that the dams had killed them off, and the hatcheries were a good working solution. But later I learned even the hatchery runs had natural cyclical fluctuations caused by shifting currents called by names such as *"El Nino"* and *"Pacific Decadal Oscillation"*, and fishermen try to predict what that will mean for the season in weather, water temperature, and run strength. If that information was available in the 1960s, it never made it to my boat. And each fisherman probably had his own ideas about conservation, but it wasn't at the top of the agenda. The larger continuing story of salmon fishing is a lot about what we didn't know, and what we still don't know, regarding salmons' lives, or how to manage them. And the management part is too much about political power.

Four decades since I quit commercial fishing, I think of the Marshall gillnetters (remember them?) and their protest fish-in that got nowhere. Back in the 1960s-1970s in Southeast it was Russian trawlers we saw out on the 12-mile line off Sitka that we wanted stopped. Now, it's our own industrial fleets that are scooping up prohibited (for them unsellable) salmon as bycatch.

Such an amazing lack of progress is partly due to missing scientific information that is in turn partly caused by short funding and partly by a convoluted managerial system. But partly the trouble is purely political power issues. The peaceful hardworking life I had imagined when I went commercial fishing was an illusion. Fishery politics were never calm. They were explosive during territorial days in Alaska and they still are in most regions. French fishermen regularly conduct huge protests against EU fishery management. New England fisheries have been in an uproar for years, and the turmoil spreads. West coast Africans are losing out to European industrial fishing firms. New Zealand fisheries are owned mainly by four huge companies. The State's goal of closing off of new entries into the salmon fleets began only a year after I left.

Chapter 9

LIMITING THE FLEETS

Whenever fish stocks are in short supply it is a given that the commercial fleets will be blamed. There are too many of them, too loaded with technology, they're addicted to overfishing, they're greedy bastards--every way to place the blame on fishermen. Why? It's easy. They are such a simple place to look for blames and solutions. George comments that they were blamed for what evolved on the Columbia, on the Frazer, and probably on every salmon river showing stress from the 1960s on, and actually earlier too. During that decade, "overcapitalization", i.e. too many boats chasing too few fish, began to be the diagnosis of government economists for many troubled US fisheries. One solution would be to stop new entries into a fleet. O course the growing fleets were an influence, but probably not the major one, just the easy one.

When the idea of placing a moratorium on new salmon licenses was floated by state fisheries departments it caught most fishermen by surprise. The managers already used methods to control the catch: the seine boats in Alaska waters were limited to 58 feet, the gillnetters to 32 feet. Their nets had limitations in size and mesh. Trollers were the least managed in the 1970s but even they had catch limiting rules to follow: no keeping of small kings, a limit of four gurdies in state waters. In most state fisheries the managers also began to favor a fleet quota based on allowing reasonable escapement for sustainability, and when the quota was reached the boats would have to stop fishing. The gradually increasing state regulations kept the fleets from being too efficient and allowed a somewhat equal chance to the licensed fishermen.

However, commercial fishing kept growing in popularity to the point where in some districts fishermen began to feel they were overcrowded and that they would catch more with fewer competitors. One can see that could be the case especially with net boats hovering at the mouth of a river. Another argument used for limiting permits was for better conservation, and another, that at least fish managers thought worth considering, was that stopping fleet growth through limiting of permits would cut costs and bother of ever-increasing management.

American cities were also getting crowded, and in some areas small businesses were crowding each other. Yet it would have been considered un-American to say that people could not move into a city or could not start a small business in a business zone. Passing legislation to halt new fishing licenses meant going against the age-old institution of the open ocean commons. But to the government economists, the growing fleets of small fishermen competing for the available fish were inefficient and inefficiency was almost a crime. To small fishing communities the traditional understanding of the fishing commons was that everyone gets fair chance at the fish. But they, too, can get nervous when too many "outsiders" show up on the nearby grounds.

This is happening today in Nome's winter crabbing, where a rising price and favorable winter conditions have tripled the number of people out on the ice and the "oldtimers" say it can be difficult to find a safe spot to drill a hole for a crab pot. They talk about being forced to go to limited permitting. Environmentalists always worry about overfishing on the available stocks, which is a genuine possibility in any fishery. The proposal in the 1970s to limit salmon permits had broad appeal, but it also had its opposers. Why couldn't the managers simply set a sustainable season quota and close down the boats when it was reached? That argument didn't fly. Too many boats, said the economists. Not efficient.

British Columbia's salmon fleet was the first to experience limited permitting and its effects. The federal Dept. of Fisheries and Oceans (DFO), the manager of all Canadian fisheries, closed the books on new licenses in 1969. At the time BC salmon runs were sliding into a down-cycle, and DFO hoped that cutting the number of boats would help. It was politically and financially easier than tasks like cleaning up the trash that powerful clear-cutting timber companies had left on the salmon streams. The situation for the Northwest states was more complicated. There, the federal dam system had to share the blame for a salmon shortage. Even before Bonneville Dam opened there had been many privately funded dams on the Columbia that had done their part to wipe out salmon spawning grounds and for the turbines to suck up smolt on their way down river. The federal dams, however, had huge political as well as hydraulic power.

Other influences on salmon runs were even harder yet to control. In the early-1970s a shift in currents brought warmer water up the coast. Salmon find more feed in colder water upwellings and stayed longer up north. Boats fishing in Alaska thus ended up with a double advantage for a time in both sea changes and run interception. The Alaska troll fleet was already a good portion "outsiders" and now that increased. Following Canada's lead, Alaska decided it must stop the north-pointed armada and institute a moratorium on new licenses.

George told me that when he heard of the plan he never believed it could happen.

"How could they close off a public commons like that, a very old part of democratic tradition? I believed they'd never get away with it--I'd always be able to fish in Alaska if I wanted to. I was wrong. I'd never bought an Alaska license again, and so I didn't get included in the limited permit free handout. Then it was too late--the price of a permit shot out of reach."

When the Alaska Legislature in 1973 passed the first "Limited Entry Permit" law it was hit with several lawsuits. The Alaska Constitution stated that all residents would be treated equally. Therefore, every Alaskan resident had a right to buy a license and go commercial fishing in the commons--or none did. People outraged by the attack on traditional freedoms sued the state. While limited permitting was delayed in court, the flood of fishermen north swelled. To be left out could be disastrous. The Legislature found a legal solution when it declared that the state's natural resources were in danger, and the Constitution said that they must be protected for the state residents. The alarmed voters agreed to a constitutional amendment allowing a moratorium on new commercial licenses at the next election. By 1975 the whole coast was closed to new salmon permits. The concept spread to other commercial fleet management. By 2010 Alaska would have 79 fleets under limited permitting. Almost all of the other states' commercial fleets would also become limited entry, and eventually most west coast federal (outside three miles) fleets as well.

This argument--our public resources in danger--would be used again and again in the future to close off the open fishing commons and is still alive on every coast and in other countries, with courts maintaining that a government does have the right to limit the number of permits fishing the commons. An important difference in the states' new laws from that in BC was that the states awarded the limited permits to the boat owners, not to the boats. That prevented the accumulation of boat permits by fishing corporations, processors, or speculators, which is what soon took place in BC. But either way, the nature of the Pacific fleets was again changed forever, this time not through technology but politics.

For the original Alaska salmon limited permit awards, if you could prove respectable harvest activity in the right years and other qualifiers such as evident economic need, you were in free. The state would give you a permit for your area and for your specific gear: gillnet, seine, or troll. You could not lease it out but you could sell it, will it, or give it away. In Alaska you would need to keep your permit active, but you could have a three-year leave for medical reasons. Newcomers to a fishery could only buy or be gifted existing permits. Whether the law improved things for fishermen overall depended partly on your political philosophy, but for most people the issue was a practical one: were you eligible or not for a limited permit? Or could you afford to buy one that someone wished to sell? Limited permits became commodities on an open market, and predictably their monetary value soared.

The Northwest harvests were so poor in the 1970s down cycle that many salmon fishermen saw it as a time to get out. Those that stayed often saw the new law as a gain. Unless they had fished in Alaska they wouldn't picture how it would affect its small off-road coastal villages, most of them under 1,000 in population, many under 300, places with no road access and with fishing and support businesses the only stable local economy. Many lawmakers apparently didn't have a clue as to what would happen to those places when the valuable and coveted permits began to transfer outside to individuals with deep pockets.

I asked George how limited permitting had affected him.

"At first it seemed okay to me--I'd have no trouble getting a permit. But later I saw it differently when I realized that what kept people coming to the docks looking for work as deckhands was the lure that someday they could have their own boat. There's not enough just in the money alone to hold a deckhand on a troller. Even your own kids will leave you. They have to have hope for their own future boat. Now they'd also need a permit. That closed off fishing to so many young people when the price shot up. None of my kids decided to go into fishing, but they would have had a hard time if they'd wished to. I was still fishing when they needed to start making a living, and we couldn't have raised the money for a permit. So the people I was now able to hire as deckhands often weren't really interested in the work. It was just a temporary job now, nothing to put your heart into learning, and it showed.

"And here is another thing. When no new people could get into the troll fishery, we lost the political allies we had, like the businessmen, and teachers that liked to commercial troll during their summer vacation. Some trollers were glad to see them go, didn't think of them as real fishermen, but that's when a lot of our political support disappeared. It made it easier for the state to increase restrictions on us all the more."

Limited Entry was especially hard on the young people in the smallest ports. Children that were born into a fishing tradition and expected to make it their livelihood were pushed out. Their parents would have to retire and pass down their permit--but to just one of the children. The rest could fish only as crew unless they could afford to buy a permit from someone retiring. Families also began to sell their permits during slow seasons to cover tax bills and other backlogged needs. Youth with no fishing future started leaving their home communities. The Alaska Legislature, if not the others, should have foreseen such an effect. The small fishing ports were out of sight from

Juneau but should never have been out of mind. Yet Limited Entry would have gone through eventually in the other states regardless of what Alaska did, and I don't know what other way Alaska could have protected itself from the hordes bound to steam north. Ten years earlier I had been one of them. New people were as eager as ever to get into fishing, so an inflationary market in permits was assured. Today if you want to go fishing you will be looking at five, even six figures for Alaska salmon permits.

As a personal example of how this affected young small fishermen, I was not eligible for a free permit award. In nine years fishing, and now 35, I had never signed a fish ticket, and therefore like other deckhands or unofficial partners I had no proven history even though I had worked as hard as anyone. There was no way I could have afforded to purchase a permit. Limited Entry was just the beginning of an era of new laws and regulations that would cut off younger fishermen from a life they had planned in the commercial fleets. For the classic economic advisors to fisheries, these changes did not present a problem, and if social scientists were concerned, it didn't make a difference.

Small ports like the ones we had sometime sold at--Pelican, Tenakee-- were affected in many ways by limited permits. If a local fleet shrank too much, a small local processor was likely to close down, while places that depended on a traveling fish-buying boat could find they were no longer on a route. Fewer boats were buying fuel and supplies locally. Another wide-reaching effect was the loss of flexibility through limited permitting. For small fishermen it was important to be able to fish in many fisheries, moving from one to another depending on the season, market price, and what species was producing best. Families, boats, and gear were also involved in both commercial and subsistence harvests, with the commercial fishing paying for the fuel and other costs for subsistence activities. That important freedom of movement between opportunities began to dry up. Fishery managers must have believed the inevitable social disruption for communities was the price paid for a needed change. But they may not have foreseen, though their economists should have, how the monetary value of permits would shoot up beyond what new buyers could afford in small economies. Many of those coastal towns, both Native and mixed, though probably not luxurious living, had been economically viable until hit with limited permitting.

Alaska Commercial Entry Commission statistics show that for years the Limited Entry Permit law, though it did keep fleets from growing, didn't shrink the salmon fleets significantly. Even after thirty years only 700 salmon permits had been deep-sixed in addition to the 1000 hand-troll permits that the state cancelled outright after it determined they were not a significant part of a family's economy. The real effect of limited permitting was to transfer the active boats from smaller places to larger ones, and to some extent, to non-Alaskans. The Bristol Bay fleet, for example, saw 500 permits transfer by 2010, and over 60% of those transferred to non-Alaskans. Statewide, all fisheries, the percent owned by non-residents in 1975 had been 18.5; as of 2012 it was 23.2 percent, though some of the shift was from Alaskans moving outside the state. The overall rural flight was significant: in 1975, 8200 people lived near their fishery, in 2012 6,000 did. That doesn't seem serious until you consider the size of the communities they mainly were leaving.

Social and economic disruption to small rural ports only became generally acknowledged when social scientists discovered that the changes created a ripe field for study. They discovered that the permits tended to transfer from Native and mixed villages like Angoon, Kake and Old Harbor to larger fishing towns like Sitka, Petersburg, and Kodiak, or out of state. Once these permits left the small communities, they tended to stay gone. Four decades after the passage of Limited Entry the state contracted a University of Alaska study by Gunnar Knapp, who came to the same conclusions, in economists' jargon, that George had:

"When fishery managers create tradable fishery access privileges, regardless of the initial allocation of those privileges, markets become powerful forces for the reallocation of privileges over time, both among individuals and among geographic regions. Great care should be taken in creating tradable fishery access privileges and in considering potential restrictions on how and to whom they may be traded."

The states did need to do something about fisheries becoming crowded, but great care had indeed *not* been taken. Yet it could hardly have been a surprise to anyone paying attention, and one might wonder why the state bothered at that point (2009) to contract for Knapp's insights, true as they were. The loss of local fleets would spread beyond salmon, described in Part II of this history. If the state had engaged social scientists earlier and listened more to the village fishermen themselves, the effects of the new law could have been at least partly mitigated. Different rules could have protected local fleets. For example, licenses would not have left the remote coastal communities if they'd had no market value beyond that district and their price had stayed within what the local market could afford. Yes, there would have been cheating and enforcement costs as there always is in any industry. As my friend John at Kodiak commented to me, there will always be some people trying to get around regulations but that is no reason to have bad regulations.

For years Alaska permits transferred from boat to boat. For every person wanting to sell there was a hopeful buyer with the cash walking the dock, and Alaska held onto its special draw. When the salmon outlook was poor, a fisherman could reasonably believe that the safest place to invest and to wait it out for better seasons would be some place in Alaska--Kodiak, Bristol Bay, Ketchikan. The fish had always come back. Meanwhile fishermen continued to upgrade to new technology and become more efficient. As the value of the permits climbed, those who benefited were people who could pay cash and/or get through a loan process, generally not younger people. Permits available on an open market today range from $9,000 for a modest hand-troll permit to $36,000 for power troll, to $303,000 for a Southeast seine permit--costs in addition to those for licenses and equipped boat.

Today limiting of permits still hasn't had much effect on salmon stocks. Fish managers, especially in the Northwest and BC, continued to rely on more restrictions. Fishermen like George who didn't own Alaska permits regretted it. Some of the lucky had other options: licenses for crab, halibut, black cod, or groundfish. George comments, "You know, limited permitting by the states was actually a forewarning of even more trouble coming from the feds, from NOAA." He was worried enough that he decided to go for another upgrade to a boat that would allow him to chase albacore tuna, giving him another option. Many of the larger trollers already fished tuna when it made sense.

George acknowledges that his venture into albacore had the feeling of the classic trap that fishermen fall into: "*If only I had the boat to...*"

Sea Ranching

"I wasn't forced into a bigger boat. I could have just kept trolling salmon with the *Rocket*. But there were those worrisome signs on the horizon--the big corporations' franchises to start salmon sea-ranching. I mentioned that article telling about the franchises for corporations, the smell of bad times coming. The feds especially liked the idea, as they saw they could get out of expensive programs like hatcheries if fishermen were replaced by self-supporting private fish farms. About six big corporations got their franchises. Weyerhauser was one that did get started, based right at Newport. And just about that time the feds began to see our government-funded fishery management as such an economic and conservation problem. Remember, this was long before any salmon were listed as endangered species.

"Later some of us went to a meeting where those companies told it to our faces, that salmon would all be farmed in ten years. All of them. Think about it! If they could drive us out of fishing, the government could get out of hatcheries. We walked away from that meeting thinking the future definitely looked poor for salmon trolling. Karlene and I thought albacore might be our only way to keep fishing. But that meant going much farther offshore, not possible with the *Rocket*. So we began looking for a bigger boat that wouldn't saddle me with a huge loan. We heard there were many for sale on the east coast due to depressed harvests. I sold the *Rocket* and we went to New England searching."

Another Limit: The "200-Mile Limit"

I never heard of plans for limited permitting during the time I commercial fished. We had other problems, we thought--the foreign trawl fleets fishing just a few miles outside us, right on our 12-mile line. We believed they were scooping up salmon as bycatch. Closing off those fleets had been a subject of debate at Congress for some time. I have a 1970 newspaper photo of my oldest son Dan, fourteen, marching along with a big grin carrying his sign: TWO-HUNDRED MILE LIMIT. We were part of a parade down Sitka's main street, followed up by yet another parade of bannered, tooting boats. We had never seen trollers organize for anything, but we were suffering another poor coho season. We were sure it wasn't we trollers to blame, but the huge Russian freezer-factory ships and their catcher fleets, visible to us on clear days.

Other fleets far more influential than small trollers were harassing their congressmen to do something. Pacific Rim countries had jumped far ahead of us in industrializing their fishing fleets, and the same was true on the east coast with several European nations fishing right up to the US boundary. Though our fleets, too, had grown, they remained small boats by comparison. Salmon fishing, crabbing, jigging, long-lining for cod and halibut, and small-scale trawling could all be carried out inshore by modest craft at reasonable cost and provide a living for families. But they couldn't compete with industrialized foreign fleets.

The same conflict took place across the Atlantic. Iceland had commercial fleets still operated by oars and sail while just outside them British trawlers were raking up the fish on the continental shelf. Iceland was one of the first to move its national boundary to 200 miles. Many in our own Congress were afraid that such a boundary could work against us politically when every country went for its own slice of ocean. We had already moved our US boundary from three to twelve miles in the 1960s and that was far enough. But the frustration and anger of US fishermen kept building. At Sitka we heard many frustrated trollers crying; they were sure of it; those silvery shapes going up through the trawl nets onto the great Russian factory ships had to be salmon.

One older fellow who fished with his wife told us later on the dock how he had run out to the line to see just what was in those nets. He said he completely lost his senses as he drove up until a giant trawler loomed directly over his small vessel.

"The captain comes out and stares down at us like who do you think you are? I gave the wheel to Alice, and I grab my rifle and step out on deck and she goes hysterical, "Don't you dare, bring that gun back in here." I was just going to fire over their head, but she's screaming at me 'They're going to ram us, they're going to kill us!' And I caught my breath and got thinking, she's right, no one would ever know. So I came back inside and told her to shut up, and she took the gun away from me. And we drove off. I had to let her drive. I'm so mad I'm still just shaking. Right off our shores, our cohos. That's why we can't get any, now you know."

Such stories may have made no difference to a reluctant Congress but in 1976 it enacted the 200-mile Exclusive Economic Zone (EEZ) for fisheries, a jubilant day for the troll fleets on the Fairwater Grounds especially. Our other small-boat fleets that fished in the new federal waters for halibut, sablefish, crab, and groundfish would soon face a very different kind of fishery management, a form that favored industrial fleets, covered in Part II.

The 1970s produced one big fish story after another: limited permitting, the Magnuson Act, the EEZ, vessel buy-backs, hatchery problems, salmon sea-ranching experiments, ever-increasing restrictions on the fleets, and in the Northwest, Indian Treaty rights. Eventually all this did succeed in doing what government desired: a shrinking of the small-boat fleets. But cutting out boats and licenses failed to calm the political waters or solve conservation problems. Salmon runs and the people interested in them were more complicated than that.

Chapter 10

GOOD TIMES AND GRIM FOR THE NORTHWEST

The problems for the Washington-Oregon salmon stocks were not on my mind as my family fished near Nome in the 1980s, commercial salmon fishing no longer our focus. I was thinking about the wild salmon runs we fished for subsistence--the chums, pinks, and cohos. But in the late 1980s shrinking of our local chum runs forced me to open my eyes to the bigger world of fisheries. First, commercial fishing and then subsistence nets in the rivers near Nome were restricted, and before long even ocean gillnetting near Nome was closed during a few years. Some time before my visit with my cousin George at Newport, I happened to be at a meeting at Sitka, a favorite town of mine I hadn't seen in a long time. It was June, about 1993, and I walked down to the floats, not expecting to see much of a fleet on a calm sunny day when every troller should have been out with hooks in the water. But here the trollers were, crowded at the floats, and most of them with gillnet or halibut reels also mounted on back. They looked strange to me with all that gear aboard. I guessed that it meant there was a closure, but also that they could no longer make a full living just from trolling. I saw for myself how Southeast salmon fishing had changed in two decades. Later I wondered if there was a connection between the Southeast troubles and our Norton Sound chum decline, but I didn't pursue it.

I was destined to get better educated yet. In late April 2002, again at a conference, this time in Juneau, I naturally took a stroll on the floats in the evening. This time the few lonesome trollers tied there made me wonder where the fleet was, and I hoped they were out fishing spring kings. Later that evening, looking for airplane reading, I picked up Jim

Lichatowich's *Salmon Without Rivers* about the crisis in the Columbia-Snake River runs. I recalled the despair of my last year trolling off Sitka, when we were down to a handful of cohos a day and didn't know why. I wanted to learn more, and Lichatowich turned out to be a good source as one of the early proponents of an ecosystem approach to fisheries management. The message got through and the amended MSA 2006 mandated ecosystem management. The total habitat has to be addressed when planning and regulating for all important stocks, and for salmon that means both ocean and river habitat.

The salmon problem didn't start with the dams, Lichatowich reminds us. Many Northwest residents know that the Columbia-Snake salmon runs of millions were the basis of a rich and sustainable economy for eons for the Native Americans along the huge river system. What many haven't realized is that after thousands of years of sustainability, a series of waves of destruction hit those runs: Euro-American beaver trappers, followed by miners and cattle herders, followed by sheep drovers, then the timber industry--all were bound to tear down the riparian borders of hundreds of tributary streams and destroy spawning areas. Even before the dams were built the runs were already well on their way down.

Many people may also not know that when the Indian tribes were pushed by treaty onto reservations to make way for land companies that then made good profit off settlers, the treaties included many traditional Indian fishing grounds off reservation. These fishing sites were gradually were lost to land companies, who promoted irrigated farming to families along rivers like the Columbia. Millions of salmon smolt were soon sucked into irrigators' diversion dams, pumps, and ditches. My great-grandparents and grandparents were among the early irrigators, and I hope their pump was screened. That's the first of many personal disclosures I make regarding the development of the Northwest on the backs of salmon. We all have played our part.

Scientists, aware of the destruction of Atlantic salmon, observed the Pacific salmon runs shrinking as early as the late 1880s. Photos tell stories best, and museums along the Columbia are full of graphic stories. On a road trip down the Columbia Gorge in 2008, my husband Perry and I stopped at a museum at The Dalles. We viewed a large photo display that recognized a leading pioneer family who at the peak of their success owned 37 fish wheels. Photos also showed the several huge horse seines this one family owned, each one using up to 40 draft horses and a crew of 30 men to haul the plugged nets. They also owned setnets and traps. Just one family's holdings. The display has no comment as to how this empire might have affected the salmon runs or the Indian fishing culture. That lack of comment reflects the politics that historically have affected salmon management.

Later, back down the river at Astoria, Perry and I walked a waterfront where signs told us that at one time almost 100 canneries had operated on the lower river. They processed the catch from thousands of wheels, traps, gillnets, horse seines, everything but the fresh market troll fish that came later. Today the only remaining cannery buildings are arty restaurants and shops. Those canneries did nothing to protect the salmon runs, but for several decades they provided employment to thousands, including my uncles who arrived at Astoria from northern Norway about 1906 and began fishing and sending funds home to bring the rest of the family.

At one museum I was delighted to see the life-size model of a two-man "row-sail" gillnetter of the era. I pictured my uncles as they first set out, learning to handle the butterfly-wing sails, but not so different for them from a Lofoten Island cod boat. Learning to navigate the Columbia's bars and channels among hundreds of competitive night-fishing gillnetters surely took longer. I was viewing an early version of the salmon feasting that big canners soon transported to Alaska.

The fishing industry was just one part of the economic engine at the turn of the century. The timber industry filled the Washington and Oregon salmon streams with splash dams, slash, and sawdust that choked fish. More disclosure: my maternal great-grandfather and his sons worked in those woods before he moved on to irrigated farming on the mid-Columbia. Battles between union and non-union gillnet groups, and between different ethnic groups, like Finns versus Norwegians, added to the hectic scene. My oldest uncle soon decided a more peaceful form of fishing was for him. I was delighted to find in my father's old album a photo of uncle Conrad's small troller, the *Frida*, and I brought it to show George. His comment impressed me:

"I know that boat. See the narrow horseshoe stern? That's the Norwegian style. The Finns were the big majority in building boats in Astoria. They built what's called the Alaska double-ender. I'm sure the *Frida* was built at Warrenton, across from Astoria. Must have been in the 1920s-30s."

That's the kind of knowledge that fifty years fishing earns you.

Hydropower and Hatcheries

With the fisheries and canneries on the Columbia supporting thousands of families, and the pack dropping, concern about the Northwest economy grew. By the 1920s the federal government had followed the practice on the east coast and instituted a system of western hatcheries to keep the industry alive. But despite increasing smolt production, the hatchery managers couldn't overcome the run declines. The solution was to finance more hatcheries to produce more eggs and smolt. Nothing was done about river conditions. Lichatowich (p. 76) believes that an incredible 30 to 50% of the natural habitat along the Columbia was already destroyed before any full channel dam was built.

British Columbia had a head start in learning the risks of experimentation with natural rivers. In 1914 a railroad construction project along the salmon-rich Frazer River canyon caused a rockslide that almost closed off passage in the river. It crippled for years a major sockeye run that had fed thousands of people, both the commercial fishermen at the mouth and the Aboriginal (Native) bands along the river. From that episode, scientists were able to convince the Canadian government to keep Frazer dams off the main stem of the river.

In 1938 the Roosevelt Administration New Deal installed the first main channel dam, Bonneville, on the Columbia as the start of a vast federal system, Bonneville Power Authority (BPA), to electrify and industrialize the Northwest. Next came Grand Coulee, then many more full-channel dams including up the Snake River, putting thousands to work in WWII shipyards and aircraft plants, my father and other relatives among them, unknowing of the salmon connection. Sacrifice of salmon runs was a conscious decision on the federal level to forfeit salmon for power on streams all along the west coast and to replace the salmon through an expanded hatchery system with funding through the 1938 Mitchell Act.

The government knew of the importance of fish ladders, and Bonneville Dam did have them, but Grand Coulee did not, as it was too tall for ladders to work and yet had to be that tall to produce the power planned. The Columbia and its tributaries' salmon runs above Grand Coulee were finished, including all but one in British Columbia. Soon ladderless dams on the Snake blocked many more upstream runs. The necessary reservoirs created behind the dams destroyed spawning areas. A great portion of the juvenile salmon headed downstream was sucked into the dam turbines. My uncle told me that his college class scoffed when the biology instructor told them smolt would get the "bends" going downstream through the turbines. In 1883 the Columbia spring Chinook commercial harvest had been 43 million; by the 1940s the entire Chinook and sockeye harvest was down to 16 million.

Lichatowich (p.117) calls the philosophy of the hatchery programs during the 1950s through 1980s an agricultural approach to fish management. Since agriculture during this period was moving rapidly toward an industrialized model through feedlots, tight pens, and antibiotics, it is not surprising that we would try to manage fish the same way. (And soon we would, through fish farms, creating a whole new controversy.) But meanwhile Washington's Fish and Wildlife director in the 1960s, Milo Moore, was totally dedicated to producing hatchery salmon and much admired by George and other commercial fishermen. Moore was confident that hatcheries could overcome the destructive impact of the dams, and during the mid-1960s it seemed he was right, as both smolt releases and harvests shot up, in Oregon to surpass the level of the 1920s.

Canada's Dept. of Fisheries and Ocean (DFO) followed the US lead and started its own hatchery system for BC. But after a few years scientists observed that despite hatcheries the BC salmon runs did not increase, and Canada closed down the BC hatcheries for years. The runs continued downhill. Then by the mid-1970s, Ottawa, observing that the US Northwest runs were at the time increasing again, restarted and expanded the BC hatchery system. Since the commercial fleets were fishing mixed US and BC stocks, everything Canada did in BC affected us and vice versa. During that decade and the next, much of both BC and US fishery managers' attention was on hatcheries and on keeping up smolt production because that was what everyone wanted. Given the commitment to dams, what other choice was there? Would anyone, at that time, have given up the industrialization of the Northwest for the sake of healthy wild salmon runs? Did anyone object to the huge federal subsidy the dams and their wholesale power customers continued to have? The Indians along the rivers objected, along with a few scientists, and a growing number of environmentalists, but not nearly enough public to halt the provision of cheap hydropower for industrial and agri-business interests. It made more political sense to keep expanding the hatchery programs. The fishermen didn't object.

Fishermen did object when hatchery releases failed to keep the harvests up with the growth of the fleets. One can find great variation in government statistics, but the expansion in salmon fleets is clear in all. Probably every working person on the west coast knew someone who had at least talked about getting into fishing. Lichatowich quotes (P.217) government figures that Oregon's total commercial salmon licenses (all gears), in 1960 at 2565, had climbed by1978 to 8566. When the Limited Entry issues were settled the fleet growth finally stopped, but fleets continued to improve technically and increase their harvest capability. For managers the next fairly easy solution, which should have come first, was to set or reduce season quotas and increase fishing restrictions. But for most commercial fishermen, the true answer was to produce still more hatchery smolt. Yet it didn't seem to work as it should have.

In Alaska the trollers weren't yet hit with restrictions, but by 1970, we were fishing off Sitka for a handful of mainly Columbia cohos a day, when a few years earlier "not how many you can catch but how many you can get cleaned" was the joke. We heard trollers blaming the dams, the timber companies, the Russians, the hatchery funding. But the Fairweather trollers that fished albacore later in the year could sometimes be heard talking on the radio about the current and temperature changes they measured. It turned out that they were onto a lot of the problem. At the time the state managers couldn't come up with a solid answer as to why the smolt production wasn't returning the numbers of spawners expected. Some people gave up and quit fishing for the season. Jack and I simply quit two years later. George kept on fishing.

Politicians continued to assure the public that with salmon hatcheries we could have power and fish both. Keeping the hatchery production high was essential with all Northwest coastal towns built on fisheries, timber, or both. Salmon production fed the shipyards, suppliers, crews, processors, fuelers, and the states' fish taxes and licenses. One group, however, that was not happy with the emphasis on hatchery production, was a growing number of "environmental non-governmental organizations" (ENGOs) that were building a power base in the Northwest. Hatchery salmon weren't their interest, wild fish were. For them it was an ethical or even spiritual mission. They joined many fish biologists and all salmon-fishing Indians in recognizing the extent of the BPA's threats to the wild salmon, and that the solution wasn't just a matter of installing ladders and hatcheries. A whole river ecosystem was being affected. But that couldn't be fixed without funds. Nor would taking on BPA be an easy fight. Most pre-dam laws to protect salmons' river environment had already been difficult enough to legislate. Now BPA's customers threatened to sue if they didn't get the subsidized power they had contracted for. The hatchery system struggled on, the various factions continuing to argue and blame each other.

After the early 1970s slump that drove so many fishermen out, the hatchery-based coho runs seemed to rebuild. But then in 1977 the Oregon "coast coho" harvest suddenly dropped from 3.9 million to 1 million. Dams were not an issue in that area, so what was the problem? Was it overfishing? Northwest fishermen blamed poor hatchery management, but by now there were scientists, too, who worried about hatchery policy and its effect on runs. They were too small a group to make major change but their influence would grow. And so would the ENGOs' influence.

Washington's total smolt release for all species of salmon and (unsellable) steelhead (Lichatowich, p.113), which in 1960 was at 125 million, would triple to a 1980 peak of 380 million, drop briefly, then level for the rest of the 1980s above 350 million. For a time, George says, trolling was quite good, so was other salmon fishing, and the processors and towns that depended on it were happy. The salmon worries faded for a time.

The *Helen McColl* and Albacore Fishing

George by now was fishing a much larger boat, one that he was able to take offshore for albacore. Wise fishermen always had other charts, other gear stored away for when they needed them. Albacore tuna was an opportunistic fishery for many salmon trollers when the tuna were plentiful, the price good, and the weather mild enough to stay out far offshore.

"I'd sold the *Rocket* and up in Maine we'd found a big old wooden boat, the *Helen McColl*, built in 1911. Her hull was in good shape, sixty-four feet, sixteen-foot beam, drawing eight feet. She'd been retired from herring tendering when they fished out all their local herring, and they were trucking them down from Canada. Their groundfish, herring, cod, haddock and other fisheries all seemed to be in bad shape, or were already gone. No better, really, up in Canada.

"I got the *McColl* for $15,000. She maybe wasn't worth that back in Maine, but she was worth that to me here. I've fished her for about forty years. But Karlene and I had other plans too. I had the idea I could rig the *McColl* with sails and do some traveling around the world. I never planned to spend my entire life trolling, you know! There were very few boats outfitted for both trolling and sailing, but it had been done."

Albacore trolling was one way the salmon trollers with larger boats survived the next years. "Your sons were very lucky to get into any fishery these days," George told me. "If they are interested, they should come down and try albacore while they still can--it's still open access."

Because tuna are classified as "highly migratory" fish that cross international boundaries, our federal limited permitting hadn't reached them. My sons were too busy with their own fishing ventures, but albacore fishing was something they probably would have liked to try--I would have myself at one time. I doubted I'd ever seen a tuna except in a can, so George gave me a view that made me understand why he liked it.

"The albacore run up the coast, up as far as the outside of Vancouver Island, sometimes close to the shore, sometimes way out, depending on where the warm current is. Quite a few of the larger Fairweather trollers come down the coast to jig albacore in August. It saved my season a few times, and it's fun in a big school, more fun than salmon trolling once you're on them. You troll your gear fast on the surface, about five knots or more--feathered, weighted jigs on barbless hooks. They're supposed to look like little squid skipping along. That's exciting, seeing the fish hit, pulling them aboard as fast as you can. You get the hooks back out in the water fast or you may lose the school. You can't take time to clean them, so you really need only one strong crewman that can stay with it."

It definitely sounded like fun. I was jealous.

"Yeah, too bad you missed it. The hard part sometimes is finding the fish. They migrate for thousands of miles in less than a year. The fleet can be out there 100 miles, even more, looking for that warm current, and when they find it, they go along it, looking for schools headed north. When someone finds the school they put their boat on autopilot and start making big circles to stay on it, and they call their buddies over. They stay out there, weather permitting, chasing those fish until they run out of ice, or fuel. And these days they have freezers, so they stay out till they're filled up.

"Karlene and I brought the *McColl* around through the Canal--quite an adventure, the kind we wanted to do more of. As soon as we got to Oregon I got a mast and trolling poles installed, rigged for albacore, and went fishing that summer. I'd gone half a trip with someone before, so I didn't go out there completely green. From then on I went for albacore whenever the salmon price was poor or the fish scarce. Or I went for albacore later in the summer and into the fall. Right now the price is poor for albacore and salmon both, but at least there's that option. For a time I sold at Ilwaco, but the Columbia Bar is a good place to avoid, and it was only thirty miles more down to Garibaldi, so I ended up with that my home port, and became part of the Oregon fleet. I even did well selling right off the dock there. I fished out of Coos Bay for a time, and finally ended up here in Newport.

"It was a darn good thing to be able to switch to albacore. But there are drawbacks. To make it pay you have to stay out there. You can't be running back and forth burning fuel and wasting time when you might be fishing a hundred or more miles offshore. And this was before everyone had freezers, so you had to be able to carry enough ice. You have to ride out most storms and only run in for the worst. Mainly you just stay out there, you close down and drift at night. So you've got to have confidence in your boat.

"It's a gamble just finding the schools unless they're running really close to shore. And it's not really good for a family operation, I don't think. Sometimes I did take one of my kids along, but you wouldn't want to be stuck out there in a bad storm with your kids aboard. For them it was better when we were salmon trolling and could anchor for the evening. I'm sure Alaska is better that way.

"Albacore fishing is definitely a specialized existence, and I don't do it anymore. When I first started, quantity, not quality was the goal. You could sell any tuna in the round. But nowadays the marketers want their fish frozen fresh. Almost all the boats now have freezers and stay out there till they're full, and I never did refrigerate the *McColl*. Remember, I wasn't planning on doing this forever. My goal was a few more years fishing, then sailing the world. But there's a lot of upkeep on an old sixty-four-foot boat I found out, more than I wanted to do. So I fished her as she was, mainly what the Coast Guard required. The price of fuel, too, will always be a big issue. Salmon trollers burn a lot of fuel, albacore trollers a whole lot more. That was another part of my idea, that we could save fuel with sails as a backup. I never did do it because that particular oil crisis in the 1980s went away, but that option was there."

George paused a bit, maybe thinking about those good times albacore trolling. With fifty years fishing you have a lot to recall.

"But you know, I think just going forever for bigger and better boats and equipment has its limits. I guess that's one reason I didn't go any further with my search for the ideal boat. Right now the albacore price is poor, and those boats are stuck with big loans for the investment they made for refrigeration, and they're overpowered with those big fuel-eating engines. Still, it's one of the last open fisheries. Tell your boys to think about it."

I nodded, but was pretty sure albacore was one fishery my "boys" would probably let go by. They most definitely didn't have the boats for it.

He again commented that despite shrinking coho harvests the early 1980s were his best years fishing. He didn't need to depend on coho as he had mastered trolling for Chinook, then albacore. He and Karlene could support themselves for the entire year and take winter vacations to Mexico. But he didn't get the *McColl* rigged with sails. There would be no big ocean trip, and instead the next adventure was ashore.

"Karlene became ever more pessimistic about the future for salmon fishing and for what we had planned with the sailing adventure. We did live on the boat a few winters, up in Friday Harbor. But by then we had a child, and that made a difference for her. She thought we ought to get a place of our own ashore. We saw cheap land advertised in northeast Washington near Republic, a wild area. We bought forty acres and built a big timbered house. Karlene really got into the romance of the place. We both enjoyed it, but I still had to make a living, I was only in my thirties! I couldn't retire to a remote country place with no living possible. But Karlene decided she was through with fishing, and I was going to fish, what else could I do? So I went back to the coast and she stayed in northeast Washington."

It was that familiar story again that I didn't want to hear--the fishermen's families that stay ashore and end up finding other lives.

George went on, "A few years later I was still fishing the *Helen McColl* and not thinking anymore about round-the-world voyages when I found Pat, who's a professional farmer, and that's what she wanted to do. She was fine with me going fishing, and she did fish with me awhile. The years we did the very best were when we fished salmon, then albacore later in the season, and sold fresh albacore on the dock. Now Pat has a prize goat farm and cheese house and gets international awards for her cheeses. Our daughter is grown and is her partner; she's taken over much of the cheese-making. So, a project that really succeeded! Well, the *McColl* helped all that get started."

Salmon Debates Grow

By the early 1980s the finger-pointing between the various government agencies working on the Columbia had become so chronic that Congress authorized funding for the Northwest Power Planning Council with the hope of getting everyone to work together. A big part of the council's charge was to find agreement on ways to mitigate the damage done to salmon by the BPA system. The focus was on improved dam passage for smolt and on hatchery management. Yet another group with generous funding, the "Hatchery Review Group", had a similar assignment. A growing number of the fisheries eco-scientists believed we had focused too much on smolt production and that more had to be done to restore the natural river habitat if we wanted to re-build the runs. But that would mean shifting funds from production, or winning new funds, an ever more difficult task.

The conservation movement to save wild salmon kept gaining strength in the Northwest and took a similar position to the eco-scientists, that we needed to protect and rebuild the whole salmon ecosystem. Many conservation-minded people who took on the cause tended to see the issues in simple black-and-white: wild salmon survival versus commercial fishermen, hatcheries, and dams. And since hatcheries then were not specifically organized to save the diminishing runs of wild salmon, many pro-wild salmon activists became anti-hatchery activists, adding more fuel to the debate.

It seems like common sense today that the original river had been essential for salmon survival, and that its metamorphosis into an industrial canal would have consequences, but that idea went against the general American philosophy that anything could be accomplished with the right technology. Yet even the pro-technology arguments were faulty. Hatcheries of the

day, says Lichatowich, tended to measure success by their smolt production rather than the number of adult salmon returning, which made no
sense. Meanwhile in the background of the debates hovered the Endangered
Species Act (ESA). If the managers couldn't stop the decline of wild salmon,
the ESA, already raising havoc in the forest industry, could come into play,
an event no one except the most committed of environmentalists wanted
to see.

Chapter 11

Rights to Salmon: Indian Treaties

Troubles for Pacific salmon and the people fishing them spread in all directions during the 1970s. It was also a season for a few victories, such as the establishment of the EEZ. But in the main the arguments kept sharpening over the rights to the fishing commons and limited permitting, subsistence rights, economic rights, grandfathered rights, tribal rights, citizen rights, and ethical rights--all offered as justification for claiming a share of salmon runs.

By the time the fleets had accommodated to the transfer of fishing power through limited permits, the recreational fishing industry was sharpening its spears for war on the commercial gillnetters who fished the mouths of rivers. The dispute between the British Columbia and the US fleets over fish passing through both jurisdictions was a never-to-be completely resolved on-again off-again war. To add to the floating opera was the growing political influence of the fish farming industry.

Right in the middle of the passage of states' limited permitting laws and the Magnuson Act came a major decision on Indian treaty fishing rights in Washington. Many salmon fishermen believe that the recognition of treaty rights caused at least as much upheaval in the fleets as limited permitting--positive for the Indians, negative for the non-Native fishermen, at least for a duration.

The Puget Sound Fish-ins and the Boldt Decision

Federal Judge Boldt's decision of 1975 was based on Indian rights originating with the mid-1800 treaties, and was so controversial that almost every

Washington resident had an opinion on it. The case was a product of two decades of protests by Indians, many staged on the Nisqually and Puyallup Rivers near Tacoma, others on the lower Columbia, and involving the smallest of small fishermen. In the middle of decades of political protests across the country in the 1960-1970s, Northwest Indians realized the time was right to go to court and gain back the ignored lost fishing rights in those federal treaties. The issues presented in court appeared to boil down to these: Did the Indian tribes of Washington have treaty rights that allowed them to ignore state fishing regulations at their traditional fishing sites and to regulate themselves, or did they have rights to regulate themselves but only inside reservations, or did they have no special rights at all, just those of other citizens?

Most Indians assumed they had treaty rights to traditional fishing sites outside reservations, as stated in the treaties, and some thought they should regulate themselves there. Many individual families had always chosen to ignore state regulations and fish at those sites, but with no intent to make it a public issue. Some also sold fish without a business license. State police at times confiscated fishnets and boats, and local judges levied fines that rarely stopped the fishing.

Billy Frank, a Nisqually, and respected voice for protesting Puget Sound Indians, was arrested the first time in 1945 at age fourteen for what was then called illegal fishing, and would be arrested 40 times more as he gradually became a spokesman for the region's tribes on treaties and fishing. In the mid-1950s Robert Satiacum led a fish-in near Tacoma with the deliberate hope of getting arrested and forcing a federal case. The treaties had been ignored by state and federal authorities for years; what apparently caused the issue to ignite was tribal concern that sport fishing in such an urban area was growing in popularity and becoming serious competition for the annual salmon runs.

Satiacum's fish-ins and others to follow got little serious police attention until about 1964 when Puget Sound Indians decided it was time to modernize their protesting, broaden their support, and get more media coverage. A report by Gabriel Chrisman for the "Seattle Civil Rights and Labor History Project" describes how individuals from the Nisqually and Puyallup tribes organized the Survival of American Indians Association and welcomed civil liberties support groups like the American Civil Liberties Union and the American Friends Service Committee to join them. More press coverage was guaranteed when Marlon Brando, Jane Fonda, and Dick Gregory arrived on the scene. That began a tradition of media celebrities lending their famous faces to underdog causes. The rate of fish-ins, net confiscations, and arrests picked up. The official tribal groups did not approve at first, believing large fish-ins were just making trouble, which was of course what they intended until the treaty issue was decided. A main opponent of the protesters, the Washington Sportfishing Association, did its share of building publicity with its own fish-ins. The police were fair and ticketed them too. The protests' increased media attention succeeded, and both public opinion and tribal officials' opinion gradually swung in favor of the activist Indians.

The protesters decided to file a lawsuit to regain their treaty rights and got the case heard in 1969 before federal Judge Belloni. He started a whole new chapter in the salmon wars with his ruling that the Indians had to have "fair rights" to the fish. Yet what were "fair rights"? The protesters decided to put it to a field test and initiated their own fishery management on the Nisqually River near Tacoma. Hundreds of tribal members came to support the fishermen, state authorities then moved in, and battles erupted with boat rammings, paddles, stones, and fists. As usual, people were arrested, fined, their nets confiscated. They soon had new nets in the river. The war heated up. In September, 1970, clubs went swinging on both sides when, as the *Seattle Times* reported, the Tacoma chief of police raided the Indian encampment at the river even though he had no jurisdiction there. Police used tear gas this time to break up a force of about thirty variously armed Indians. It became clear that exactly what were "fair rights" had to be settled. That fell to Judge Boldt, to gain him everlasting fame or infamy depending which side you were on. He, like Belloni, favored the Indian arguments.

Sohappy's Battles

I had special interest in the lower Columbia River battles because of a family, the Sohappys, which filed many of the complaints for the "River Indians", as they came to be called. Sohappys had been seasonal neighbors of my grandparents from April to October at White Bluffs, right near pre-atomic Hanford, the Wanapum camp being about a fourth of a mile upriver from my grandparents' fruit farm. My uncle recalled to me how envious he was watching from the bank as David Sohappy, a boy about his own age, learned how to spear salmon out of a canoe. My grandmother picked neighbors' strawberries for cash alongside the Sohappy family, and David's grandmother and mine traded salmon and handcrafts for fruit. Then in 1943 everyone was evicted for the Hanford Atomic Works construction. But Wanapum baskets passed down to me from my grandmother made a lasting connection for me.

In the 1960s David Sohappy and his wife Myra and children were living year round at Cook's Landing, a traditional fishing site on the lower Columbia, along with several other families Others came just for the seasonal salmon runs. Prior to the flooding from Bonneville Dam at least 2,000 Indians had fished each season at Celilo Falls alone. Roberta Ulrich (p. 29) describes how the river Indians had been trying to get "in lieu" sites awarded to them for sites lost to Bonneville Dam construction. She reports how in 1939 the Indians had submitted their claim for damages listing 23 locations that included fishing sites.

Thus began a parlay that would go on for 51 years. The BIA asked the families living year-round at the *in lieu* fish sites to move out, that they were allowed only as seasonal camps. The Sohappys and other river Indians refused, stating they had a right to be there as long as there was some kind of subsistence or ceremonial fishing to be done. The argument over residing at the fishing sites went on at the same time as the battle over treaty fishing rights.

The river Indians had yet another related issue with the state, (Ulrich p.38) "...their desire to make money in a money-based economy" by selling salmon, again running them afoul of state regulations. The position of the

River Indians, like the Puget Sound Indians, was that the ignored treaty had awarded them customary and traditional use at traditional sites, which included barter. An argument that has held up in some courts is that today's forms of barter can include cash, an interpretation Washington Fish and Wildlife did not agree with.

The Sohappys leadership included strict personal habits, but not obedience to laws that they believed their rights superseded. On our trip to the Columbia fifty years later we saw that the fishing sites still active looked very much like our subsistence camps on Norton Sound and nearby rivers, though our camps have floors above the permafrost. Sohappys were also at Cook's Landing on a related spiritual mission as leaders of the Washat religious sect. The Washat argument for the importance of salmon in their spirituality reminds me much of the role of salmon in the spirituality and culture of Western Alaska as I understand it.

The same year that Sohappys moved to the river, 1965, the salmon runs were in that downward cycle I described that continued into the 1970s. Fomented by the Washington Sportsmen's Association, some of the public began to blame the Indian fishery in the river for the small salmon runs, thus providing another useful scapegoat, along with commercial fishermen and dams. An official estimate was quite different--that the Columbia Indians took only two percent of the harvest, while for the whole state the Washington Dept. of Fish and Wildlife estimated Native harvest at twelve percent, hardly the cause of poor runs.

During the same period that the fish-ins were taking place at Puget Sound the police issued about 600 citations on the lower Columbia, most resulting in fines, and always in confiscation of gillnets. In 1968 David Sohappy, after spending a few nights in jail over a fishing violation, decided to ramp up his protest and go to court himself as a plaintiff, with Oregon and Washington fish and game officials the defendants. Like the Puget Sound Indians, he found white sympathizers and soon won a Ford Foundation grant for legal support. He was evolving from a conservative religious leader to one of the most consistent fighters for Indian fishing rights, but stayed with non-violent resistance, gradually learning how to work the media and the courts. At one point his wife, Myra Sohappy, took their argument to the Human Rights branch of the United Nations.

In 1974 federal Judge Boldt clarified Judge Belloni's decision on what was a "fair share" of the salmon: a fair share was half the salmon harvest (later extended to other fish) for the Puget Sound area tribes. Judge Belloni then extended the ruling to cover the Columbia River and Oregon tribes. The ruling outraged many if not most non-Native fishermen. But Boldt argued that the way the salmon were disappearing, if he didn't give them half the fish they might not get any at all. The Boldt Decision was appealed but upheld in the US Supreme Court. However, Justice Black found the states could regulate fishing in the interest of conservation, which the states tended to interpret broadly.

Ulrich (p. 146) says the Northwest states tried every way to get around the "half the harvest" rule. Oregon ruled that the concept only applied above Bonneville Dam, so that non-tribal fishermen still had a great advantage. Washington went even further; when non-Indian commercial fishermen on Puget Sound ignored the court's rules, Washington Fish and Wildlife refused to enforce them.

George recalls that the non-Native fishermen he knew were full of gloom, some seeing the decision as anti-white racism, others as another insidious government attack on small fishermen on the heels of the states' limited permit laws just passed.

"The Indians at the time were taking only a small percent of the fish and couldn't gear up soon to harvest fifty percent. So we that fished out in the ocean would get closed earlier than normal when they figured we had caught our half the fish, and the Indians wouldn't be able catch their half as the fish moved up river. So no one made their quota. I was fishing out of Ilwaco [Washington] at the time. Some third generation Finnish trollers I knew there were devastated. Some of the fleet moved on down to Oregon to fish. I did myself, and found out Garibaldi was an easier port to get into, so I became an Oregon resident.

"But what was going on wasn't just about Indian rights. At the time I saw the whole Boldt case as a red herring. The reduction of our quota in the ocean meant the big corporations planning for private fish farms thought they could gain a foothold--that with the 50% rule there would be quota left over for them. Anyway it worked out that way--we saw them move in, Weyerhaeuser, BP. and the rest."

That byproduct of the Boldt decision was probably unintended--being fair to Indians was Boldt's interest, not big corporations, but in fact that is what happened. "Leftover" salmon did become the captives of private ocean fish farms in the late 1970s. Later Oregon outlawed ocean farming, but farms still operate in Washington.

The river Indians' battle with the state of Washington continued as the state ignored federal orders and continued to ticket Indians for fishing out of season or selling. Sohappys and other river Indians also continued to refuse to follow BIA orders to leave their fulltime camps on the river. Eventually lobbyists opposed to the Indian sales of salmon were able to get the federal Lacey Act passed, which made selling of unauthorized salmon across state lines a federal crime. Ulrich (p. 126) describes the next phase in the battle: "In 1982 their permanence and their openness about their fishing made David Sohappy the focus of ...a several-hundred-thousand-dollar government sting operation known as Salmonscam. Salmonscam sent him to prison and hastened his death."

The Sohappys and about 70 others apparently participated in the obvious sting as they assumed they would win another advantageous federal case. They were wrong. This time the Supreme Court refused to hear their appeal, (Ulrich (p. 162-163) and David Sohappy and his son David Jr. "went to federal prison for five years for the sale of 153 salmon and 25 steelhead." She points out that the sting also involved a group of non-Native poachers who illegally marketed 300 tons but received only civil charges and fines. The point seemed to be to teach the Indians another lesson. But the governor of Washington and others saw it differently and worked to get the Sohappys released based on the father's diabetic condition and a stroke. Sohappy senior retired from fishing but died soon after from his illness.

Evie Hansen, a gillnetter quoted by Allison, et al. (p.90) speaking as a frustrated gillnetter, said, "I can honestly say I see both sides of the picture.

I've lived on a reservation and I know where the majority of Indians are coming from…The power of the media is something we are all struggling with…The media tried to make it a racial issue."

Was that the case, an example of the media's common goal to try to create hot stories where possible? In any case the racial focus made a bad situation worse. Hansen went on to describe the effects on her family that to her were economically unfair. "[Before Boldt] …we could fish way out in Puget Sound or way inside…it could be our own decision…! But now there were…maybe seven days of sockeye fishing this year [1980] compared with forty, fifty days before, and in those seven days a real small area to fish in too. So we've been just regulated down to no fishing at all."

Non-Native fishing groups fought enforcement of the Boldt decision for years. Their anger had at least four causes. Racism was certain present. No understanding of the reason for Indian treaties was the case among most of them, and also among most, no recognition of socio-economic conditions on reservations and how they came to be. But there was for many fishermen a less racially-colored wrong in the Boldt decision. What other US business would go down quietly, regardless of history, if told that the government would arrange that its source of potential goods would be cut 50%? The non-Native fishermen furthermore believed that they were especially penalized because in other cases of settling Indian treaty rights, the public at large had paid whatever the cost of the awards through federal taxes. In this case, the non-Native fishermen had taken the economic hit on their fish tickets. A polarizing situation was made so much worse.

Many fishery enforcement officers probably didn't understand why treaties mattered either, though they should have if they had taken an in-depth Washington State history course and paid attention. They did know that Boldt had made their job harder as they faced the logistics of how to cut 50% of the season quota from the non-Native fishermen, allocate 50% of it to the tribes, and then assure that the various tribes up the river got their share of that 50%, all the while besieged by alarmed, angry white fishermen, both commercial and sport groups. Ulrich (p. 147) comments [The Supreme Court later wrote it was] " 'the most concerted effort to frustrate a decree of a federal court witnessed in this century.' " Suits and counter suits, in county and state courts in addition to the federal courts, continued. When the states proved unable to enforce the Boldt decision, indeed refused to in many cases, the federal court took over management of Indian treaty salmon, and later, management of all Columbia salmon.

Though many Northwest residents complained that non-Indian fishing income was severely reduced largely because of the Boldt decision, Ulrich (p. 148) quotes the U.S. Civil Rights Commission, which countered that " 'Even those most vocally opposing the Boldt decision knew other causes were more reason for the commercial fishing industry's economic problems...' ". Ulrich points out that four lower Snake dams had opened in 1975, the size of the commercial fleets had doubled in the previous ten years, Canadian fleets' interception of south-running salmon outside Vancouver Island was significant, and that Washington and Oregon fleets faced increasing restrictions as the runs continued to shrink. There were many reasons why fishing got harder out on the ocean, but at the time it was easy to target "Boldt" and the Indians.

Wishing to ease the economic and political pressure, Washington State used emergency federal funds to buy out many non-Indian gillnet boats. Other white gillnetters packed up and headed for Alaska. They had to hurry as the price of limited permits was climbing.

Though I believe it was a hard time for many non-Native salmon fishermen, in my traveling jobs I saw too much the other side in the economic conditions in Alaska Native villages during that period, and I'm sure Washington reservations endured much of the same. Those families needed a massive shot of economic development to bring them to the level non-treaty commercial fishermen would call bare basics. Fishing was one of the few ways rural Natives--Alaskan or Northwest--could earn cash, over or under the counter, to buy such basics as piped water. Several Northwest tribes later did get a one-time settlement for loss of traditional fishing sites--the Yakamas received $3750 per head. That aside, I can't imagine any small-scale fishermen with more need, let alone rights, for a continuing, reliable salmon fishery.

Later the river Indians won another victory. Ulrich (p. 179) reports: "In 1990 the Court of Appeals finally ruled that the Bureau of Indian Affairs had exceeded its authority when it ordered [the Sohappys and others] out of their "in lieu" homes." They would be allowed to stay on the river. Congress also approved about two-dozen Columbia River traditional fishing sites that Indians could continue to use. But the federal stonewalling had certainly cost the taxpayers. Ulrich (p.221) comments, "The Corps could have spent $50,000 [for the sites] in 1945 and settled the whole issue. By 1995 the agency estimated project costs of more than $67 million..."

When you walk into the Yakama Heritage museum the first photo display you see has the strong face of David Sohappy at one of his arrests. Whatever embarrassment he may have caused some less radical Native leaders at the time, they give him and the other fish-in leaders full recognition now. The treaty Indians' battles helped me better understand today's issues on the rivers in Alaska, as Chinook runs flounder and Alaska Native people demand more control of fishery management at their traditional locations. Although I wasn't on the scene during the Indian treaty battles in the Northwest, I'm very aware of the movement to preserve subsistence fishing in Alaska and the conflicts over it today. The Alaska Native Claims Settlement Act of 1971 returned land and awarded cash settlement to new Alaska Native corporations with the potential to make money for their tribes. However, it didn't award any special fishing rights, its focus being the business of profit-making corporations, and with the current shortage of Alaskan salmon, we see similar conflicts boil up.

BC's Aboriginal Fishing Rights

Two decades after the Boldt Decision, Canada's First Nations people also used the courts to improve their economic condition and recover a share of cultural losses. In 1992, First Nations won a federal suit that allotted a specific share of the harvest to each of the bands with fishing history, ranging from 30 to 50%. The Aboriginal Fisheries Strategy (AFS) gave all formally recognized Native communities rights to harvest for food, social, or ceremonial purposes only, but not special commercial fishing rights. For that they would follow the same regulations as all commercial fishermen.

As non-Native licenses came back to the government through buybacks, the Dept. of Fisheries and Oceans (DFO) transferred them to bands to become collective community licenses used for the AFS fishing. It was a definite improvement for the bands, but not a complete solution in that they still couldn't sell any of the fish, and many had no access to boats, gear, licenses, and limited permits for commercial fishing.

First Nation harvest rights also became first priority, after conservation needs, in allocating quota, with salmon sport-fishing second priority, and non-Native commercial fishing last. Being bottom of the list didn't help build good will among BC's fishermen toward federal government. They complained about the cost of the AFS to tax payers. Another complaint was that under-the-table sales of Aboriginal fish took place. The First Nations maintained that they needed to be able to legally sell AFS fish when there was so little economic opportunity. DFO began slowly to allow salmon sales by specific bands as pilot programs and to encourage them to look at other ways to build income through fishing-related activities. But to the First Nations these improvements have come very slowly, hampered by bureaucracy and a slowness to allow Native people more involvement in fishery management.

History of previous government injustices to Native people didn't rank high among BC's non-Native fishermen any more than in the Northwest states when fishermen feared for their own livelihood. Critics like Dennis Brown observes that the government policy split small fishermen into hostile camps that prevented them from uniting against common adversaries, probably meaning DFO. Divide-and-conquer strategies work as well in the fisheries as anywhere, although that wasn't how the AFS originated; it came as part of a settlement of national Aboriginal claims. But certainly divisive issues have worked to federal government advantage, including our own, when it comes time to introduce any unpopular regulations or to deflect attention from the run losses that dams or other industrialization have caused.

It is possible that the Boldt decision and BC's AFS delayed non-Native fishermen's wish to collaborate when they needed to, but there were other conclusions. Again quoting Ulrich, p. 151, "The Belloni and Boldt rulings...brought belated and reluctant admission that the Indians must have a say in management of fisheries on the Columbia and elsewhere in the Northwest...". Both federal and state governments had to look more closely at the economic conditions in the Native communities as well as at the failures of salmon management on the rivers. NOAA mandated collaborative planning and management that included the tribes. On the Columbia system the tribes now have their own biologists on payroll and play a major role in salmon run restoration, including taking the position that much more attention be given to river habitat restoration and ecosystems in general--something the eco-scientists had been saying for some time.

Chapter 12

More Rights to Salmon: Canadians v. US Fleets

British Columbia's salmon fishermen haven't escaped any of the problems I've described. Through the 1950s a British Columbian gillnetter, like one in Alaska, could make a basic living from a 30-foot boat and nets for around $600--a start no different from my cousin George's. But by the 1960s the numbers in the BC salmon fleets had also surged, and every problem our fishermen and managers faced was experienced in BC, often with solutions and failures that have pointed the way for ours. BC began canning salmon a decade before us, began limited permitting a short time before us, and recently has even introduced individual quotas to the salmon fleet.

Impact from fishery interceptions due to geography has been worse in BC than for the US, as BC fishermen get it from both directions, with US fishermen, especially Alaskans, intercepting more BC fish than the other way around. There are exceptions: Northwest state managers estimated that until BC species closures took place that over 80% of Chinook caught off West Vancouver Island were of Northwest origin. But there were also Washington trollers fishing there.

The BC fishermen are resentful that in their view American fishermen, especially Alaskans, are held in more respect by their government than Canadians are by their fisheries boss, DFO. As small fisherman supporter Dennis Brown describes it, they often express disgust that their federal government won't stand up to ours and believe that Canadians almost always get the short end in fish treaty negotiations. But Canada's federal government has ample economic and political reasons to keep good relations with the US a higher priority than salmon harvests.

Though many of BC's salmon stocks are worse off than ours, and their fleets restricted more, it's not through DFO unwillingness to experiment. It has spent significant money on research and stock and habitat rebuilding, just not nearly enough. But like our fishery policies, in the past Canada's BC policies largely benefited the processors. The government then had to give aid to fishermen thrown out of work, reduce the fleets, and try to figure out how to rebuild collapsed runs. Not surprisingly, fleet shrinking didn't restore the runs in BC anymore than in the Northwest states.

BC fishermen have not gone down without a battle, and often a colorful one. They come across as feistier and better organized than ours in their fight to survive. They speak for the record, demonstrate, and strike more, expressing outright disdain for their federal management. Brown, writing of BC salmon fishery history, gives an opinion widespread in BC (p. 26), "Ottawa claimed it had acted in the name of conservation, but it can be argued that the subsequent dismantling of the West Coast commercial salmon fishery had more to do with the interests of powerful commercial fishing companies than with the health of the resource."

Though BC fishermen's activism makes the news, it doesn't mean that they can claim success in building enough small-fisherman unity to win battles. But their history has aided much my own understanding of North Pacific fishery issues because BC citizens have written books about their struggles, often addressing political issues from the point of view of small fishermen. Until the advent of the Internet you couldn't find that on our side of the border. Possibly our own management is simply more complicated to try to write about with our endless stream of agencies, boards, and managers involved, and both state and federal management. In Canada, the DFO runs all fisheries. Dennis Brown, in "Salmon Wars", for example, speaks as the son of a fisherman and as a person who has been involved with salmon politics his whole working life. He doesn't worry about being politically correct. He finds Alaskan fishermen, and the government that protects them, as unreasonably greedy when it comes to interception of BC fish. I was once unwittingly one of them, but totally ignorant as to how my hooks in the water affected the catch of a BC troller.

Brown has company. Eric Wickham and Alan Haig-Brown tell personal stories to show how in their view DFO's fishery management has almost ruined BC small salmon fishermen. All three writers tell of how their families had a good living from salmon fishing until the government, like ours, decided to solve salmon problems by manipulating the fleets. Again it was the simpler, cheaper solution.

The Davis Plan

Canada's mistakes with its eastern fisheries are a familiar history. But we, too, had ample opportunity to learn early on from DFO mistakes on the east coast, and chose not to. Brown writes (P.91), that BC, already worried about overfishing by 1889, "...limited the commercial salmon licences to 500, of which 350 went to the canners, giving them an overwhelming advantage both in controlling the harvest and setting the price." That philosophy toward the fisheries was probably the common one for the times, but BC has never gotten past it entirely. (I'm not sure we truly have; Part II here explores that more.) During and after WWII the province had the same huge influx of new fishermen that we did, resulting in the same processor complaints of "too many boats chasing too few fish." Of course those small boats supported families, even if in a modest way. But DFO, buying into the economists' charge of "overcapitalization", decided in 1969 to deliberately shrink the BC salmon fleets to half through the "Davis Plan", and again managed it so that a large percent of remaining licenses went to the canners.

Canada had recognized for some time that the salmon runs were heading for crisis, but was overwhelmed with its cod fishery problems on the east coast. Newfoundland and New Brunswick had fishing constituencies many times the size of BC's and far more significant in the national economy. But now it was time for BC's fleets to come under the knife. The Davis Plan cancelled all commercial salmon permits, those for the smallest boats permanently, and then issued new permits of two types--but to eligible boats, not to the fishermen. Two fishing canoes or skiffs could be turned in for one larger boat license. The fishermen saw through the scheme, but couldn't stop it, and predictably canners and other investors bought up boats and their new limited permits

Haig-Brown (p. 93) tells of the way the Davis Plan was easily manipulated to hurt the smallest fishermen: " In one case a non-Native fisherman traveled to up coast First Nation villages buying up small gillnet boats and consolidating the licences to make seine licences. He then invited doctors and others looking for tax deductions to invest in new boats for which he provided the licences."

Eric Wickham, troller turned blackcodder, describes his view of the Davis Plan (p.125-126): "Ottawa has always hungered to turn fishing over to the big companies because it simplifies their regulation of the industry. They only have to deal with a few fishing company executives who share a lot of their views about the world." He tells how the Davis Plan directly affected him and other small fishermen (p.122):

"In 1968 salmon licences cost about $10 per foot per boat. [When] I bought the *Joy II* the licence cost me about $400 out of the total $7000 price. Ten years later a licence for the same length boat (any 32 ft. boat) could be sold for $75,000! This went on until Bamfield [his village] had no little boats left. That took about fifteen years. Even Bamfield's larger commercial trollers and seiners left the village….The catch had to be taken straight to Vancouver."

Wickhams' bitterness is voiced by many fishermen in both Brown and Haig-Brown's books. But though the Davis Plan had a serious social effect on the little coastal settlements, the desired reduction in harvest didn't take place after all, as most of the remaining boats, mainly gillnetters and including those now owned by canneries, were, Brown says, refurbished to be more efficient purse seiners. Over time DFO would cut the BC salmon fleets in half, but it took decades, more plans, more buy-backs, more unemployed small fishermen, more devastated towns and villages.

The general BC public did not rise up in protest. The salmon runs were still shrinking, and though earlier it had blamed the canners for the depleted runs, now the public pointed at the fishermen. Brown comments (p.26), "In another era the problems facing BC fishermen might have aroused public sympathy, but not now on the heels of the east coast cod collapse which to many Canadians symbolized the wanton plunder of the natural environment…vilification of commercial fishermen [became] an easy substitute for changing the ways society as a whole threatens the salmon resource." That statement says a great deal about what took place for the next decades along the whole North Pacific coast.

BC fishermen, like our own, believe that the government pays fishery managers, partly from fish taxes collected from every load landed, to care for fish, just as the government pays police and educators to care for people. If stocks are poorly managed it is the government's responsibility to correct the problem, and not just by reducing fleets. As BC struggled with its shrinking salmon runs, one critic declared that DFO couldn't manage an aquarium. It was an accusation that over the next years seemed to be more and more on the mark to British Columbians.

In addition to vessel buy-backs and limited permitting, Canada also was looking at privatizing rights to fish, heavily favoring the largest producers, and was soon considering it for salmon. Another long source of trouble for the BC salmon fleets has been their weak market prices, significantly due to the province's decision to allow private salmon farming enterprises in its remote inlets, even with worries raised over disease transfer to wild fish.

British Columbia's salmon runs continued to decline on the same schedule as ours, and by the 1990s only the four largest rivers--the Taku, Stikine, Skeena, and Fraser--had marginally healthy salmon runs. Chinook and coho runs, especially, were in poor condition everywhere, and those were the fish the recreational fishermen wanted. As in the Northwest, sport fishing associations were growing in political influence. They naturally tended to point at the commercials as the blame for declining runs. Even though commercial fleets were now smaller, the boats were more advanced technically and able to catch more fish. But fishery managers recognized several additional causes of weak runs: Alaskan interception, hatchery releases not rebuilding runs, and river habitat deteriorating through logging and other industrial pressure. Still, several judicial reviews of Frazer sockeye runs' great fluctuations concluded that a variety of ocean conditions were the main cause.

Another Try: The Mifflin Plan

Canada was determined to avoid another smelly fisheries failure like the Atlantic cod disaster. The smallest BC settlements had seen their fleets decimated by the Davis Plan, but DFO decided the simplest way to save the runs was through another radical trimming of commercial licenses. Under the 1996 "Mifflin Plan", DFO proposed to cut by half the over 4000 salmon licenses (seiners, gillnetters, and trollers) remaining, and half of that through license buybacks. It imposed higher license fees on the remaining fleet to help pay for the buybacks. The Mifflin Plan also divided up BC's waters, with boats to fish only in their assigned areas. But "stacking" of licenses, another strategy, would allow a fisherman to buy or lease licenses out of his assigned area. Stacking favored larger operators that could afford to buy those added licenses, and keep the freedom to move around as needed.

The *Vancouver Sun*, April 8, 1998 editorial commented,"...the Mifflin Plan puts virtually the entire burden on the small owner-operators...." Dennis Brown (P. 264) reprints a colorful photo showing how fisherman protests provided great spectacle, such as the "Full Mifflin" dance troupe, their own version of the popular comic movie "The Full Monty", performed in front of the BC Legislature in 1998. But talent and humor couldn't stop DFO, as it declared that there was a crisis and nothing short of fleet slashing could turn it around.

Dividing BC waters into fishing areas with the licenses attached--a strategy also tried in the Northwest states--almost guaranteed that the fishermen who couldn't afford to "stack" would have poor seasons. Most of them expected to move following the runs; to stay all season in one area was possible only in peak run years. A group of fishermen sued to stop stacking as unfair to smaller boats, but the court rejected it. The fairest part of the Mifflin Plan, most fishermen agreed, was license buybacks. Even so, that was harder for small operators as it left them with the legal and physical problem of disposing of a boat that the government would not deal with and that could no longer be licensed to fish.

Mifflin, like the Davis Plan, especially hurt the small ports as many of the surviving fishermen relocated to Vancouver and other cities. The BC jobs commissioner himself stated that 7800 crew jobs were eliminated through Mifflin, and this time only a few qualified for unemployment compensation. Clearly DFO needed to do something different in salmon management, but it continued to choose strategies that moved fishing power toward corporate firms and other larger operators. It rejected recommendations from Canadian social scientists that unused licenses be turned over to local community entities to control, thus preventing so much urban flight. By 2004 Canfisco, BC's biggest processor, owned half the seiners and almost 250 commercial boat licenses in various fisheries. Social scientists estimated that small rural communities lost over 50% of their gillnet and troll licenses and almost 70% of their seine licenses. DFO conducted public hearings but didn't change course.

Though Mifflin did succeed in cutting the fleet to half, it didn't bring about salmon stock recovery. In 1998 DFO tried again with the Anderson Act. $400 million more was dedicated to what was essentially a continuation of the Mifflin Act policies. This act promised more economic aid to "retired" members of the fleet, but those funds in the end got chopped. Habitat restoration funds were included this time but couldn't begin to deal with the extent of the problem.

There didn't seem to be much mystery regarding coho and Chinook declines. Massive destruction of spawning areas had been going on for some time. Back in the 1960s we saw long, long miles of clear cutting along the BC portion of the Inside Passage. An American Fisheries Society study found that most coho stock extinctions were due to logging, hydropower development, and urbanization. Both coho and Chinook spend more time in the rivers than chum or pinks and therefore are bound to be more affected by damage to river habitat. Brown (p. 255-256) quotes Ron Kadowaki that the survival rate of coho during the 1980s, about 14% of the fry, had by the

1990s dropped to about 4%. Brown goes on that although DFO could not explain the losses, its own "...study conducted in the 1970s found that 70% of the original tidal lands in the Frazer River estuary had been altered as a result of human activity." Brown (p. 257) also quotes a Sierra Club report of 1997, that " '...83% of all streams were clear-cut to the banks...' ". But at the time BC had no "Spotted Owl" listing to address forests. It would be another few years (2002) until Canada passed its own "Species at Risk Act". It doesn't take much of a leap to suspect that similar impacts were affecting coho in Washington and Oregon.

Frazer River Mysteries

Canada directed millions to research, stock management, and license buy-backs, all with the goal to rebuild salmon runs and prop up the industry. The coho runs crashed in BC at the same time as they did in Oregon. From then on commercial coho fishing was regularly cut back more until today BC cohos are essentially a sport fish--or a farmed fish.

The more recent Frazer River sockeye crashes and booms should make any sockeye fishermen and their managers nervous. The Frazer feeds British Columbia just as the Columbia feeds our Northwest. Both regions' fish are chased by Alaska fleets as well as their own. Both regions have other rivers of importance, but those two are the queens. The judicial inquiries produced no solid answers. Frazer sockeye always are a challenge to manage. The returning fish on different years even take different routes through BC, confounding the managers as they try to set seasons and openers. Since the inquiries could come to no conclusion the typical solution was more restrictions on the fleets.

A inquiry after a recent crash, however, resulted in 70 recommendations for the salmon managers. Among them were many regarding possible impacts of fish farms: to consider a freeze on farms near the routes of migrating wild salmon, to fund more research on the effects of the farms, to close farms if negative effects were proven. The judge recommended removal of fish farm management from DFO because of an apparent conflict of interest between that and the DFO role of managing sustainable wild fish runs. He was also critical of DFO's management of First Nation subsistence and ceremonial use of river salmon. Far too many Aboriginal Fishery sockeye were found in cold storage, appearing to be illegally destined for the commercial market. DFO wasn't monitoring this, or it chose to look the other way rather than have another battle with First Nations. Larger treaty issues were on the table.

A weakness of the Frazer inquiries was that since they were concerned with sockeye, a commercial and subsistence fish, they therefore did not take on the issue of the growing recreational fishing impact on Chinook or coho stocks, or on habitat.

Even more political and environmental issues hatched on the BC rivers. In fall 2012 the BC Parliament voted a surprising "No" to a scheme to develop hydropower for a private industrial mine planned for the remote north. The project was bound to affect the Skeena River and other salmon runs, and was loudly protested by First Nations communities dependent on river fisheries that chose long-term salmon subsistence over a hope of future jobs in mining. In 2015 an upriver band refused to accept $1 billion in exchange for allowing a LNG plant to be constructed on its land. Alaskans should appreciate this conflict, so similar to their own over the proposal for the Pebble Mine project in the spawning grounds of the famous Bristol Bay sockeye run.

Interception Issues, Alaska Intransigence

Interception has been a chronic problem for BC salmon fisheries. Washington and BC fishermen regularly trade interception complaints. Though it is clearly counterproductive for fishermen to fight each other, run interception is hard to ignore when the runs are at high risk. Alaska's west coast gillnetters historically have caught Yukon spawners heading for Canada, Southeast Alaska gillnetters and seiners near Dixon Entrance catch fish headed for BC's Stikine, and in Stephens Passage they intercept stocks headed for the Taku River, its headwaters in Canada. Trollers intercept just about wherever they fish. Alaska's defense of its interception is that if BC rivers empty into Alaskan waters, it has a right to these fish too. Sometimes the problem reverses: BC trollers carry on their own interception of Washington and Idaho-bound fish. Scientists estimate that over half the cohos and up to 90% of the Chinook caught in the Gulf and Southeast Alaska, most of them by trollers, are headed for BC or stateside rivers. Clearly troubles on any of the Canadian or Northwest rivers could also wreck the Alaska fleets' seasons, as we found out in the 1970s.

As salmon get scarcer, "interception" wars get hotter, yet they don't necessarily affect individual fishermen's freedom of movement. During the years that I trolled and we traveled the Inside Passage between Washington and Alaska, we never had a problem with Canadian Customs. They ignored us as one of hundreds of boats passing through. A Canadian skipper cheerfully towed us into Campbell River one fall when our starter went out. This kind of courtesy happens all the time on a personal level. But in years when DFO and BC suffered from public outrage over the salmon fisheries, they would both try to deflect the anger toward the behavior of US fleets and our Congress.

The BC Public and the ENGOs

The BC environmental movement grew in strength about the same time that it did in the Northwest states. During the 1970s Canadian ENGOs insisted that serious attention to rebuilding salmon runs was more important than millions being spent for license buybacks and emergency aid to fishermen. Just as in the states, the public's building interest in salmon conservation didn't translate to concern for the families who depended for a livelihood on those fish. The voters were afraid of a repeat of the cod disaster when much of east coast Canada was forced to resort to seemingly permanent dole. The Canadian public, like the American, couldn't argue well about ocean ecology or fisheries management, but could point its finger at a visible target: modern salmon boats sporting all the latest bells and whistles of new technology as they unloaded at the cold storage docks and canneries.

To people who didn't know what a life's investment in fishing meant to a family, small-boat fleets looked more and more like subsidized rich men's fleets, a perception often fanned by the BC media.

The Pacific Salmon Treaty

BC, under the gun to save salmon runs, tried every strategy to make the boats fishing in Alaska cut back on their harvest of BC coho and Chinook. It began to impose heavy fees on US boats passing up the Inside Passage on their way north, and refused to allow US boats to anchor in west Vancouver Island coves during bad weather, and so forth. We in turn closed our grounds to BC fishermen. Finally worrisome conservation issues caused the two countries to form the Pacific Salmon Commission and sign a treaty in 1985. It addressed fair sharing of the harvest, stock conservation, hatcheries, and habitat protection. The negotiators settled on a fifty-fifty split for the allocation of fish between countries, and then, as with the Boldt Decision, had to figure out just how to manage the split.

For years the Treaty Commission had stormy waters to work in. BC fishermen felt doubly abused--by both their government and big sister US, both in general and in particular regarding salmon. The BC fleets regularly erupted into demonstrations, with their target usually DFO. During most of the 1990s the Treaty Commission was too much in conflict, mainly over interception, to even hold a meeting. But in 1995 BC found an unexpected legal ally when the Washington tribes covered by the Boldt decision sued Alaska over the interception of their federally authorized 50% of the harvest. That summer Alaska's fleets were ordered to stop fishing, for the first time support for BC complaints. Meanwhile, just as in the Northwest states, restrictions on salmon fleets continued to expand, and little was done about salmon habitat. Let fishermen fight fishermen.

BC fishermen hoped for more action in their favor when in the 1990s they elected a left-wing Premier, Glen Clark. He made stronger threats against the US for its salmon interception. Ottawa thought he went too far for neighborliness. But his leadership must have injected spirit into the

BC fleets, as the Northern BC gillnetters turned to direct action to get attention. At first it was attempts to block packers bringing in Alaska-caught fish to Prince Rupert processor that the canners had bought at a lower price than unionized Canadians had bargained for. Then the Canadians decided to stage a more newsworthy event. In July 1997 two hundred gillnetters and a few seiners organized a blockade to hold the Alaskan ferry *Malaspina* hostage in Prince Rupert for three days.

Again, the issue was intercept, but this time included sport fishermen who objected loudly that their quota had been cut. Canadian gillnetters at the same time suffered sockeye closures in order to allow steelhead up the Skeena for DFO's higher prioritized sports fishery. The commercials held up the ferry and its load of tourists to Alaska while the two governments danced around the international embarrassment. Politicians arrived on scene and argued that Pacific Salmon Treaty negotiations with the US were at risk. The blockade leaders were arrested. The fishermen decided they had done what they could with the event and dispersed. Alaska's Governor Knowles sued over the ferry blockade and eventually settled out of court for almost $3 million's worth of Canadian services to aid Alaskan tourism.

Perhaps the ferry blockade did help break the stalemate with Treaty negotiations, as by 1999 the stalled meetings were back in session. Though DFO told Premier Clark and BC fishermen to back off any further radical actions against the US, the fleets continued to protest against their own management with strikes and demonstrations whenever more restrictions came down. Eventually the US conceded that it needed to compensate BC for US fleets regularly taking more than their 50% share and agreed to pay a sum of $75 million over four years to be used for BC habitat restoration. It would continue to pay, as needed, for equity.

In 2012 the US paid as usual for the imbalance in harvest, but Canada decided not to use the funds for habitat or stock rebuilding as agreed, using it instead for more vessel buy-backs, this time for its troll fleet. Many trollers complained that they didn't want to be bought out, they wanted compensation for fish quota once theirs that had been given to sport fishermen. The lines of battle had shifted a little. The new treaty continued the trend in both countries of allocating more of the harvest quota to sport fishing. Most of BC's commercial fishermen continue to suffer small harvests and are required to share them ever more with recreational fishers. Today the very same issues are debated in the small ports of the Gulf of Alaska that have lost boats--even fleets--due to fishery policy, and fear they may be ghost villages before long. BC small fishermen, hearing of their troubles, may shrug: yes, we know what that feels like; now it's you Alaskans' turn.

The ferry blockade gave an energy surge to the fleets feeling abused and unable to get attention to their woes, but is a good example of small fishermen's inability to break through to the bastions where the big decisions are made about fisheries. Nothing really changed for them, even with a sympathetic premier. Nationalism and sense of impending crisis rule the fishermen's actions when they need to work together on common adversaries like industrial pollution and untargeted fish dumping.

Even top leaders don't get it. When Canada and the US couldn't agree in 2012-2013 on an albacore tuna treaty, Canada announced that US vessels once more would have to have a special license to enter its waters, which they need to do in order to travel to Alaska or to find refuge in storms. One person at least had more sense, a BC Tuna Fishermen's leader who commented, "There is no point in retribution…. What are you going to affect? A bunch of American halibut and salmon fishermen? We are all fishermen. Those guys have no influence on the tuna treaty. Why would we want to interfere with them making a living?" Why indeed.

The lessons from BC's salmon troubles would be repeated along the coast. Eventually even Alaska, the last stronghold for wild salmon, would begin to suffer like BC as its own west coast Chinook started to disappear.

Chapter 13

SAVING SALMON; SINKING FLEETS

I was half-watching a T.V. fishing channel scene of two fishermen in a skiff trying to wrestle a king salmon into a dip net. My heart wobbled as I recognized in the background the familiar cliffs of Point Amelia, near Sitka. Memories of our best days, circling off that point, pulling in the ancestral cousins of that fish, were soon crowded out by memories of a few years later, trolling for hatchery coho off those same majestic cliffs, but for nothing at all. It was the last place we fished before we pulled our lines and called it quits. Though there were short upswings in harvest, the Northwest hatchery coho runs couldn't regain their old strength. They still haven't.

David Montgomery describes a depressing history of earlier salmon run destruction in both Europe and the North American east coast that was impossible to ignore, but was. Legislation and regulation to protect Pacific salmon from similar disaster didn't have high priority among enough US politicians or publics, and the same inexcusable destruction would be repeated. Salmon fishermen I knew by the 1970s believed urban expansion and hydropower were too powerful to overcome and that habitat was going--or gone.

Montgomery goes on to describe the unproductive agency wars over solutions for salmon. Lichatowich agrees in his most recent book, "Salmon, People, and Places" on the influence that bureaucratic passivity and agency protectionism had in driving down the runs. For the fishermen, the political reality was that hatcheries were by the 1990s the only answer. Nothing else could work. Public acceptance of this made things worse, as it allowed managers and their funders to discount habitat rebuilding.

The other factor not mentioned so often, though by Montgomery, and by George and other Oregon salmon fishermen, was the federal movement to privatize fisheries. As George argued for the government, "Why pay for hatcheries or support commercial fishermen if you can get private business to take over?" This movement to privatize, far larger than salmon issues, is a main theme in Part II here, as it played out in our federal fisheries.

Salmon Byzantine

The specific blames for the loss of Pacific salmon runs make a long list, depending on who's talking. I've already mentioned most of them. Everyone agreed that the amount of trawler bycatch waste was unacceptable, but stopping it required federal government action on a large scale. Almost everyone also agreed that dams and their reservoirs were destructive to both juveniles and spawners, but fishery scientists' arguments weren't going to win over the hydropower lobby. Agreement broke down on the rest of the blame for salmon declines. Was it overfishing? Absolutely the main cause, said the dam managers. Absolutely not said the fishermen, and especially by the mid-1990s with the fleets and their harvests so reduced. Environmental groups and sports fishermen blamed both commercial fleets and dams, and some were beginning to suspect hatchery policy.

Jim Lichatowich's two books are Northwest salmon history as many eco-scientists came to understand it, and their view continued to gain ground. Actually, as Montgomery points out, many scientists had recognized the importance of habitat since the 1920s at least. Regardless of all the other impacts, you can't have salmon without rivers, and by the 1950s the Columbia was no longer a natural river by anyone's definition. But any steps that could bring back salmon their normal habitat would be a direct challenge to agribusiness, timbering, that powerful agency, the Army Corps that operated the federal dams, and even to hatchery programs. And all this would be in addition to the impact of industrialization in general. Though Lichatowich blames the salmon managers for their passivity--and where that is the case he is on target--how do you fight adversaries like those that lined up for the salmon wars?

Salmon activists among the ENGOs, fondly called "fish nazis" by some of their adversaries, focused specifically on the survival of wild runs, which in many streams seemed to be virtually extinguished. For many people that focus was irrelevant as a salmon was a salmon--interbreeding proved that. For others, perhaps there was some intrinsic value to wild fish, but they believed it was too late to save them. For the commercial fishermen, including Alaskans, the big policy question was not over wild versus raised salmon, but rather why were the hatchery budgets shrinking when commercial salmon was such an important part of the Pacific Northwest economy? I don't know if the fishermen had a fair chance to discuss this puzzle with managers and chose not to, or if they weren't invited to, but among fishermen along the entire coast the consensus continued to grow that powerful forces were working against the commercial fleets, and that they were at least as threatening as their traditional enemies.

The fisheries were not alone in their funding problems. In the 1980s-1990s a chopping of the federal budget was due to the growing number of conservative voters who favored more limited government. They objected to what they saw as a subsidy of commercial fishing, pointing especially to the large outlay for hatcheries. They could have had the same objection to the federal dam subsidies, and to oil, cotton, soybean, rice, and many other subsidies, but the agribusiness and oil lobbies were among the country's most powerful. Politicians, as usual, chose their fights. Those not dependent on coastal votes had other priorities than salmon runs.

Ocean versus River Blames

With so much pressure on departments of fisheries to better manage their salmon, researchers fell somewhat into two groups, one concentrating on the river life of salmon, the other on their ocean life, and no doubt some researchers tried to follow both. Both groups discovered how heavily environmental changes impacted the stocks. "What? It wasn't us?" I picture fishermen snorting. Scientists studying the ocean milieu discovered that the strength of salmon runs correlated strongly with changes in ocean climate and currents. Salmon increases were already known to be associated with upwellings of plankton that come with cold currents. Now they suspected that not only El Nino/La Nina currents, but also larger cyclical influences like the Pacific Decadal Oscillation appeared to affect salmons' food availability. Their forage fish, like sardines and pollock, had their own cycles dependent on ocean changes. Even spawner returns at the far inland Snake dams seemed to be affected as La Nina/ El Nino patterns influenced the rise and fall of snowfall affecting stream conditions.

Changing ocean currents could also bring in unusual numbers of sea mammals to finish off the fish in the estuaries that predator birds didn't snatch. With all this accumulated information, many scientists came to believe that the years of strong Northwest runs had probably been attributed too much to hatchery success, when actually favorable ocean conditions in certain years could have been more important than any other factor.

Trollers like George who also fished albacore had long known of the relationships of ocean currents to fish runs. The fleet could have insisted the scientists investigate this earlier if the communication between them had been better, but it is a common complaint still today that research often fails to take into account the anecdotal information fishermen have.

The North Pacific harbored man-made problems as well: illegal high seas drift nets, wasted trawler bycatch, agricultural run-off and other toxins, salmon farms in BC's inlets, and overfishing of forage fish for farmed fish feed. Of these human influences, the controlling of illegal high seas drift nets was the only problem in the ocean that we had made progress on, or even seriously addressed, by the 1990s. Another question scientists raised was over the number of hatchery salmon now entering the North Pacific from at least four countries. Could there now be too much competition for food in places where salmon congregated, such as the international waters of the "donut hole" in the Bering Sea? But the experts reported that the amount of hatchery salmon in the Bering Sea was about the same as a decade before.

The eco-scientists studying rivers found another wealth of probable and possible causes for weak returns. They had little trouble showing the effects of dam turbines on downstream passage of juvenile salmon, or the effects of spawning area destruction, and rising water temperature in slower current caused by drought was also easy to show. But the number of scientists grew who believed hatcheries could be more detrimental to runs than realized. Spawner "straying" was one area they looked at more. Normal straying by wild fish into new streams would aid diversification. But increased inter-breeding through hatchery fish straying into wild fish runs could eventually weaken wild stocks adaptation to their local streams and possibly shrink runs over time. This could have occurred especially in the 1980s when there were such massive releases of hatchery smolt. As a young person I had gazed at the bountiful scene in several Northwest hatcheries, and all those fish swimming around looked great to me. What could possibly be wrong with it? This was the general public consensus I'm sure.

Another more broadly recognized effect of hatcheries was that spawn-ers returning to their home hatcheries instead of to a natural spawning area didn't leave their carcasses in a stream. An essential part of the natural annu-al fertilizing of the food chain in the river was lost, and without it juveniles were short-changed in their diet.

Both groups of eco-scientists--those concentrating on rivers and those on the oceans--saw habitat as being the essential factor in salmon survival, neither group now pointing to commercial fishing as being, if it ever were, the major cause of shrinking or fluctuating runs. But the ocean regime was beyond human control unless governments could legislate huge changes that would slow phenomena like ocean warming, and soon a related con-cern, growing acidification caused by CO_2 releases. Negative changes in the rivers, however, could respond to many less drastic fixes. Concerned publics could even make some changes without politicians through such projects as trash cleanups. Fishery researchers and managers concentrating on rivers now focused on two goals: habitat recovery as the real answer for salmon restoration and improved hatchery programs, both of them costly undertakings.

The Northwest Power Planning Council eventually was able to win mil-lions over the next decades for river habitat corrections, most of it contrib-uted from federally controlled BPA revenues. But river restoration was still hugely expensive even if everyone did agree. Most of the hatcheries, due to budget cuts, were no longer maintained well for routine smolt production, much less for special projects. Soon debates over hatchery changes almost equaled those over dams, and the resistance to government spending on salmon, in an era of general budget-cutting, kept growing.

Lichatowich (p. 195-196) quotes General Accounting Office figures that surely aggravated budget-cutters. "From 1949 to 1981 the average annual cost of salmon restoration was $15.4 million per year; from 1982 to 1991 it was $122 million per year; and it could reach $425 million per year in the near future….About $3 billion has been spent attempting to restore salmon on the Columbia River over the past fifty years." In 1995 politicians complained that the Oregon salmon harvest value (after closure of the coast coho fishery) was about one-third its management costs. The message the public got was that commercial fishermen were getting an undeserved free ride. Oregon fishermen objected that these figures omitted the fact that during those years they were paying the dockside-value fee into the state fishery fund. George says their common belief was that the fees had actually built up a surplus in the fund until the money began to be siphoned off to the state's general fund. Such a move is not unlikely where states have that prerogative.

Enter the ESA

The first Pacific Northwest salmon, a Snake River sockeye stock, finally made the ESA list in 1991. All the combined planning energy, from NOAA and U.S. Fish and Wildlife on down, all failed to come up with solutions that could avoid ESA listings. By 1998, thirteen more Columbia-Snake salmon stocks were on the list. That April the *Alaska Fishermen's Journal* commented, "The spotted owl controversy was mosquito-like compared to the specter cast by the proposed listing of Pacific Northwest salmon stocks." The thousands of commercial salmon fishermen were now caught between two federal acts, both with positive intents: the MSA, intended to save fish and develop American fisheries, and the ESA, which looked past human communities to save wild species. Oregon's Dept. of Fish and Wildlife, faced with a threatened ESA listing, had in 1994 closed the coast coho fishery entirely, including for the many hundreds of charter boats. Even the recreational fishing industry's growing political power had not been enough to stop the closure. Saving the tourist economy of communities like Garibaldi or Depoe Bay was not an ESA concern; saving wild salmon was. An ESA listing required the managers to do whatever they could to stop any federal program from threatening the listed stock. A state ESA law had the same goal. Next, the government, worried that the coast coho hatcheries could be implicated, closed them too.

George recalled to me the effects on the small Oregon ports he sold at:

"The coho charter boats and the dory coho fleet and other small boats were hugely important. You know I had been a coho fisherman to start with. I remember there used to be so many charters out of ports like Garibaldi that you could hardly find a parking place for your truck at the harbor. Now, with no coho season, several small towns' packing plants closed down, and all the fishing supply businesses were hurt. There was a fleet out of Tillamook that was completely closed down. Tourism was hurt because that's why they come to the little ports--to fish and to walk on the docks and take photos of small boats and people working. They aren't interested in boats as big as freighters. If you take away the small-boat fleet, tourists aren't going to spend any time in places like Garibaldi past a lunch break."

The coho trollers thought the closure might last a season or two while the runs recovered. But even though there were no dams to blame, and after 1994 no commercial coho fleets, the rebuilding didn't proceed well. Two decades later, except for a corner of coast north of Cape Falcon, Oregon coast coho fishing stayed closed. As I listened to George, I wondered, what was going on in those rivers with hardly a dam? The spotted owl ESA listing gave a possible answer, the same as in BC: much of the loss could have been due to logging damage to spawning streams, especially affecting coho and Chinook. But restoring forest stream habitat would require major federal projects.

The ESA had a huge reach. For listed Pacific salmon, it pointed at the already much criticized Bonneville Power Administration (BPA), the Bureau of Land Management (BLM), Army Corps of Engineers (the Corps), Bureau of Reclamation (BOR), Bureau of Indian Affairs (BIA), National Environmental Protection Agency (NEPA), any hatcheries that received federal Mitchell Act funds, and probably others. NOAA and US Fish and Wildlife (USDFW) were the agencies charged with the unenviable job of seeing that all those federal agencies, and any others, complied with the ESA. Storms were sure to come. An environmental group like the Center for Biological Diversity could fill a petition that a species be listed. NOAA had to assure that the Corps, or other involved federal agency, developed a comprehensive plan for protecting every listed Columbia salmon stock and its critical habitat that the agency might impact. If no acceptable plan were forthcoming, NOAA itself, or the USDFW, would have to prepare one, or if the ocean outside three miles was involved, a regional fishery council could. A federal judge then had to accept or reject the plan.

Thus the ESA moved the balance of power away from the states and more to the federal level, but also forced the development of more consortiums of fishery interests, wild nature preservers, and federal agencies to work on these plans. For the commercial fishermen and their managers that trend could be positive, but the planning process under the ESA was certainly even more complicated.

The Eco-Scientists Gain

The Power Planning Council, as a response to eco-scientist pressure, meanwhile had created another group, the Independent Scientific Group (ISG), funded through Bonneville power user fees and charged it to provide "unbiased" opinion. The group focused on the Columbia Chinook and their river habitat but also covered other species and soon expanded to include ocean habitat. Its book, *Return to the River,* is not exactly unbiased as it has a strong ecological perspective. One ISG contributor stated that though overfishing and habitat destruction were both culprits, a fish might survive overfishing if it had its natural habitat, but a fish that had lost its essential habitat would not survive even if it were not fished at all. That was a gloomy announcement--but there was still hope. Eco-scientists believed that although it was unlikely that the main stem of the industrial Columbia would ever be a natural river again, it was possible for tributaries to be rehabilitated if the funds and the will were there. Not all fisheries staffs agreed with the eco-scientists, but budget priorities did shift a little toward habitat restoration.

Commercial fishermen saw the reduction in their harvests came from cuts in smolt production as funds were shifted, perhaps to habitat projects, and they complained. However, many like George, as I described, also suspected cuts in smolt release were promoted by the fish-farming lobby that wanted space for its own fish to grow. Such influences are complicated to trace, but it is true that decreased smolt production did take place concurrently with the increased interest by the government in private salmon farming.

Though the political interests that had supported huge hatchery production had lost ground, Northwest biologists would try simpler, cheaper, more "natural" ways to enhance fish runs such as incubation boxes and egg-planting directly in rivers. Incubator-raised fish return to their natural river to spawn, die, and leave carcasses to complete the natural cycle. Projects like this were also good for building public interest in stock and habitat rebuilding. Watershed volunteer associations began to be popular. George saw great success from one privately supported incubation project--his fleet caught the marked fish--and wondered why the method never expanded. Yet such projects are labor intensive and require on-going commitment over years from local volunteers. I am acquainted with a few sport fishermen who are happy to give their time to such work, but volunteers need staff to coordinate them and provide support services. Though projects like incubation boxes can work on a small scale, they can't cure the huge damage done to streams by industry or solve the ESA compliance problems of mega-agencies like BPA.

Disagreements among fishery managers were deep as to what direction to take. Some scientists believed that regardless of the ESA it was no longer possible to restore many of the wild runs on the lower river, that there were not enough purely wild fish left to protect. Others believed it was physically possible, but would be more costly than Congress, or as some put it, "the voters" would support. But whatever the beliefs about the future of wild stocks, or about the effects of hatcheries, or the loss of habitat, the ESA would rule what went on with federal projects. BPA would raise its subsidized rates to pay for projects directly involved with its dams. Other funds were much harder to come by.

George's troll fleet did not go under--not all of them. Though hundreds of the smallest Oregon boats were wiped out with the Oregon coast coho closure, the trollers that, like George, were able to fish more offshore could continue fishing for Chinook as long as they turned loose the forbidden coho. George recalls this grim period:

"After 1994 we were fishing just kings, even though there might be so many cohos swarming around us that they kept us from landing the kings. But still, we could make a living at twenty-thirty kings a day when the price was good. We were ordered to use barbless hooks, but that was okay, we already did for albacore and knew we could still catch fish. But then some of the Chinook stocks got ESA listed, so we were also ordered to stay out of areas where schools of wild salmon were passing. Then the managers limited us to only four spreads on a line. That means you're not covering as much water with your gear, so you have less chance to grab onto a school. Then they also instituted corridors. You could fish one period on one end of the coast and then that would close and they would open up another section. The bigger trollers could run up and down the coast, but that was hard for the smaller boats still trying to fish kings. Then they not only kept shortening the seasons, they instituted "days at sea", so we were fishing four days and closed down three days. Every time they added another restriction a lot more of the smaller boats gave up."

The only thing keeping the Oregon and Washington fleets afloat during this period was a new loophole in the ESA that allowed a commercial pastime to continue--in this case, catching fish from federally funded hatcheries--if it would not increase the endangerment of the wild listed stocks. Fishermen could keep harvesting if they released the ESA listed salmon alive. This was workable with trollers using barbless hooks, less so with seines and gillnets.

Gillnet Crisis

With the increased pressure of the ESA, recreational fishing lobbies like the Coast Conservation Association (CCA), always competitive with the commercials, now turned their national focus to getting gillnets outlawed. The Oregon coast coho closure made them especially determined to see that happen on the lower Columbia. The Columbia gillnet fleet, like those at Bristol Bay, is a small-boat fishery fishing close to shore with the crews of two or three, frequently family operations. Their fishing conditions at the mouth of the river mean fishing the tiderips at night, often in fog, while listening and looking out for not just drifting trash but passing freighters. But that fleet had picked nets in that dangerous estuary for three or more generations. With the increasing coho restrictions, those with Alaska permits spent at least part of the summer in the north, usually at Bristol Bay. Many who couldn't head to Alaska tried switching to other species like crab even though they didn't have the proper boats for it. Only ten percent of the Columbia gillnetters said they gave up fishing. I know the wrenching that would cause, and I commercial fished only ten years. Like trollers, gillnetters saw growing opposition to their livelihood, but when an initiative made the ballot in Oregon, the voters refused to close down the gillnetting. For once the public was activated in favor of small commercials--their neighbors. The CCA wasn't giving up however.

The state still had ESA compliance, not the CCA, driving its actions. Counting on the simplest solutions, it continued to offer voluntary license buybacks at intervals, especially for the seine fleets. In all fleets for the older fishermen worries about retirement were always hovering, and a buy-back/buy-out was a means. When I asked George about that for his troll fleet, he laughed.

"No need for that! There was nothing left of us. And when less of the hatchery budget was put into smolt production, there weren't so many fish for us to catch, so we weren't contributing as much fish tax, so then less revenues went to the hatcheries budget, so less smolt releases, so less fish. The downward spiral! As for the federal government, it really wanted to privatize everything that it could, and that included fisheries."

During the next down-cycle the poor runs extended clear into Alaska. Even Bristol Bay's "never fail" sockeye harvest did so poorly in 1997 that the governor declared it an official disaster. That failure had nothing to do with impacts on the rivers like dams or logging or urban pollution, or trollers--the Bay was free of all those. The culprit had to be out in the Bering Sea. But the Bristol Bay runs would always recover while the Columbia-Snake runs did not, not for years, and then the percent of wild fish was low except for runs that got expensive special attention.

ENGO Power

When the truth came out about the radical depletion of the wild salmon runs the growing conservation public was outraged. Salmon management came under ever more scrutiny as the environmental movement announced it would not allow the wild fish to be lost, whatever the interests of the industry. More petitions for ESA listing came down. Today over twenty-five biologically separate Columbia/Snake Chinook, sockeye, coho, and steelhead stocks are on the ESA list and many more from other western rivers. After 2002 the listings also began in BC through their version of the ESA, the Species at Risk Act.

The Chinook had grown in special stature, like the spotted owl, the wolf, and the whale, to become a symbol for how we were greedily destroying nature at its finest. But federal resources for expensive projects were now harder to win, even for politically powerful ENGOs like the Sierra Club and the Environmental Defense Fund. The political position of the ENGOs and the question of how much they influence federal and state policies are complex. They are not just "covers" for industrial interests, as some salmon fishermen believed at the time, though there was certainly that influence hovering. ENGOs include a wide variety of supporters, and the organizations' agendas don't always agree. Some ENGOs favor small commercial and subsistence fishing, choosing industrial fishing as their target.

Alaskan fishermen, perhaps more than others, appreciate the true power of the ENGOs and what they stand for partly because of the state's history starting with President Theodore Roosevelt and early conservation forces. The battle over the Alaska National Wildlife Refuge (ANWR) in recent decades has ENGOs fighting against big oil over the proposal to open up 19 million acres of wildlife reserve. Earlier, conservation forces helped kill the state's huge Rampart dam project, and today, along with fishermen, are in major battle with the proposed Pebble Mine near Bristol Bay and with a good chance of winning. The Center for Biological Diversity, the group that submits most of the requests for ESA listing, regularly petitions one Alaskan species after another, some successfully, and none of its victories appear to be to the advantage of industrial interests, except, perhaps, the "propaganda t-shirt" industry.

Dam Breachers

A special breed of ENGO became famous in the Northwest for its demand for dam breeching. Though after the Oregon coho crash, BPA could no longer be blamed so much for salmon losses, dams obviously did have a much to do with the decline on several important rivers. Dam breachers believed that for wild salmon to recover, many dams had to be removed or at least breached, to allow rivers to flow unobstructed. Breaching grew to a national movement, has waxed and waned, but has never been abandoned. One Northwest focus has been the breeching of four lower dams on the Snake River with locks built for barge passage, turning it into a canal. The four dams furnish irrigation to a few farms as a secondary benefit. The breachers helped to stop a planned fifth Snake dam from being built, so believe they can succeed again.

Breachers bring out interesting facts, such as that the lower Columbia is not essential for transport, and hasn't been since around 1900 when a busy railroad line ran to the coast, taking farm products into Portland. Now a four-lane highway runs alongside it. Breachers point out that though hatcheries are attacked for being subsidized, the Snake River's Port of Lewiston, the river barging, and the maintenance of the entire river transport system, like the dam system itself, are all heavily subsidized. Every business and household hooked to Bonneville power is subsidized.

As of 2010, over 500 non-federal dam owners in the US have had dams breached or removed, usually when the cost of bringing old dams up to stricter modern standards or replacing them would cost more than breaching them. It is a different story with the lower Snake dams. A federal scientist working on the river told me that the government would never agree to breaching the four, even if they are "less essential" ones, as it would set a precedent and (he said) the dam-breaching forces would then go after "essential" dams.

Dismantling of the two dams on Washington's Elwha River, the focus of Bruce Brown's book "Mountain in the Clouds", ended in victory. It took thirty years, but the message is not to give up on salmon restoration. The dams had been constructed in the 1890s to provide power to sawmills and growing towns, with hatcheries installed to supplant the wild runs. The Clallam tribe in collaboration with ENGOs won the legal battle to tear down the dams, and by 2014 restoration of Chinook runs on the Elwha was well underway.

While the breaching battles go on, BPA continues its ESA-required salmon mitigations. The main debates and court battles are over how to move the most smolt safely down river past the turbines. The dam operators prefer trucking smolt downriver. The other option, regulating spill at the dams, could work better for salmon needs but means the power wouldn't be produced at optimum for important corporate customers. Technical arguments are thus, as usual, also political arguments. The hatchery managers continue to try, with reduced funding, to produce smolt and keep all the breeds of fishermen happy, but also to protect and even rebuild the wild runs and their habitat, keeping the ESA watchdogs happy too.

Going for Other Fish

I have only told part of the Northwest fishing story. Some Oregon fisher-men were doing well enough in the stressful 1990s. Back visiting Newport again, forever drawn to fishing ports, I was lucky to catch George ashore. He steered me down the dock to point out to me the recent winners. Among them were big, well-maintained hake (whiting) trawlers, some of them also Bering Sea pollock chasers in season. Hake, like pollock, was another scrap fish that had become respectable for Americans. George commented that the humble hake had saved the town of Newport and others from going under when the salmon fishery sank. I recognized the name of one trawler that was under fire up north for hauling up excessive salmon bycatch. Oh, these tangled webs.

George smiled, "Pretty fine boat, eh? Well, NOAA financed fifteen or so of them to gear up for hake and to come into the area where we trollers usually fished. They said it was 'to develop an underutilized fishery'. Salmon trollers, a fishery already utilized, would be in their way, but we're mostly closed down now, conveniently for them. The hake boats are out there drag-ging on our grounds, while we're tied up. Our salmon eat hake so we're doubly injured. Their nets are two football fields long! Now we fish three days a week in a short season if we're lucky, and our salmon, several listed as endangered, are taken as hake boat bycatch. Somehow that's okay."

It was the old gear wars resurfacing again, but with the winners already called. Of course NOAA wasn't putting its entire attention to hake; funds were pouring into salmon research, thanks to Senator Ted Stevens' earlier ef-forts. But what the trollers saw before their eyes was what counted, and that was a monster harvest being unloaded at the Newport Cold Storage where they had for generations been the glory fleet.

"Lately, Newport's been bragging about its fish landings, but it's the hake, nothing else. They buy tons of hake for 3-5 cents/ lb and make it into surimi and sell it for 35 cents /lb. It's like Goodwill Industries! And we have to sit out there and wait for hours to unload if we happen to come in behind a hake boat. One of those hake skippers was heard to say, 'Hey, my boat can supply all the fish this processor needs. Why have these other boats?' Meaning us."

It was hard to see one fishing fleet pushed out for another, hard for me to hear of salmon being traded for hake. I recalled the MSA's Standard 8: "to provide for the sustained participation of traditional fishing communities...to the extent practicable." Saving the west coast salmon fleet apparently wasn't practicable anymore. Newport was doing better with hake--for a time. But ocean warming was picking up its pace.

Clearly the adversaries of the salmon fleets are not always so easy to prove. Their declines are not always just that we were/are "too much on the fish", or that "technology is to blame". Neither can trawler bycatch take the whole blame, or dams, or logging. Much is still a mystery as one learns from the recent debates over Alaskan rivers' Chinook losses. The disappearance of Yukon and Kuskokwim stocks can't be blamed on drought, dams, and very little on timbering or agricultural toxins. That is why so many people, myself among them, have joined the cause of reducing trawler bycatch.

There are so many pieces to this story that it is difficult to hold to the focus for this book, the loss of our small-boat fisheries and the causes. The court battles on the rivers, and their effect on the fishing fleets, could be their own book. A look at four rivers will be a start.

Chapter 14

RIVER BLINDNESS:
FROM THE YUKON TO THE SACRAMENTO

In 1999 I stood on the bank of the mid-Columbia looking at what restored spawning areas could look like. But it was almost unique, for the Hanford Reach had never been ruined for spawners. A month later we would have seen a successful return of fall Chinook spawners, over two-thirds of them wild salmon, the others from the Priest Rapids hatchery a few miles upstream. The Reach had special meaning for me as just a few hundred feet away my family had started their fruit orchards in 1907. David Sohappy's ancestors had spent a good part of each year just a quarter of a mile upriver. Scientists said in the 1990s that it was the only place on the main stem of the river that was still in semi-natural condition. I hope that today there are many more like it.

This sanctuary for salmon is normally closed off to the public on the west side of the river where the decommissioned nuclear plants are. Built in 1943 at Hanford and at twin hamlet White Bluffs, the site was where the plutonium was manufactured that was used to wipe out the town of Nagasaki in WWII. Later, as the plants produced peacetime energy, tons of radioactive waste were dumped into tanks in the ground, and more let to the winds, to the detriment of human down-winders' health. Hanford waste is still a major disposal problem, the worst in the country. No area wants the poison, apparently, for any price we can offer. But in a strange twist this part of the Reach is now a wildlife sanctuary and national monument, and wild Chinook and other salmon and steelhead thrive there. The Reach is not an entirely "natural river", but close. Because of the Vernita Bar Agreement, the Priest Rapids dam controls water flow through the spawning beds at a rate right for salmon to carry on their natural cycle, an exception in a river system now so sluggish.

My uncle, who grew up at White Bluffs, his daughter, and I were visiting the old White Bluffs site at the Reach on the one day each year when the 1943 evicted residents and their descendants were allowed into that no man's land on the west side. As we walked along the river to the farm site, the place where my cousin George also spent his early years, we could see ominous smoke from one of the plants upriver. But the bank where my family used to picnic was now a wild park with reeds, willow, and wild roses drooping down over the cut bank of the slough. Ducks, deer, and a porcupine watched us. The view of the tall, bright bluffs across was magnificent, big fish hovered in view in the clear water, while the noxious smoke drifted just around the bend, an ironic portion of the story I'd been following. Fisheries history is full of ironies, most of them not as gruesome as Hanford Reach, but not with such a positive byproduct either. For people without the whole history, the Reach is simply a great model for habitat and salmon restoration.

It was a good trip for me, being so immersed in George's gloomy fishing news, to have a positive salmon story to hold onto. But lately I have to hold on hard with the Chinook runs on the Yukon still getting worse, and the news of the problem spreading to other big Alaskan rivers. Chinook have continued in long-term trouble almost everywhere to the south, too, but for other reasons. On the Sacramento River and other California streams 2013 was declared the worst drought year in decades after many years of on-and-off drought conditions, 2014 was no better, and 2015 was projected to be the same. The drought was spreading north into Oregon and Washington with there being so little snowfall in the previous winter. The west coast wasn't the only region in trouble either; wetlands along most of our coasts, essential habitats for juvenile fish, were drying up or being paved over for parking lots and malls. National ENGOs focused their attention increasingly on the alarming conditions of western rivers and the restoration of disappearing habitat. But what could humans do about such long-term drought except reduce water allotments to irrigators and cities? How could the federal government, for the sake of salmon, cut off essential water to our national food basket and to huge California cities? The extended drought compounds all the other problems on the rivers.

ENGOs early on had concluded that to save wild fish they must do as the ESA provided: take the federal water users such as irrigation districts and their defender, the Bureau of Land Management (BLM), to court. From 2002 on, court cases regarding west coast rivers surged. Former adversaries saw the advantage of collaboration, and finally commercial fishermen began to join plaintiff conservation groups in lawsuits. Suits were entered even against the NMFS and the Pacific Fisheries Management Council. It's too much to summarize all the court battles over salmon and their rivers, so a few will do to reveal the salmon river dramas and the crisis their management faces.

Oregon Coho to Court

The stream of lawsuits over Oregon coast coho is a rich, sometimes bewildering example. It began with the NMFS petition and the ESA listing of the wild coast coho in 1998, four years after that harvest had been closed. The coast coho still hadn't rebuilt. Commercial fishermen had turned to Chinook or given up on salmon entirely. Coast coho spend their first year in many small rivers: the Newhalem, Trask, Siletz, Alsea, Coquille, and others and the fish are especially susceptible to whatever is taking place in the river, such as drought or trashed tributaries from clear cutting. In 2001 the Alsea Valley Alliance, a property rights group of businesses and farmers, took the NMFS to court to get the coho delisted since the listing put restrictions on land use. The arguments were the classic ones about wild salmon. One side argued that the wild fish needed the ESA protection to survive, while the other argued that since the coho, wild and hatchery, were the same species and interbred, the listing made no difference to species survival, so wild salmon protection wasn't merited under the ESA.

Alsea won their case. NMFS appealed to the Ninth Court in 2003, but Alsea was upheld in 2004, and Oregon coast coho were de-listed. The government chose not to appeal further. Instead NMFS decided to reconsider all of its current 27 western salmon listings. In 2006 it approved them all--except for Oregon coast coho. However, fishing and conservation groups joined to appeal for re-listing, and in 2008 coast coho were again listed, this time not as endangered but as "threatened". But that wasn't the end of it. The Pacific Legal Foundation filed a countersuit to keep the coho unlisted. The next year the federal court upheld the coast coho listing. By then the coast was virtually closed to commercial fishing anyway due to the Klamath Chinook crash due to water shortage. Widespread weak coho runs continued, and by 2010 the NMFS count for returning wild Northern California coho was 400, a number one could assume very close to extinction. Clearly the Oregon and Washington based scientists didn't need to blame themselves entirely for the tragic state of affairs.

If this court battles history seems extreme--it does to me--it's not that rare in ESA cases where industry or corporate agriculture is involved. Eventually NMFS allowed a small amount of commercial coho fishing targeting hatchery stock off the northern corner of Oregon and in Washington to compensate for loss of Klamath and Sacramento Rivers Chinook fishing. The wild coho caught would be considered "incidental". A similar run of court battles went on over the Lower Columbia coho (Common Sense Salmon Recovery v. NMFS). Then, for years, the remains of the lower river wild coho stayed unlisted, as 'indistinguishable" from hatchery stock.

A tremendous amount of time and money poured into these coho battles, and though the cases may seem frivolous when read in such a condensed form as this, they aren't when you consider that they were precedent-setting for many western mixed wild and hatchery salmon stocks lawsuits to come. They were also a warning of more political action to come when Washington joined Oregon in declaring recreational fishing to be higher priority than non-treaty commercial.

The Snake's Wild Sockeye Runs and Judge Redden

By 2000 the dam breachers decided it was time for more action on the Snake River, and a consortium of fishermen, tribal, and conservation groups entered a suit in federal court against NMFS, BOR, and the Army Corps demanding that the Snake's four lower dams be breached. The governors of Washington and Idaho didn't support the lawsuit, while the Oregon governor did. The plaintiffs argued that NOAA's annual plan for listed Snake stocks did not deal enough with actual recovery of wild stock and habitat. But meanwhile the ESA had been modified to state that "recovery" was not required of the federal agencies. They had only to prove they were not causing a stock to go extinct. But as one tribal representative put it, if you have one salmon of a stock left in the river, it's not extinct, but even if you do no harm to it, that stock is still going to disappear.

The dam opponents pointed out that the four dams were not essential for irrigation but were essential navigation locks for barge traffic (such as wheat) from the interior to Portland. But BPA and the Army Corps would fight with all they had to save the dams. They claimed that annual salmon enhancements on the Columbia/Snake cost $13 million, and of that, barging or trucking juvenile fish down past the dams was only $3 million, a much cheaper solution than breaching four dams, a cost the Corps estimated in 2002 at $267 million. Thus began a case that thirteen years later was still alive.

The 2006 re-enactment of the MSA strengthened the plaintiffs' case since it required more attention to protecting and rebuilding "essential habitat" for protected stocks, a far more expensive process than just providing salmon passage past the dams. Economists posed the question: How much did "society" want to pay to restore wild salmon runs? Of course there never would be an answer that society would agree on; the federal agencies themselves couldn't agree, especially during a period of growing polarization at Congress, with a strong faction determined to cut government spending at all cost. Yet there were also positive changes for salmon through the Columbia consortium alliances created. The Indians had beliefs, historical knowledge and practices, and treaty law on their side, and had begun to hire their own biologists. The conservationists had the ESA, eco-scientists, the growing body of research, more attention from the public, and sources of funds for court cases. Sport fishing groups, with significant influence on the states' budgets, saw habitat as more important than BPA's profit margin, and they too had money. The commercial salmon fishermen, what was left of them, had a shrinking piece of the economy, and no money to speak of, but they had an activist coast-wide association that now joined other groups in court: the Pacific Coast Federation of Fishermen's Associations.

On the other side were defending federal agencies lined up according to their missions, each with its army of biologists and economists ready to testify. They had powerful lobbying groups, such as Northwest River Partners and Pacific Northwest Waterway Association, whose goals were to keep the river "as a servant of the industrial economy". Much of the research presented by the pro-dam forces was funded by BPA, either through its support of science centers or through direct fees. This battle of one group of funded scientists against another, research scientists' recommendations versus fisheries managers' policies, and one government agency against another would continue for years. NOAA/NMFS states on its website that it has been in almost continuous litigation with the Army Corps and Bureau of Reclamation since the advent of the ESA.

Federal Judge Redden, who had taken on Columbia/Snake salmon as his special cause, continued, because of the ESA, to find in favor of the fishing/conservation forces. He regularly gave NOAA time to present realistic plans for protecting wild stocks, and if the agency could not, the judge said the only acceptable answer was indeed to breach the four lower dams. Over the next years, NOAA submitted one plan after another for Snake salmon protection, all rejected by Judge Redden, who determined that none of them would satisfy the ESA. The hatchery fish returns did improve, however. A Corps fisheries biologist told me in 2008 that officially over ninety percent of juveniles (mixed wild and hatchery), now passed downstream through all the dams in the Columbia-Snake system and made the estuary. Another more recent government study showed that the overall Snake River sockeye spawner returns were up to the level of 1936, pre-dam system. About the same time, another piece of good news was that almost 20,000 mid-Columbia coho spawners had passed the Rock Island dam near Wenatchee, the seventh dam from the ocean. Only twelve coho had been counted there in 1999. Would those successes satisfy the judge? No. The ESA's concern was saving wild salmon, not the size of mixed runs that were mainly hatchery fish.

It seemed as if the case would go on indefinitely, but then about 2009 four of the five tribes involved in the suit decided to settle out of court. They would drop the suit to breach the four lower Snake dams and in return would receive $90 million, though only half of it new money, from BPA. It would be paid out over the next decade to cover their research and management in restoring Columbia mid and upper river ESA-listed stocks and improving their essential habitat. One hundred fifty projects were selected, mostly for Chinook. The tribes' explanation for settling out of court was that they had done well winning in the courts but had not seen an increase in wild fish. It was time now to shift to actually changing the river habitat, rebuilding more areas like the Hanford Reach. A fifth tribe did not settle, nor did the conservation groups in the suit.

In 2011 Judge Redden again rejected NOAA's latest plan for the original reasons. He gave NOAA a new deadline of 2014 but stated he would be stepping down from the case before then. In 2012 he did so. Was he simply exhausted with it all? But ESA was still the law of the land, and a few remaining beaten down commercial fishermen were among those waiting to hear the outcome of the next NOAA plan and the decision of the next federal judge.

The Klamath and Sacramento Rivers Disaster

The Columbia/Snake case was equaled by the Klamath Chinook crisis that drew more cases, fresh soldiers, and media coverage on the effects of drought. The Klamath River, once the third top producer of Pacific salmon, begins in Oregon, empties into Northern California, and has old private dams with no ladders on the lower river. Upriver it has some of the most pristine streams and lakes in the far west. Battles between irrigators and fishery managers have been going on since 2001 over water draw downs, Some years the farmers win, sometimes fish and fishermen do. In 2002 BOR gave the irrigators more water and did not leave enough in the already drying river for spawning salmon to make their way up. Seventy thousand Klamath Chinook salmon died before spawning.

George reported, "It was a major scandal on the coast. A fisherman I know went down to the Klamath, and in one area he saw just puddles in the riverbed. In another place the river was running upstream." It was actually a common sight, and the problem was bigger than low water. The build-up of toxic blue-green algae due to the slowing and warming of the river was 4,000 times the limits set by the World Health Organization. A virus endemic to the river was also able to propagate as never before. By 2005 scientists estimated that 80% of the Chinook juveniles on the Klamath were infected, with a mortality rate of 100%. Faced with such a huge loss, the Oregon Dept. of Fisheries responded with more restrictions on the troll fleets. The PCFFA joined conservation and tribal groups to file a motion to reduce the Klamath irrigation diversions in the future. They lost and appealed. The drought continued and spread.

In 2006 the Ninth Court found in favor of the pro-fish plaintiffs and ordered the BOR to limit irrigation withdrawals. NMFS called for breaching of six lower dams owned by private Pacific Power. But the irrigators also filed a suit against the government for losses due to inadequate water, and won $16 million. The Pacific Council had closed the commercial Chinook fishery below Cape Falcon (most of Oregon) until the Klamath run could rebuild, while the coho closure was still in effect for the same area. After huge fishermen protests the Council agreed to a limited Chinook harvest. But in 2009 the dressed poundage of salmon for Oregon, all species, was roughly 500,000, one-tenth of 2005. The Klamath was still almost unusable by spawners. Trollers in some areas below Cape Falcon were fishing for a two Chinook per day limit--hardly enough to pay for their fuel.

Twenty-nine parties signed an agreement to breach four of the old Klamath dams. Removing the dams (part of Warren Buffet's empire) was going to cost less than bringing them up to modern fish passage standards.

The dismantling would take about ten years; meanwhile the salmon would have to hold on. The customers would pay the costs of the dam removals through a surcharge, but they would have had to pay for any dam upgrades anyway. Signers to the agreement included water districts, commercial fishermen's organizations, government managers, and conservation groups. Congress would have to approve and appropriate funds. The Center for Biological Diversity (CBD) and three other conservation groups sued to get some of the Upper Klamath Chinook ESA listed and to have fall Chinook considered for listing.

Events on the Sacramento River were of more concern to what remained of George's troll fleet, as that river was where most of their catch came from, and was the largest of many California rivers in drought trouble. As far back as 1987, 80 to 90% of the water had been directed to irrigators of the huge crop-producing Central Valley, an area focused on water-demanding corn, rice, cotton, and especially almonds. However, if one looks at the per capita farmer involvement, not volume of crops produced, only 9% of California farmers were receiving federal subsidies. So the situation became for a time big agri-business power against almost powerless small salmon fishermen. By 2013 the drought would democratize the situation as all farmers saw their water cut back, or even off, and their crops in death throes. Almond trees in crisis made big news. The state was slow to order mandatory water cutbacks for the cities, but in 2015 felt forced to.

Historically the Sacramento-San Joaquin river system had produced a run of two million Chinook. By 2009 it had already dropped to 41,000. Fishery managers were soon forced to truck the smolt to the estuaries. As I write, it seems most likely that rivers will keep drying for a while and California salmon will continue to collapse.

Probably these are enough examples to reveal how tedious and polarizing salmon court battles and salmon management in general can be. But as frustrating as it is, we have to wonder where would the wild salmon be, at this point, especially the Columbia salmon, without the ESA? Nothing else seemed to have worked to save them. Few of the successful corrections at the federal dams would have taken place without the ESA. And would

Oregon have dared to close down the coast coho fishery without the ESA threat hanging over it? Would a major review of hatchery policies have taken place? Would the Chinook bycatch reduction programs for pollock trawlers have gone through without the ESA listed Chinook involved? And would any serious habitat restoration have ever take place?

NOAA Throws a Bone

One evening in about 2005 George called with a more upbeat tone in his voice. "Well, guess what, NOAA threw us a bone! A little open-access black cod pot fishery. I think about half a dozen of us are taking advantage of it, and more at Coos Bay. I can't make any real money at it--it's only a part-time fishery--but it helps. You remember those fine looking boats I pointed out to you, tied up next to the *McColl*? Those are the real black-codders. They can gross $500,000 a season. But it's one of those privatized fisheries with their own personal quotas to fish, and for a black cod tier you'd have to pay $80,000-$100,000. (He laughs). Some even have three tiers stacked! What ordinary fisherman can do that?" (I thought it sounded too much like NOAA's brainstorms for the Alaska federal waters fisheries.) "Yes, that's what it is. Of course they don't want us to really become black cod fisher-men. But the Magnuson-Stevens Act says NOAA can't just let us sink, so they've opened up this little part-time fishery for us and I sort of like it. It's only three days a week and I'm allowed only 1000 lbs.--enough to buy the fuel, ice, and pay some bills on the farm, but at least I'm out fishing."

For the next couple years George's calls were more cheerful.

"You know, I really could like blackcodding. I can do it by myself, don't have to hire any crew. I run five or six pots, baited with hake. I head out to 150-200 fathoms, throw out my pots, and fish till I have my quota. It takes me three days of the week, a day to get out there, fish a day, pick up my pots, and run home. I can even do it in the winter in decent weather. It would be a good fishery if we were allowed a larger quota. There are lots of black cod and the Japanese pay big bucks for it so there's a sure market. But no, NOAA doesn't want the small boats to actually make a living because that would keep us in the fishing business. I hear the big boats are already complaining about us (he laughs) cutting into their profits."

I observed that the middleman was doing all right. George was being paid $2/lb, and in Anchorage that winter blackcod fillets were going for $19/lb. retail. But NOAA's open access fishery was intended only as a transitional program, an aid while the salmon fishermen tried to find new ways to survive. The next year the trollers-turned-blackcodders were cut to 800 lb. a week, and by 2008 they were cut again, being weaned. NOAA had done its duty by the MSA's sorely neglected clause that ordered federal managers to do "as little harm as practicable" to the traditional fisheries. A small amount of open access fishing had to be available. George said that another open-access fishery was also available: slime eels for a dollar a pound.

I thought back on the long journey my cousin had taken, starting with that first trip as a kid to Southeast Alaska, then to become a successful salmon troller, and now a part-time blackcodder, saying, well, he could do that. After the initial high from blackcodding wore off, George's calls with me slipped to a less cheerful tone and more cynical humor. I could identify with it all, yet be free of it, fishing subsistence only with no worries about money.

The Oregon troll fleet scattered along the coast and across the fisheries. Some went north of Cape Falcon for wandering Washington hatchery coho, others were able to chase albacore. The lucky ones with Dungeness crab permits switched gear and ended up with some good seasons, but for many, there was little reason to untie from the float. For years hake would probably be the main harvest keeping Newport and Westport processors operating.

Re-toolings and Recoveries

After two decades of technical efforts, though there has been success with special projects in mixed-run restoration on the upper Columbia, as I write there is no news that wild salmon runs on the lower river have been rebuilt. Rehabilitating streams takes time, and is not only expensive and labor intensive, it usually involves getting agreement of private landholders. As for hatchery runs, the state managers, probably for both scientific and financial reasons, dropped the policy of ever-increasing smolt releases. According to Washington's Dept. of Fish and Wildlife, its annual releases (all salmon and steelhead) dropped fairly steadily from 1990 on, and that was also the case in Oregon and California. However, the public still demands a healthy return of spawners. For every commercial salmon fisherman that drops out, five new sports fishermen will probably head for a river--or maybe twenty-five. The love of fishing isn't going to go away.

The Northwest rivers haven't yet suffered from drought like the California rivers even though drought conditions have crept north. Many salmon runs were increasing again by 2010. But according to the Washington Dept. of Fisheries, over 90% of Columbia salmon and 75% of Puget Sound salmon returning were hatchery fish. Will this small percentage of wild fish satisfy the ESA? A federal judge will decide. To do better than this we will have to what the eco-scientists had been saying for at least two decades: to rebuild wild runs we have to concentrate more on river habitat restoration.

The theory about Bristol Bay runs' consistent success also has gained more supporters. Breakthroughs in DNA tracing have given scientists proof of what they had suspected as to why certain salmon stocks had so much more resiliency than others. Unusual genetic diversity is what has kept the Bay producing and producing. No matter how many stocks are pummeled by disease, flooding, ocean conditions, and bycatch, as long as they have scientifically set quotas there are so many distinct stocks in the Bay that most will survive and through natural straying will re-propagate streams that have had losses. Could that keep happening with global warming? No one knows. But aside from that, the implications are clear from that theory, that the hatchery systems need reorganizing. And a growing number of managers do believe in the priority of habitat restoration. The ESA forces the others to listen.

An article in the Seattle Times (Feb. 28, 2015) gave hope in that direction, reporting that there were 350 restoration projects going on the upper Columbia and some showing real progress. The year before, a run of wild sockeye salmon arrived at the Okanogan River, a tributary of the upper Columbia. They were greeted by a group of jubilant folks, members of a tribal consortium that had used some of their settlement money to undertake a restoration project, and it was their first return on that brood stock. A run of 900,000 of up- and mid-river fall Chinook was forecast for 2015, most of them wild, from the Hanford Reach area and above the Priest Rapids Dam. The promise of the Reach, the place where I had watched wild Chinook spawners back in the 1990s, has proved out. Does that mean that the crisis now in about its sixth year on the western Alaskan rivers is temporary, with no dams, no pollution, no overfishing, (though some level of a bycatch problem), and that those stocks, too, will come back?

The Northwest commercial salmon fishermen are now in worse shape than the fish if you accept stock rebuilding as including hatchery fish. For several decades the "treaty tribes", with their traditional concern for community and their clout through the federal court, have been the only Northwest groups to win action from the government with the argument that the welfare of fishing families, not just fish, must be recognized. The exception, NOAA's disaster relief and retraining programs, are actions that do help fishermen, but are no true solution for people who have spent their lives fishing and want to continue to, or for whole communities based on the fishing industry.

Salmon fishermen, bitter in their lack of broad public support, need to open their eyes and see that to win anything at this point in history they need the political clout and energy of groups like the treaty tribes, sport fishermen, and ENGOs on their side. Without that, can fishing ever get better? I ask myself, why are people still trying to fish salmon commercially at all? It must be because there is always Bristol Bay with its promise. Fishermen and managers must believe that if wild salmon runs can occur at the Bay year after year, no matter the obstacles, they can recover again somewhere else, sometime, along this long coast.

Chapter 15

The Beacon Bright

The words "a beacon bright" are from the official Alaska State Song, and reflect the hopes of the many fishermen who come north each summer that Alaska will again treat them well. Of the over 30,000 people who were licensed to fish commercially (all species) in 2012, a significant percent were from "outside". They included about half the Bristol Bay gillnetters, for example. The large Bering Sea and Gulf of Alaska industrial fleets and floating processors that hail from Seattle play a big part in keeping that city's waterfront services profitable. For those chasing salmon--commercial or recreational--Alaska overall has treated them very well. The 2013 statewide salmon harvest was a record breaker, with Southeast and Prince William Sound harvests both beating famous Bristol Bay. 2014 was another record breaker, a harvest of 157 million worth at dockside $113 million, and more yet are forecast for 2015.

Alaska historically has done much better than the Northwest by its salmon runs and its salmon fishermen. Unlike some regions, fisheries big and small, and including commercial, subsistence, and recreational, still hold a strong place in the Alaskan economy. The value of the state's commercial salmon fisheries is still number three after oil revenues and military spending--in 2014, $407 million in value-added revenues just to Alaska; $1.1 billion to non-Alaskan fishermen, brokers, and processors. Almost half the nation's seafood exports are sent from Alaska--two-thirds of what Alaska produces. The state counts 136 Alaskan communities as dependent on recreational and/or commercial fishing for cash income. That doesn't count the many more dependent on subsistence fishing where no cash is exchanged.

The biggest population of commercial fishermen, like the rest of the residents, lives in the Anchorage-Matsu bowl. However, hundreds of fishing communities with fewer than 1500 residents still exist and are dependent on their small-boat fleets for their economic survival. And here is where a problem lies. Sometimes it seems they are almost forgotten when it comes to state and federal legislation. But then groups with political power--the Alaska Native organizations, or the recreational fishing associations, or sometimes a conservation group, will remind the State of its duties to its small-fishing constituents.

What the general public--and some politicians that should know better--seems to miss is that Alaska's resident boats, including those fishing in the federal waters outside three miles, are largely small boats owned by families and small enterprises that are the concern of this book. In addition to the salmon fleets, which constitute about half of the commercial licenses, virtually all of the Alaskan-owned boats, even trawlers, longliners, pot, and jig boats--are less than seventy feet, and are operated by families or small combinations, not large firms. As an example, Petersburg, a small but important Southeast fishing town, has 75 registered commercial boats, with only thirteen of them longer than 58.5 feet. This local fleet character is true among all of the coastal communities. The huge catcher-processors and factory freezer vessels that fish the Bering Sea and the Gulf of Alaska may unload and fuel up in Alaska, but they are based in places like Seattle where there are shipyards and experts that can service them. This fleet and community character is important to recognize when you analyze the effects of the management changes that started in the 1970s. Salmon boats, however, have stayed small boats, wherever they hail from.

Though Alaska's fish stocks are in much better shape than in the Northwest or in British Columbia, this isn't particularly because the managers are brainier or more dedicated. More influential are geography, history, and unique political influences such as the Alaska Constitution and the Alaska National Interest Land Conservation Act. Alaska's fishery managers, especially after the oil revenues began flowing, definitely have enjoyed advantages. The State has so much invested in the fishery business that "Fish" has to have a separate state board from "Game". Regional advisory councils of citizens, both state and federal, have regular involvement in management and to ongoing lively debates. Recently one of the stormiest battles, described more in the next chapter, has been the recreational industry versus the gillnetter fleets, a northern version of a conflict that has spread across all of our coasts.

Alaska's Salmon Hatcheries

With all the recent distress over declining salmon runs--the Chinook in western Alaska, the sockeye on the Frazer, the Chinook and coho in the Northwest--one gets the impression that the Pacific coasts are running out of salmon. Yet overall, the Alaskan salmon harvest, has hit records, or been close in recent years. The problem is, they weren't the Chinook that many west coast and river communities depend on and the environmental groups focus on, nor the Chinook and coho that the sportsmen want. The great harvests included 34% salmon from Alaskan hatcheries: over half the chums caught, almost half the pinks, 23% of the Alaskan coho, and 12% of the Chinook. If you also count all of the Washington and BC fish passing through, hatchery-raised salmon are now likely over half the total commercial harvest in Alaska. They are 70% of the total hatchery releases from the west coast US. The humble pink is the biggest salmon harvest by volume, and second after sockeye (mainly wild) in value. Southeast Alaska chums, largely hatchery bred and a fish that trollers once ignored unless desperate, are now so numerous and bring in such a price that a troll fleet out of Sitka targets them. If this move into hatchery produced runs sounds familiar, I hope it is not a forewarning of the trouble for Alaska that the Northwest and BC have seen.

Alaska has also made exceptional advances in marketing these big runs, with the State putting millions to marketing research and development. Alaska's chums and pinks have far better markets that they did when I commercial fished--I recall dressed pinks were $.50 apiece. Frozen fillets can now be sent to China to be thawed, have pin bones removed by hand, then refrozen, then returned to us for sale profitably to Europe. Canned fish are still available, but frozen fillets are the favorite fast-food fish dinner grabbed up around the world. Still, the market has its limits; during some peak years the Alaskan pink harvest is much more than the market can absorb, despite the popularity of the fillets and government purchases for food banks, schools, and humanitarian aid.

Alaskan hatchery history is not just a rerun of the Northwest's however. It began much later, in the 1970s, with the salmon runs just coming out of a cyclical downturn, and with every other western state providing lessons from their own hatchery systems. The state decided it would not supplement wild runs for commercial fishing but rather would stabilize them by evening out the natural cycles. (This didn't succeed as we can see by the pinks continuing their every-other-year pattern.) In 1992 the state strengthened a regulation that made wild salmon protection highest priority, with the hatchery fish--mainly cannery-bound chums, pinks, and sockeye--to be kept separate from the wild runs, thus avoiding the risks of mixed stocks. Alaska prohibited hatchery practices like transferring of smolt or eggs from other regions, and discouraged installing of hatcheries upriver. But today in Prince William Sound over 80% of the returning salmon are now from five hatcheries. The Sound's wild runs, whatever they once were, are not the important haul there. But it is important for the future of all salmon runs to recognize that 95% of the remaining wild North Pacific salmon are hatched in Alaska, with few still coming from the Northwest and BC rivers.

For years Alaska operated only about fourteen hatcheries, and instead promoted private and non-profit coop hatcheries. If commercial fishermen wanted hatcheries, they could get some technical assistance from the state, and oversight on sustainability issues, but they had to operate the hatcheries themselves. They did this through private hatcheries or by creating coops and charging members a percent of the catch for the right to fish those runs, or contracted boats to fish for them with the profit going to hatchery cost-recovery. Each hatchery opened and closed based on its own successes. If problems developed with the hatchery runs and the fleets that depended on them, those fleets could get advice from biologists, but unless they were violating standard conservation practices, the decisions were up to the coop. Probably federal and state governments regret that they didn't early on create something similar in other regions. Today the state still owns only eleven of the 26 salmon hatcheries.

Alaska had many other advantages to start with, not only the Northwest hatchery successes and failures to learn from, but runs that were not badly declining when it began its hatchery program. Despite logging and mining impacts on many streams, the state had river systems still in good condition to select for hatcheries. Making the task easier, too, is that the troll-caught Chinook and coho are mainly Washington and BC fish passing through, with those regions largely responsible for their sustainability.

The opportunity to raise, catch, and sell great numbers of fish will always attract people, and one of the Fish Board's duties is to say yea or nay to new proposals. Curiously, so far, no ENGO has raised a loud cry about the effects of unintended mixing of hatchery stocks with wild stocks. Is it because chums and pinks don't have the iconic status of Chinook, and therefore we can produce as many as we like? Biologists are split on the harmful effects of hatcheries, as described earlier. But it is also relevant that no native Alaskan salmon has yet to be ESA listed. If that should take place, there would probably be immediate concern about the fish pouring out of hatcheries.

Bossing the Fisheries

The Alaska Board of Fish has the job of setting policy and quotas not just for salmon but for every species in the state waters and it shares shellfish with the federal council. With all the complex ethnic, cultural, geographic, and fleet issues, often at least one group is going to be unhappy with a proposal before the Fish Board. One proof that Alaska still cares strongly for its commercial salmon fleets, not just for the significant urban sport fishing constituency, is that the state has yet to rule recreational use a higher priority than commercial. Sportsmen have considerable influence on the Board but so do the commercials. Subsistence use is highest priority.

The Board has wisely spread responsibility and deflected cronyism by keeping for itself all the heat over annual allocations of quotas for districts or fleets, while seasonal management decisions like emergency openers and closures are in the hands of regionally-based biologists. The biologists are thus freed of much pressure, which is always to the benefit of the smaller, local, least politically powerful users. Fish and Game's local biologists manage seriously for adequate salmon escapement and don't hesitate to open and close fishing times based on how a run is doing. Nome people's grievances get listened to through the regional advisory councils, both the state's and the federal Fish and Wildlife's. Though Nome district management is probably an easy assignment compared to what the managers deal with on the Kenai Peninsula, we have had our share of hot meetings over the reasons for poor runs, badly timed closures, interceptions, and destabilizing experiments like the Board's chum Tier II project.

The regional advisory councils are important access for the public to fisheries decision-making, and where the smallest fishermen may succeed in getting changes initiated--or stopped. For the State Board, any resident is free to submit a proposal for a new program or regulation change. All of them go into an annual book. In 2008, the last time I saw a paper copy, it was a typical 283 pages, containing 367 proposals of concern to citizens somewhere. Now you can read this report on the web--what a paper saving. The staff reviews the proposals before the Board takes them up, but it forwards all of them whether it agrees with them or not. In 2010 the Board spent a week reviewing dozens of proposals submitted on just one issue alone--personal-use dipnetting. But democracy was never efficient; that has never been the point.

The current distress over the long-term poor Chinook runs has formed a gloomy cloud over Alaska's Dept. of Fish and Game. Not only are a growing number of fishermen of all types demanding that the state do something about missing Alaskan Chinook, the Canadians through treaty expect a certain percent of the fish to make it up river to them. When run failures extend over years, as they have with Chinook, normally quiet groups like the Marshall fishermen become vocal.

In 2012 the subsistence chum fishermen on the Kuskokwim River, once more closed down to allow Chinook up the river, launched a chum fish-in and 33 were ticketed. It triggered high politics as US Senator Lisa Murkowski said, "I think this is an embarrassment for how our government and federal agencies ignore tradition and culture of our first people." US Representative Don Young called for a "much, much, much lower" Chinook bycatch cap for the Bering Sea pollock fleet. He went on to pledge support for representatives of Yukon and Kuskokwim River villages to have more involvement in salmon management in order to get decisions with more "in the field" fishing knowledge, and to win support for fishery management decisions. If anyone cares about the long-term survival of the salmon runs it is the Alaska Native families that depend on them. But the court found against the illegal chum fishermen and it remains to be seen if they will appeal.

As each spring the Chinook problem seems to spread to more Alaskan rivers--the Copper, the Taku--the situation recalls the Northwest where in the 1990s wild coho were disappearing, and Frazer sockeye fluctuated wildly. What or who's to blame again becomes another shouting match over salmon, this time in Alaska. There are no big dams, logging companies, or huge urban pollution sites on the Kuskokwim or Copper Rivers. Was it the commercial gillnetters on the lower rivers as the sport fishermen claimed? Or was it the sport fishermen with their boat traffic on the Kenai, but yet no factor on the Kuskokwim? Or was it trawler bycatch, as so many west coast fishermen suspect? Or, as many scientists suspected, was the Chinook problem mainly caused by changing ocean currents and temperatures? But answering such questions takes research funds, and a great amount has already been spent on very similar questions in the Northwest.

In 2014 Alaska Fish and Game decided to expand an experimental dipnet-only subsistence chum fishery on the Yukon, which had seemed promising, over to the Kuskokwim River. That way, the at-risk Chinook caught during a chum fishery could be released and chum fishing would not have to be closed. The Kusko fishermen who tried it were not happy with the results. Alaska Native organizations again asked for more involvement in salmon management on the rivers. Coastal Alaska Natives had already proven years ago that that co-management could be successful with marine mammal management, so why not fish? By spring of 2015 Alaska Fish and Game had made steps toward more involvement of the fishermen in discussions over what should be done next. The federal managers of a reserve area on the Kuskokwim announced that they would allow a greatly expanded community subsistence fishery of 7,000 for Chinook, and with set amounts for selected villages and selected families, with local leaders coordinating the catch. It was the beginning of co-management with the feds.

The Alaska Board of Fish is only one of the governmental entities involved with fishery management, and salmon is only one of the stocks embroiled in controversy. Another recent problem was how to allocate a shrinking seasonal halibut quota between the commercials and the charter halibut fleets in the Gulf and Southeast, with the International Pacific Halibut Commission (IPHC), North Pacific Council, NMFS, as well as various fishermen's associations all drawn into the issue. The IPHC demanded Council action to better protect halibut stocks due to an unusual percent of discarded small illegal fish. Mature breeding eleven-year-old females in the Gulf, which in the 1970s averaged 50 pounds, from unknown causes in 2012 averaged 20 pounds. To overcome this the IPHC had continuously dropped the commercial fleet quota until in the Southeast it was less than half that of fifteen years earlier. But meanwhile the halibut charter fleets had regularly gone over their share. The government ordered the charter fleets to limited permitting, a change that fleet had fought for years, beaching a good number.

Several coastal communities like Homer that depend on both charter and commercial fleets were drawn into the dispute. All of the classic fishery issues emerged: MSA requirements, fish stock problems, fleets' growing competition, cheating, community economy, and politics. Some of the charter boat skippers were former commercial fishermen whose neighbors said they shouldn't be thrown to the sharks another time, especially after all the years the North Pacific Council had delayed mandating a halibut bycatch program for its trawlers. The economic plight of small fishing communities that once were healthy fishing ports, but now relied much on chartering, was a situation that the Council had helped create through introducing individual quotas to the commercial fleet (more on that in Chapter 28). Now the Council needed to deal with what it had created.

The final plan, after two years of study and negotiation, made few people happy. It increased the commercial quota and cut the charter fleet quota. Through a "catch share" program the charter boats would be able to buy or lease more quota from commercial boats if they could find willing sellers or leasers. The plan also set aside quota shares for local use by small communities that had lost fleet though the Council's halibut program. But because of the declining size in caught halibut, charter boat clients already limited to only two fish per day each would have to take one of them at eight to ten pounds, a rather small halibut. Would charter clients pay $200 a day to come aboard for two fish, one of them small? Apparently, yes.

Dams Again and Now Mines

Such resource controversies emerge regularly in Alaska. A huge Rampart Dam project that the state planned decades ago was killed due to environmentalist and Native concerns over the region's ecosystem. Now the state has embarked on another similar huge dam plan to provide hydropower, temporarily on hold because of the state's financial picture due to oil prices. The proposed giant Pebble Mine development at the headwaters of streams that flow into Bristol Bay is the project generating great heat. These streams, as most people know by now with all the media exposure, support the largest sockeye run in the world. Three years of constant video messages pro and con "Pebble" had the frenzied feel of election propaganda. The battle forced everyone's attention to an eternal issue in the West: local job opportunities (in the mine) versus a renewable resource and communities' long-term sustainability through fishing. Too often, potential dangers to fishing families have not been obvious until it was too late. This time they had ample warning. The EPA also entered the fray early to try to stop the mine. Alaskan commercial fishermen and sport fishing businesses combined their energies to fight Pebble, proving that when people feel a real crisis in the making they can put aside differences. One major Pebble partner, worried over bad publicity and great expense, withdrew from the project, and soon another. The Pebble project may be dead--or it may not be. Other potential partners may be out there.

A similar battle is now building over BC's plans to industrialize its remote north through development of multiple mines. One mine's huge pond recently breached its dam--not a good event for fish. Salmon producing streams from that region flow into Alaskan waters. Canadian Indians opposed to mining development on their subsistence streams are already planning to bring in federal law that protects their rights. Alaskan salmon fishermen again should see that stopping big projects--mines or fish farms--will demand a combination of powerful voices.

These northern examples are no doubt a picture of how things get done, or don't get done, in fisheries everywhere. The difference is that today fishermen have better means to keep up with events and connect with each other. That doesn't mean they can afford to stop fishing during an opener to go to a meeting. But even if dead-tired, fishermen can use the Internet. Last week I was reading a fisherman's blog from the east coast when another person broke in with his comment. It was a groundfish skipper off England. I switched to his conversation. He went back to talking to another of his fleet, commenting about "that Spaniard [boat] over there". Today everyone is linked, and certainly it is more difficult to make secret deals.

The Board of Fish will not find its job easier as each decade it deals with a larger and more informed often disputing constituency that knows how to work the media and the legislature for causes. Today there are a score of websites and blogs dedicated to US fishery issues. But the small commercial fishermen can't just communicate more; they must become more politically sophisticated.

State and Federal Management Intersects

Alaska is home to many situations where state and federal fish management must find common ground. One long-term struggle is over state versus federal policy on subsistence fishing. Usually it stays smoldering embers, but the flames tend to re-ignite whenever there is a salmon shortage and the board feels forced to reduce harvest quotas, such as during the recent Chinook decline. Because of the Alaska Native Claims Settlement Act, the Native regional corporations may choose to make their private selected lands available only to shareholders' use of resources, but the state has only one Indian reservation, and no special tribal or Native fisheries. Instead, in Alaska the federal lands are covered in the Alaska National Interest Land Conservation Act (ANILCA), passed in 1979, which has a section preserving rural subsistence use of fish and game as first priority over commercial and sport uses. The state, meanwhile, enacted a law that also made subsistence a priority over commercial and sport uses, but for all residents, not just rural.

An Athabascan elder, Katie John, played a role in the long legal battle over subsistence rights, reminding me of David Sohappy's battle on the Columbia River for Indian fishing rights. She was an active subsistence fisherman on a tributary of the Copper River when the state closed the tributary to subsistence fishing in the 1960s, before ANILCA was passed. Meanwhile, it allowed sport and commercial fishing on the lower reaches. After years of being denied traditional fishing, even after ANILCA applied, in 1985 Katie John and the Native American Rights Fund entered a class action suit that that eventually lead to a federal court decision in 1999 that ANILCA applied to the majority of navigable rivers, not just the adjacent federal lands, and therefore the rural subsistence priority applied to fishing in those streams. There was great objection to the rural preference from some groups who saw this as a threat to their activities.

The rural priority also put the federal regulations in conflict with the state's subsistence priority for all citizens, not just rural. After struggling over this difference for years, with some Alaskans even trying a failed effort at a state constitutional amendment, the state entered a lawsuit to remove the rural preference and lost. It was in the process of appealing when the governor at the time decided to drop the state's suit in 2001 and adjust to the concept of dual management. The court found his action legal. But the issue didn't stay dead. A decade later another Alaskan governor entered a petition to the US Supreme Court to have the rural preference dropped. In 2013 the Court rejected it and Alaska is left with a unique but manageable fish and game dual system. The issue of who manages came to the conference hall again during the ongoing Chinook crisis. Most Alaskans agree with the rural preference but will be glad if the issue is put to rest. The salmon issues they want attention to are the rebuilding of their Chinook runs, and as part of that, serious reduction of salmon bycatch by the industrial fleets.

West coast Alaska's fishery management is an even more mixed brew because of the six regional Community Development Quota groups based in communities traditionally dependent on subsistence salmon, and for decades also relying on commercial salmon fishing. These communities now have a significant interest in their CDQ royalties and ownerships in federal industrial fleets, a main one being the pollock fleet with its salmon bycatch problem. As I described in Chapter 1, families may have a foot on each deck--one on the small salmon boat that feeds them, and the other on the trawler that pays their community royalties and may make salmon fleet services more available but hauls in thousands of salmon as bycatch. I hear traditional subsistence families raging at the waste of salmon bycatch, yet at the same time, some of the regional CDQs have invested much in local fishing support. With the western Alaska Chinook fishery practically closed down for years, those communities and families wonder which fishery is really important to them in the long run. Not every CDQ has the same answer; not every family does.

The issue of the state's role with the federal fisheries is a knotty one, and not just because of salmon bycatch. The most serious mistakes in fisheries management, from the point of view of small-boat commercial fishermen, have not occurred in fisheries managed by Alaska Fish and Game but in the federal fisheries outside three miles. Since the passage of the Magnuson Act and the formation of our federal waters management, these fleets have been managed by NOAA/NMFS and the regional fishery councils. The small fishermen out in the federal waters have had a difficult time under administrations that favor the industrial fleets with vessels three times their size, as described much more in the second section here. Since these small fishermen are resident Alaskans, they feel that the state had a duty to protect their and their communities' interests at the time federal fishery regulations were being developed that would end up favoring the industrial fleets.

My sons that fish crab and halibut with small boats in the federal waters off Nome, were spared this discrimination because the region won the designation of "super-exclusive fishery" and before long became part of a CDQ. But many small communities outside the CDQ regions lost their local boats, and even most of their local fleet, through the federal decision to change to privatized fishing quota. Today, although the majority of the boats fishing waters off Alaska are still small boats, 60 feet and under, the big volume of catch from the federal waters is from industrial-sized vessels resident in other states, mainly Washington. The "outsiders" know the waters they fish, those regulations, and the port they unload at, but maybe nothing more about Alaska. When they go home at the end of the season, they go on to other concerns, not the problems of little fishing villages.

A great amount of federal support has gone into Alaska, and some of it, such as marketing research, eventually benefits all fishermen, including the smallest. Alaskan fishermen have also gotten their share of federal disaster aid. But the philosophy and practice of NOAA, its fisheries branch NMFS, and the North Pacific Council have caused a small-fisherman dropout rate in the non-CDQ regions that is probably higher than in any other area of the country except, probably, in the Gulf of Maine. The socio/economic loss to their communities is detailed more in Part II.

One could understandably be puzzled as to how Alaskans allowed this to take place. When federal fleet restructuring was first introduced in 1995 for waters off Alaska, some state political leaders did express concern as to how it could hurt small Alaskan fishermen. But there was no protest of the news-making kind we see now in New England over similar federal fleet restructuring. Neither the Board of Fish, nor Legislature, nor the Alaskans on the North Pacific Council put up enough of a fight for places like King Cove or Craig. Yet the state had experienced how its own Limited Entry had negatively affected its coastal communities twenty years earlier, as we have seen, so it couldn't claim innocence. The days when the state would butt heads with Washington D.C. seemed forgotten, while "outside" fishing corporations, foreign canners/processors, and Washington DC had their political way, and many small fishing communities suffered yet again.

The political challenge for a state fishery board to attempt to influence federal fisheries is obviously a tough one. The Alaskans on the North Pacific Council at the time of the re-structuring of the federal fleets either didn't see the economic implications for the small, remote communities or didn't have the voting power to stop it. The long-term passivity of Alaska concerning small commercial fishermen in federal waters is more understandable if we recall that the state gets a huge amount of federal funding, much of it through NOAA. We in Nome have had our share in the development of our port. What group wants to bite the hand feeding it? The federal fisheries may also not have seemed the place for the state to meddle when, after all, most of the state's fishery wealth comes from state-managed salmon. If NOAA and the Secretary of Commerce signed off on any fisheries program, Alaska's Senator Stevens, in his long tenure as Chair of Appropriations, made sure it was funded. His political energy addressed everything from fish stock research to marketing to vessel buy-backs. Now it would be major fleet restructuring.

Perhaps state officials also rationalized that the state didn't create the problem; let the feds straighten up their mess. If so, the obligation to their fellow citizens got lost somewhere. For years both Alaska's executive and legislative branches put aside direct requests for help from the Gulf of Alaska fishing towns that, starting in 1995, began to lose their local boats due to the federal fleet programs. When the North Pacific Council finally did create a halibut quota buy-back program for Gulf communities, described more in Part II, the government didn't fund a loan program to make quota purchases possible. The state then took years to fund one.

Those towns' plight was not much different than Alaskans faced during the days of the salmon fish traps: outsiders versus Alaskans, and big versus small. But this time the Alaskan public didn't take on the cause of its small fishermen. I doubt very many people outside the fishing towns knew what was taking place. I only found out because of people I knew in Kodiak and Old Harbor.

All that is history. The "Beacon Bright" didn't put up much of a fight, and today the industrial fleets in the federal waters of the Bering Sea and the Gulf have far fewer little fishing boats to get in their way. An urgent question for today is how tall will Alaskans stand for the state's small fishermen as NOAA proposes more major changes in their lives, changes that claim to be what the US needs for a modern fishing industry. Recently, the Alaskans currently on the North Pacific Council have spoken up more for their neighbors, and it has clearly made a difference in the debates. But they still have to win a majority vote on a change. Many times that has not been possible. Obviously much depends on the fishermen themselves, who need to study the political strategies used by the environmental organizations and recreational fishing industry. They and the organized Alaska Native groups like the Alaskan Federation of Natives are able to win battles. Alaska's small commercial fishermen have no spokesman with power like a Katie John. They need to change that.

The joys of fishing get lost in these political mires--fleet versus fleet, state versus federal, industrial firms versus small fishermen. I want to put them all aside and head down to my camp, set out the net, and wait to see that big splash. Here is a fisherman's comment I want to hold onto from last summer:

"One of the greatest things about living in Nome is the wild salmon we have here. In the tent on the riverbank at night, listening to the humpies coming up the river--it's just a great feeling, and it's tragic, the places that have lost their wild salmon."

Those words were not from a romantic urban "greenie" but from a dedicated subsistence hunter/fisher/gatherer who is happy for any new technology that makes life safer, who has no problem with small-scale commercial fishing, or sport fishing, and who wants everyone to have a chance at fish for the larder. But he also wants to know that a wild fish is still out there and headed home. He is one of the transplanted Alaskans who came over forty years ago and stayed because that is still possible here.

Chapter 16

AND MORE RIGHTS TO SALMON: SPORT FISHING

Each summer the Anchorage daily paper prints a photo that makes me cringe. It shows scores of sport fishermen crammed elbow to elbow, casting for salmon on one of the Alaskan rivers one can reach by the road system out of Anchorage. Behind the anglers we see their army of RV's, many of them from outside Alaska. Sometimes right among them are scavenging bears. Out on the river are the schools of powered skiffs, each trying for a big fish.

Sport fishing attracts people with a wide range of attitudes about fish stewardship and nature in general. One image we have is a Norman Rockwell painting of grandpa helping a small boy bait a hook on the bank of a stream. Another is a party boat full of weaving rowdies with fancy poles, tossing beer cans in the water, chasing trophies. Another is the quiet fly fisherman stalking steelhead on a remote, clear stream. And another I have is that herd of RVs lined up along the Kenai River and the dozens of chartered boats roaring up and down to find a spot with room to cast, their wake sometimes washing away spawning areas.

A source of worry for commercial and subsistence fishermen in Alaska and everywhere is the growing economic and political strength of the recreational fishing industry. It can't overcome the subsistence priority, but it can target commercial fishermen as competitors to demolish. Today it especially focuses on gillnetters, accusing them of bad conservation practices, taking more than their fair share, and holding up the rebuilding of prize sport fish runs. Whenever a salmon or other coveted run is lower than average, angry debates before the Board of Fish or other governmental groups are guaranteed, with ENGOs jumping into the fray as well. If the recreational fishing associations don't win there, the next step will be the legislatures or the courts.

In past decades, the identified enemies of our salmon were the foreign fleets, big canners, big irrigators, timber companies and pulp mills, ladderless dams, creek robbers, and the intercepting commercial fleets. Today some of the noisiest fish battles in the West are waged between gillnetters and sport fishing associations on the Lower Columbia and on the rivers on the road system near Anchorage. This kind of competition will not be a problem for my region unless they build a connecting road from the state highway system into Nome. But elsewhere the battles drain energy of fishermen and managers from more basic problems: where went the missing Chinook salmon? They didn't all go into the totes of gillnetters. How will we get the funds appropriated to rebuild habitat wrecked by mines and logging? What about growing acidification in the ocean? Instead the Board is besieged by people angry over gillnetting. It's an easy target.

Competitive pressure on western stocks of salmon, halibut, crab, and cod increases as the share of recreational fishing in the economy increases. The pressure is noticeable in areas around cities, but also in rural towns where the commercial fleet has disappeared and that now depend on tourist fishing. Among the charter fleets everywhere, not just in Homer, commercial fishermen have turned to skippering charter boats to survive. The recreational fishing sector's share in the national economy was about $50 billion annually, involving at least 320,000 jobs in 2012.. That may not count the jobs with fishing suppliers. States receive significant support for their fish and wildlife programs from sport licenses, a big share of that from non-residents. With the growing number of sport-fishing constituents and the revenue from sports licenses greater than from commercial licenses in certain districts, the Northwest states and BC have made sport fishing higher priority than commercial when it comes to doling out the fishing quota. But often the reason for recreational industry's lobbying success is that it is more sophisticated politically and has serious money committed to its battles.

All along the west coast, groups like the Coast Conservation Association (CCA) with the avowed goal to close down gillnetting are succeeding. Gillnets are an efficient gear for capturing salmon that are schooling to move up a river. The small boats are maneuverable, historically were affordable for smaller fishermen, and can be operated with a crew of two. But they are less than ideal for conservation in that they can't distinguish between fish of the same size, when one species is targeted and the other undesired or prohibited, and releasing an untargeted fish without injury from a gillnet is difficult. The death rate for illegal steelhead and wild (ESA listed) Columbia salmon released from gillnets is, some say, 40%, while managers estimate only 10% die from sport catch-and-release if done with care. The pressures to protect more fish for constituent sport fishermen in the rivers and to protect wild ESA-listed stocks, are what drive the spreading war. But the sport associations ignore the well-known fact that the numbers of salmon lost to gillnets are nothing compared to the death from mid-water trawlers' bycatch. The CCA goes after the most vulnerable adversary. The gillnetters are ordered to change their mesh size to protect species at risk like Chinook, but that is not enough for the sport fishermen.

Cook Inlet near Anchorage is one current site for war on gillnets. The gillnetters are mainly after a healthy run of sockeye through Cook Inlet, while sport and personal-use fishermen are in the rivers that empty into the Inlet and are after Chinook, sockeye, and coho. The Chinook have been far down in numbers lately, and the setnetters are the visible scapegoat. The true cause of weak Chinook returns may be beyond the Kenai, out in the ocean, or it could be closer to home--sport boating destroyed spawning areas. If the board is really serious, the commercials ask, why doesn't it restrict those sport boats more? But for the Board, the immediate issue, since none of the fish are ESA listed, is an allocation between the competing groups they can peacefully accept.

The anti-gillnet movement is not a new one. In 1996 a survey of lower Columbia gillnetters by Jennifer Gliden and Courtland Smith, sponsored by Oregon State Univ. Seagrant, a NOAA program, revealed the hot feelings at the time. Here are three gillnetters' anonymous written responses in the space provided for comments.

"What really would help is for those who've benefited from salmon habitat destruction (hydro, aluminum, irrigation, navigation, poor logging practices) to step up to the goddam plate and make a commitment to right past wrongs and work with, not against, salmon industries."

And another: "If you could send out a survey and ask the gillnetters if they would sit on the beach for the next five years or more to save salmon, they would say yes, but you will never find another user group, whether it be from other fishing groups, agriculture, utilities companies, or aluminum industry, ever make such a commitment." And another: "Society has made a few commercial fishing pay for the past sins of all." Most fishermen would cheer for that last remark.

Today, in both the Northwest and Alaska the battle of gillnetters versus the recreational industry has reached the courts and legislature. The commercials, believing that they were being discriminated against, sued recently to be returned to federal management. They claimed that the state's Board of Fish was not following MSA guidelines. But the Board is besieged regularly by sport fishermen who claim it discriminates against them in favor of the commercials. As one supply merchant commented, when someone spends $10,000 to come to Alaska to catch two Chinook, the state needs to take it seriously. The economic value of sport fishing is not spread evenly across the state. Probably about half was in the Cook Inlet/Kenai area accessible by highway. In 2012, with the Alaska Chinook decline going on for several years, the recreational sector in the Anchorage/Kenai area believed its political power had reached to the point where it could force legislative action. The debate shifted from board and council rooms to conferences that drew in the public. The next step was a petition to place on the ballot an initiative that would outlaw gillnets near urban areas. The Lieutenant Governor rejected such an initiative as illegal under the state constitution, and the battle moved to the courts. A group called Alaska Fisheries Conservation Alliance sued the state over his action, and another new group, Resources for All

Alaskans, formed to support his action, arguing that allocation of fish by initiative was illegal. What was missing, to balance things, was an initiative to outlaw sport boats on the rivers. That would probably have been illegal too.

Another current example of sport fishing power is right in Puget Sound where I grew up. Sport boats now dominate all of the many small-boat harbors and pay for their support. The Sound's commercial Dungeness crab fleet finds itself in a losing competition with recreational crabbers who enjoy taking their sport craft out to check their one crab pot and bring home their own dinner. It's fun--how could anyone be against it? The current commercial crabbing licenses for the Sound are now only 149, while the sport crab licenses are now around 250,000. Add to that the other voters involved with sport crabbing: boat manufacture and sales, repair shops, fuel, fishing supplies, restaurants, moorage fees, and licenses. Who will win any debate over quota? The Washington Dept. of Fish and Wildlife dealt with this in 2010 by giving the sport sector 50% of the non-tribal crab quota. The next season the sport crabbers received an open quota.

The lower Columbia gillnetters have probably lost the fight with the two governors' combined rule to outlaw the nets by 2016. If the "Salmon For All" pro-gillnetting group loses its legal appeal, the roughly 150 gillnetters remaining will be able to operate only in designated sloughs off the mainstream after a phase-in period. As an alternative gear, a pilot program using beach seining and tangle nets was tried in 2014, but so far did not prove to be a viable alternative. The CCA has already succeeded in winning gillnet bans in ten states, but none in the West, so the probable ban at the Columbia is an important victory.

The county governments on the lower Columbia strongly protested the phase-out of gillnetting for communities with five generations' history of commercial fishing. Why, they argued, should the gillnetters be targeted when a Bering Sea trawler fleet is still annually allowed bycatch of Chinook salmon of 60,000? That is where this story began. But sportfishing associations know how to organize campaigns and work the media. They have influential individuals among their members willing to speak out at a meeting or write an op-ed to the local paper--more willing than ordinary commercial fishermen. They can also be socially useful, such as in projects like habitat restoration of particular streams. These aid the state fisheries budget, but also build public awareness of the value of habitat stewardship and volunteerism. Sport fishing associations sponsor other worthy projects like cleanup of abandoned commercial nets and gear. But too many of them see commercial fishermen as easy opponents and continue to make them scapegoats for every bad news about fish.

As urban populations grow, the recreational fishing industry and its political power will grow too, but the numbers used to promote recreational fishing are deceptive. As an example, the 2013 Alaska's eastside Cook Inlet setnetters brought in a dockside total of $5.3 million, while in 2008 a state economic report found that sport fishermen in the same area spent $104 million on all of the services. These are not fair comparisons, five years apart, but point out the misconception common in the press. The value of the setnetter catch shouldn't end at the dock. It would be a more valid comparison to look at the wholesale price of the fish and on up the chain to the value to stores, restaurants, fuel docks, shipyards, and supply houses. Too often in such arguments over economic value these second tier revenues are counted in with the sports fishing value but not with the commercial.

The value of the recreational fishing industry, on the other hand, shouldn't be understated, as it has bailed out many tiny communities on this coast left broken after commercial fishing, logging, mining, fur trapping and farming all dried up. The area around Anchorage experiences the same over-impacted trend as in Puget Sound, while in more remote Alaska and BC, wealthy clients fly in on floatplanes, have fun in lovely settings, may even obey their bag limits, and leave without doing much ecological damage. Local residents love or endure them for the money spent.

When I commented to George regarding the economic value and political power of recreational fishing today compared to commercial, he shrugged,

"Of course. There are so few of us left in the Northwest. In the 1970s there were at least 4,000 active commercial salmon licenses in Oregon, and in Washington too. What would have been our economic comparison then? At that time they figured that every dollar spent on the Oregon commercial salmon fishery brought six into the economy."

Yet George doesn't have a special quarrel with the sport fishermen. Like me, he thinks sport fishing is a fine activity, far better than many leisure time diversions. If I couldn't subsistence fish I would be out on the river with a rod and reel too. The problem is one of narrow political interest subsuming the bigger environmental concerns of the oceans, rivers, and fish-- a temptation we all have to fight. George adds, "The sport fishermen will eventually be victims too."

The MSA recognizes the growth of sport fishing sector nationwide, especially charter boat businesses, and requires that all federal waters anglers must now register, and their catch figures submitted, the purpose being to find out just how large their influence is on the ocean stocks. How do salmon derbies that focus on catching the biggest fish affect spawning? Are charter fleets causing part of the worrisome trend in smaller halibut size? Still, no faction is trying to close charter boats down; their situation is not like that of small commercials. The CCA and similar groups have stubborn faith that they will win out over the commercials, that it is only a matter of a few years and a few changes in makeup of state legislatures and the election of sympathetic governors. So far, they've been right.

Gillnets are definitely not a perfect gear, and more could be done to make them less destructive, but closing them down draws attention away from the really big problems harder to cure: poorly funded research, scores of river habitats that need rebuilding, drought, agricultural run-off pollution, hatchery programs that need reorganizing, trawlers that need to install salmon-excluders now available, regional councils that need to tighten up on salmon bycatch, and dams that need breeching. Both recreational and commercial fishermen need to accept the fact that to help fish stocks stay sustainable they need to collaborate and go after today's real adversaries of healthy fisheries. Commercial, sport, subsistence fishermen, and conservationists are all people that want to protect and rebuild fish stocks and their habitats. They will come to understand they need to work together or they will all lose. Regional councils, state fish boards, and fishery managers could help them understand this.

Chapter 17

SUBSISTENCE THESE DAYS

Today you can still go down the coast a short way from Nome and see families subsistence fishing much as my husband Perry's grandmother taught it to him. I think she would be happy to see that still alive. We have a unique situation for today in that Alaska's west coast is still available for subsistence salmon gillnetting openers anywhere along the ocean shores, and in some river areas. Some areas are also designated for subsistence beach seining or subsistence hook-and-line. Halibut long-lining is also included in southeast Alaska. To be fair to residents who live outside designated subsistence areas and can't travel to them, the Board of Fish has created "personal use" net fisheries, both fresh and saltwater. Fortunately there has been no flood of people north recently to establish residency for those permits, the other economic advantages in Alaska being much less sure.

Alaskans have subsistence opportunities others yearn for because of the Alaska Constitution and coming later, the federal Alaska National Interest Lands Conservation Act (ANILCA), with its rural subsistence priority. Our extended family is one of the happy beneficiaries of the subsistence priority. By its very nature, that priority favors ordinary families and the small boats and gear they use. But what works in Alaska could never have protected subsistence resources adequately in more populated regions of the country. Though it is most important to the indigenous people whose entire traditional culture is linked to subsistence activities, both Native and non-Native rural Alaskans today put high value on subsistence and on protecting it for their children and grandchildren. My oldest son, Dan, and his wife Naggu are good examples of modern people of the work force who make the most of the subsistence life. Because they both work fulltime they are among the families that can afford all the modern equipment and gas and bring home their catch for others who don't have those resources. Dan's Inupiaq in-laws were his first mentors, right after he married. He'd first learned to fish on a commercial troller in Southeast; now he was learning a whole different kind of fishing, very much as Perry learned it as a boy. He wrote this for me:

"In the 1970 Naggu's family was commercial gillnetting near the mouth of the Ungalik River near Koyuk. During closures they would boat a little way up the river and beach seine for their subsistence chums and pinks for drying. It was my first exposure to gillnetting or seining, and to filleting and drying of fish. I jumped to help with the net and right away fell in the river as I was bringing the end of the net around, to the great amusement of my new in-laws."

Maybe it was a good way to start, assuring everyone of his properly humble status. He goes on to tell that after the women cut the fish on the bank the crew transferred them down to the main camp on the coast to be hung on their driftwood racks. As long as it wasn't rainy or foggy, the coast was ideal for drying, just as Perry had experienced as a boy.

"After we had our two kids, three summers we flew to Koyuk in July with the kids and camping equipment and supplies and set up camp for a week with Naggu's mother, Esther, up either the Koyuk or Ungalik River. They preferred the salmon that were up the river a ways as they had less oil and could dry faster. Esther showed me how to properly set up the white wall tent and build a fish rack that eventually held 300 fish--steady work until the rack was full. Things could relax a little for us then, with just the chores of keeping the fish shaded and doing the fly patrol. Even though two of those years a big storm came in from the southwest and we had generally miserable conditions for a few days, we saved the fish with smudge fires. We left most of them for the Koyuk family and brought home enough for ourselves."

When one year someone gave Dan an old gillnet, he decided we should give it a try near Nome. I was ready to join in. He tells how that venture turned out:

"I thought we could combine Perry's childhood memories with our Ungalik experience and everyone's hunger for fresh salmon. We didn't have a boat to set the net, but I thought we could use a tiny two-man rubber raft. Perry was the person I counted on to help me set up the lines and stack the net right in the raft and shout advice from the shore. He hadn't done this since he was a teenager, but I figured he'd remember. He probably thought I was nuts to try this with a little rubber boat in our typically choppy seas, but he went along with the idea as long as the water was calm enough, and I would be the one out in the boat while he directed from the shore.

"So that June we were at the Sandspit just off Nome, where no one had a net, and no one would watch me, and I was loaded up in the little raft with 100 feet of net, anchor, float, shore line, and oars, wearing a life jacket at least. He pushed me off, and there I was, madly rowing out to get past the surf and set the net. And I got it set. The next morning we had more chum salmon than we knew what to do with. But Perry had made arrangements with his relatives at Fort Davis that we could put up a rack on an unused lot. We rushed our big haul down there, and while Naggu was showing Nancy how to cut fish for drying we hurried to get the rack finished."

I was deep in my own apprenticeship; I had cleaned thousands of salmon but never filleted one. But Naggu certainly had, thousands probably, and I was bound to learn. Dan adds, "The next year that rack got expanded until it could hold well over a hundred fish, and we used it for many years, producing very decent dried salmon."

There was more to my apprenticeship. I wrecked that net the first year due to too much eagerness to fish when there was a threatened storm. The waves brought the net to shore rolled up tight around a dead, smelly walrus. I took a lot of teasing as I cut the floats and lead line off, but redeemed myself in the family's eyes by learning how to hang nets and from then on it was another subsistence task I liked. The condition of the ocean doesn't always cooperate of course. One year it was too stormy in August to get a coho net out at all, but we instead did some exceptional rod-and -reeling for kings down near Unalakleet where Dan and Naggu and their children were living for a few years.

Before long I no longer missed commercial fishing. As a 40-hour week employee, I never had to worry about, "Will we make our season?" Yet I still had all the other elements of the fishing life: challenge, physical outdoor work, pride, and beautiful, delicious product. Our monetary investment for fishing stayed modest however. After the original expense for Perry to build a cabin and a couple outbuildings, and the purchase of a small skiff, four-wheeler, freezer, and the net gear, the ongoing cost for thirty years has stayed basic: net web and rope replacements, tubs, stovepipe, bear damage repairs, 4-wheeler maintenance, and truck gas for the ten-mile ride to our camp.

There are always plenty of challenges at our camp to keep us entertained, especially as the weather gets unstable from arctic warming. It's often too surfy now to launch the skiff for an opener. It took us many trials to devise a pulley-ring set-up, copied from my friend Esther at Bristol Bay, with an anchor that would hold in the current and surf so that we could haul the net from the shore. Some years, even so, it is a challenge to fish the ocean at all in August. In 2012 wild August storms caused us to switch our net to the river near the mouth. Then, exceptional high-water lifted hundreds of dead spawned-out pinks from the upriver beaches and sent them down to our net. Lots of fish, very soft and smelly. Some years the Yukon sends down many times the normal amount of driftwood to float on up the coast to plug and rip Nome area nets. But the stove wood is a trade-off. In June of 2013, Dan and I got the net out the first time, caught some chums, and it was eleven more days before we dared put it out again. It rained all of July and few people were able to dry their fish. We wonder if this is what we face in the future.

The solution was to smoke the fish, something people here have not had to resort to often in the past, or to try to finish drying it inside by an open window, with a fire going and a strong fan blasting. It doesn't always work, and you can't dry hundreds of fish that way and choose to do it again. We also have frozen the drying fish to wait for reasonable drying weather, then thaw it and re-hang it. As the weather changed and we seemed to have more flies each year, Perry built a screened drying shed. Bears tore at the screen when we are not there, and sometimes we had to share with them more than we wished. In 2013 we finally made an expensive investment in an electric bear fence for the fish rack. It works.

Looking back in my journal I am surprised to see that although I remember all seasons as successful, actually there were summers like 1994, when I wrote: "We don't have one dry salmon. Now in August the surf is so high we can't launch the skiff for an opener. Not a coho is in the freezer either." That was the year we taught ourselves how to smoke fish and began to experiment with pulleys.

The many years of poor chum runs in the 1990s were serious for Nome families, who count on them for dry fish, especially in the pinks' alternate off-years. Most people blamed the gillnetters' and seiners' intercept fisheries at False Pass. More recently, bycatch by trawlers takes most the blame from local fishermen. Possibly it is neither as scientists are now finally able to use DNA tracking to identify the source of trawlers' bycatch, and most of their chum bycatch was from the Pacific's western shore. But about half the Chinook bycatch is from Alaskan rivers north of the Aleutians, confirming our salmon fishermen's suspicions.

During our chum run declines that had begun in the 1980s Alaska Fish and Game instituted a program called Tier II that heavily restricted even subsistence chum fishing near Nome for years. Beginning in 1999 you could receive a Tier II permit based on your years of fishing-and-using chums in the Nome sub-district. It didn't affect us personally as Perry had been part of fishing here since he was five, but it did worry us about the chums and about other families needing fish who had to stop fishing near Nome for several years. The young people and others who didn't qualify and couldn't afford trucks, riverboats, jet units, and gas to move their fishing site out of the Nome area were left out of their needed subsistence fishing. Fish and Game alleviated this somewhat by allowing hook-and-line subsistence fishing for pinks in the Nome River in the strong (even) years, and it is still allowed. However, some young people who should have been developing their fishing skills and accumulating equipment didn't do so, and a family tradition was lost. Other families who had moved their fishing to another location stayed there, and in some cases might have over-stressed those rivers with all the added gear in the water. Tier II was probably a useful experiment in management but had those unintended consequences, and is another example of how every change in the fisheries is likely to affect much more than the targets.

Last summer I was in the middle of picking a net on the shore when I thought about my cousin George's words: that subsistence fishing is the most important of all if you look at societies worldwide, from Greenland to South Asia. Furthermore, commercial and subsistence fishing are often intertwined. The crew that brings in a sockeye for dinner is probably the one that with a switch of gear brings in coho for the buyer on the next opener. The commercial season can be a failure, but families still have to eat. They will tell you that the profits from commercial are very soon gone, whereas subsistence fish properly stored can last all winter, and also be traded.

Today Perry and I are retired, but for years we were, like Dan and Naggu, weekend warriors. We didn't have to subsistence fish but we liked doing it and looked forward to it all year, then eating the product, and sharing with others who didn't have the opportunity. Dan and Naggu are tent-camping every year at a favored river site for their half-dry/smoked fish. He makes a comment that is true for many families who can afford to buy everything from the store but stay with subsistence activities: "Whenever the season was right for food collection, we were doing it, and always with the kids along from the time they could walk."

Sixty years ago subsisting meant survival. Today salmon, and other fish too, but especially salmon by volume, is still essential for the majority of rural Native families. Higher-grade hamburger can be $9/lb. in Nome. You fish and hunt, or you feed your kids hot dogs. Several household surveys have been done in Western Alaska to discover how much of the local diet is from subsistence foods, and of course this varies widely between families. But especially among Native families, if they don't collect subsistence food themselves, they have traded for it with their network of relatives and friends throughout Alaska.

Statistics collected in the 1990s revealed that the average Western Alaska family's protein supply over the year will contain 33% salmon, 22% other fish, 10% wild meat (moose, caribou, birds), and 25% store-bought meat. Across Alaska, of the subsistence foods, 60% is finfish. Yet, as we have all heard many times, subsistence life is not just about collecting good, nutritious, sometimes cheap food. It's about traditional heritage and values, and families doing things together. That heritage is still strong here. But would I fish if I were not married to an Inupiaq? Yes, berry picking and fishing are part of my northern rural heritage too.

Perry's oldest sister, Linda, also retired, outdoes us in what she produces in subsistence food for her family. She has her own camp on a tributary of the Copper River--Athabascan country. Perry and Linda exchange fish reports almost daily all summer. Like our campsite, hers is through an annual lease from the local Native Corporation. Her camp consists of a cabin, a fish wheel her son built, a fish rack, a smokehouse, a propane cook stove, and two big two-tiered canners that can each take 14 quarts. At the peak of a run she has all these operations going at once: catching, cutting for drying, cutting strips, smoking, and canning. While she's doing this she is soaking fish heads to make traditional fish oil. It plays the same important role that seal oil does on this coast.

In 2012 river flooding caused Linda's fish wheel to be swept downriver and out of commission for the summer. Replacing it would be a huge task, I thought, for her son. But she disagreed. "It's okay, he's done several now so it goes pretty easy for him." But she was blue, not only from the lost food but from lost activity. Like me, she looks forward to fish camp all winter. Three days later a friend, unannounced, brought her 96 sockeye from his fish wheel. She jumped to handle them, hardly slept, but got it done. By fall her son had the new fish wheel built. Linda has a freezer for her camp that runs off a solar panel she installed herself. She is able to manage the whole operation and provides all of her four children's, and several grandchildren's families with their year's supply of several varieties of salmon. Linda enjoys it all, "If it wasn't fun I wouldn't do it," but she also comments, "I especially like it when my grandkids come out. They need to learn how to run a river fish camp and how to take care of themselves during hard economic times."

Recently, the surge in gas prices and equipment is having its effect on poorer families. Northern Alaskans are, like everyone, eager to try new technologies. For those that have camps farther from Nome and want to use them year round, they may have not only trucks, skiffs with outboards or river boats with jet units, and trailers, but snow machines, CBs, GPS, satellite radios, and four-wheelers. Many Native families also have a large sea-going hunting skiff. All this is quite an expense to maintain, but it is hardly just for fun, rather for a serious food production system for an extended family. In 2011 a big change came to our fish camp, good news to mix with the warnings of climate change. Frank's partner, Cherilyn, and often her parents, Frank and Norma Kavairlook, became part of our setnetting team. Cher drove the 4-wheeler, helped haul the net, cut fish, and smoked strips. So although many families gave up on the ocean due to storm and surf and went to the rivers, we kept on, and in 2015 still caught what we needed.

How long Alaskans, Native and non-Native, can hold onto a productive rural lifestyle is a serious question for us. The goal of every rural family is, of course, to have the best of both worlds: economic development and protection of our natural environment and subsistence too. Some don't believe it's possible. If we in rural Alaska were on a state road system we would be overrun by people wanting these same opportunities. (I can see the ads: "Come visit the westernmost city in the US!") People here put up with long winters with high unemployment to hold onto a lifestyle that was once common throughout the continent, now hardly replicable. Many "outside" people want to hold onto Alaska's "wilderness" value. I can understand that, but for us that is not seeing the whole picture, how human communities have for a very long time been part of what makes Alaska great. We may not be able to salvage subsistence fishing for everyone, but we should at least protect it where it still exists, and the Alaska Native organizations are determined to do that. They also look at the trends in climate change. Subsistence fishing families will be surely affected by any change, as their economies are so closely tied to what is taking place in the natural world.

Chapter 18

Fish Feed Lots

While debates rumbled on over how to rebuild salmon runs, to some scientists and economists advising the government the battles over inefficient traditional fisheries versus industrial fishing were rapidly becoming irrelevant. Aquaculture--farming fish--would be the only possible means to provide enough protein for the world in the not too distant future. Aquaculture was where the US priorities had to go. I sat on our bench up on the bank at our Cape Nome camp, read that news, looked at my grandchildren, and now great grandchildren, playing down on the beach, and wondered about their lives to come.

The fish-farming issue is a serious one for our catch fishermen. At least 80% of the fish for sale in the US is imported, most of it from farms. After much effort by certain ENGOs and US fish marketers, a few grocery chains now have agreed to offer US "caught" fish only. Whether this trend will continue and will make a serious dent in that 80% is unknown. The farm interests are huge, global, and growing. They have succeeded in getting our own federal waters opened to aquaculture leases, to start by 2016 off California. Closer to home, British Columbia has for years seen factions warring over the fish farms in its inlets. These are pens where fish stay their entire lives, but which some scientists insist show large escapements that contaminate wild stock through interbreeding and pass on diseases and toxic wastes.

BC biologist Alexandra Morton, after being discounted and ridiculed for years over her insistence that the farmed fish infected with sea lice were passing them to the wild smolt, finally got acknowledgement in 2010 that there might be something in her claim. Many fish scientists, not just

Morton, believed that the Canadian government has dragged its feet over the research necessary to prove or disprove a connection between BC's sockeye run failures and the fish farms. But though as of 2014 there was still no general agreement on effects of about 20 farms active in the inlets, DFO hasn't ignored BC's salmon problems, budgeting $26 million just for the most recent judicial inquiry into the radical swings of Frazer River sockeye runs.

BC's wild fish activists and the commercial fishermen are for once on the same side. But the farms provide about 3,000 jobs along with the rents for BC inlet space, and so far the province seems to think this more than offsets the problems the farms might be creating. Fish farms, even if foreign-owned--like almost all of those in BC-- are also much easier to manage than feisty independent fishermen. Even viruses are, apparently, for the farm industry survives even when diseases worse than sea lice drive out individual companies. The farm spokesmen dismiss the whole uproar over the farms as protectionism by commercial fishermen. "It's all about money." That did have a lot to do with it on both sides, of course. In 2004 the US Congress's "Trade Adjustment Assistances" program subsidized both Washington and Alaska fishermen to offset the drop in price due to the international farmed fish competition. So much dispute continued in BC that in 2010 the Canadian government took oversight of the farms away from the province to be a duty of DFO. But the farms continued, and so did the dispute. Researchers themselves can't agree. DFO recently did authorize one Indian band to build an enclosed pen as its own pilot project. A private corporation also recently started one. Biologist Morton's stubbornness begins to get results. Putting aside projections of where our protein must come from a hundred years from now, right now the conservation-minded public doesn't wish to see wild salmon sacrificed for fish farms' profits.

For the salmon fleets the immediate threat for the fishermen is not disease but the cheap farmed fish flooding their market. George recalls years ago coming in from a trip and how disgusted he was with the price he received.

"I went up to the office to complain, and the guy that happened to be there was a little more forthcoming than some of them. He waved his hand out at the parking area, 'Well, seven truckloads just came in from Canada.' That was putting it plainly! "

The salmon fishermen by the 1990s had only to look at BC to conclude that there could be another plan afloat in favor of US farms. It was not mere paranoia, as they had already witnessed the "sea-ranching" experiments off Oregon of the 1970s. The salmon farms on the west coast were an early version of the federal government's encouragement to privatized fisheries that is the focus of the second section of this book. Weyerhaeuser, one of the timber companies losing their market, was among several private groups that had put much energy into "sea ranching" development on the Oregon coast with the assumption that the fish would return to their site of origin for the companies to harvest them. Though within a few years all of the Oregon ventures fell apart, the ones in BC are apparently thriving well.

The last Oregon sea-ranch--Weyerhaeuser's "Oreaqua Foods" in Yaquina Bay at Newport--closed in 1992, leaving its abandoned straying salmon for the state to clean up. It created a bad taste for farmed salmon with Oregonians, but Washington State has farms, as does California, and Hawaii approved them in 2009. Washington waters today produce around 20 million pounds of farmed Atlantic salmon. However, most of those fish are raised in contained enclosures preventing them from infecting what's left of the wild salmon in Puget Sound.

Southeast Alaska's waters are so like BC's that if salmon farms moved in one can see how the same worries would soon surface. But Alaska's Constitution states that the resources must be sustainable and protected for the citizens. So far, that has helped keep out "outside" aquaculture interests. Fish farming is also not for marginal firms. BC's salmon farms grew to a peak at 135 private and government farms but are now consolidated to four private firms, only one of them a Canadian corporation. Ninety-two percent of the BC fish farming today is Norwegian-owned, 55% of it by one company. When Norwegian lawmakers created stricter regulations within their country, the farm corporations moved to BC, Peru, and Chile.

Even though most states have so far rejected private fish farming, our federal government is outspoken in its support. The Marine Fisheries Advisory committee to the Secretary of Commerce said in 2007: "The development of a significant domestic, environmentally sound aquaculture industry is essential." The group went on to recommend that the start-up be federally subsidized, with a goal to have the industry operating by 2020. Thus, the Oregon trollers' conspiracy theory had reality behind it, the drama just postponed three decades. The US aquaculture movement, as it develops, should create new battle lines, as the farms' adversaries will be not only traditional fishing groups and managers but also sportsmen and conservationists. But now there is a new alien to combat: the "Frankenfish", a genetically modified (GM) creature that is part salmon and part a fast-growing eel. The concept is gruesome enough that the media coverage has created a larger public to go on attack than ever did over farmed fish. BC has recently authorized Aquabounty to begin growing GM salmon, strategically off-stage in Panama.

The Joys of Farming

Fish farming is obviously profitable but has a myriad of problems that scientists work on continuously. Many of the problems are familiar to traditional farmers, just a watery version. Unlike hatchery fish that are released to the open sea to feed themselves, farmed salmon are kept in pens throughout their lives to be harvested by their owners. The pens' close quarters encourage spread of contagious diseases, and like feedlots, produce waste. But a high rate of disease, as Paul Molyneaux observes in "Swimming in Circles", will not stop farm companies. They expect to make profits that can handle the losses from disease, just as crowded cattle feedlots do.

Overcrowding in the pens, a means of cost-saving, exacerbates pollution and disease, while chemicals and antibiotics fed to the fish to protect them are transferred to humans. Viruses and parasites carried by farms can be transferred to the wild fish when penned fish manage to escape. Interbreeding by escaped farmed fish can result in weakening of wild stock. A bad solution has been to move the pens to new, fresher locations and to continue to overcrowd and pollute even more areas. When the toxins are released to the waters they pollute the ecosystem, just as feedlot and fertilizer run-off pollute rivers. In a complicated process, ocean fish farm waste encourages an abnormal growth of phytoplankton that in turn are eaten by zooplankton that flourish, use up oxygen, die, and sink to the ocean floor. This process from the fish farms, like fertilizer runoff, is one cause of the increasing dead zones in the ocean. The fish farms also seem to concentrate chemicals. A 2005 study found ten times as many chemicals in farmed salmon as in wild salmon. Since the aquaculture leases that NOAA promotes could be as little as three miles off our coasts, they could bring their ecological, and fish health problems to states that don't want more. Another farm impact that may be affecting wild salmon right now is the huge tonnage of small forage fish like sardines that for years have been caught to feed farmed salmon. The ratio given is three to four lbs. of forage fish to produce one lb. of "cheap" farmed salmon. These ecological problems of the farms are well known.

Fighting fish farms is a battle where ENGOs can find common cause with sport, subsistence, and commercial fishermen, and a good number of scientists. Still it will be a difficult fight with so many economists, and even some biologists, speaking out about the long-term shortage of available protein supply and arguing that humanity must have the farms. But there is logic missing in that argument. If it is true that the US should take responsibility to feed the world's hungry, why does our government continue to support agricultural subsidies to wealthy US farmers along with free markets that have ruined small third world farmers, such as rice farmers in Haiti and corn farmers in Mexico? It is well known that the number of families in poverty worldwide has risen in part because of free markets and our competitive subsidized grains. Continuing with this policy, our government announced recently that it would subsidize the start-up costs for federal waters private fish farms. This will hardly help the poorer people of the world. They did far better with small-scale subsistence farming and fishing, independent experts as well as the UN now acknowledge, while poor urban people aren't helped either. They can't afford salmon, even farmed salmon. They buy fish like mackerel and carp.

I don't know how many in the federal government share the view that aquaculture development could eliminate forever the knotty problems of our hatchery system and the management of the fishing fleets, and furthermore feed the world. More likely the opposite is true, that if industrial fishing and fish farms destroy traditional small-scale fishing worldwide (subsistence and commercial), it will mean millions more humans will be suffering famine or malnutrition. On a more personal level, I imagine working as staff at a fish farm is about like work in a fish cannery. You can make good money on the slime line, but once you have commercial fished, to go back to cannery work is a definite downgrade in quality of life. It is like moving from commercial small-gardening to working in a vegetable cannery. I would rather go broke fishing, and almost all fishermen would agree I think. But this is irrelevant if you are convinced we are soon going to run out of protein and must either farm fish or resort to jellyfish.

The issue of healthy oceans is bigger than fish farms of course, but the farms debate should help enlighten the public that the whole ocean ecosystem needs more protection. Our new federal "Joint Ocean Commission Initiative" is mandated to study and protect the ocean using an ecosystem-based approach. Surely that will have to include review of the effects of fish farms. As usual with fisheries, our federal government seems to be going two directions at once.

Marketing to the Rescue

We have seen close at hand at Kotzebue and Bristol Bay how foreign farmed fish can drive down the price of domestic "caught" fish to where small-boat fleets can't make a living, just as the Oregon fishermen of the 1970s foresaw. Even our groundfish fleets see their prices drop as competitive farmed tilapia swamps the market. Cod, tuna, and halibut farms are also underway somewhere. Cost of processing, transporting and marketing fragile fresh salmon, especially from remote areas like Emmonak or Unalakleet, has always been challenge enough. Aquaculture competition adds to the problem, with the smallest, most remote fishermen bound to be hurt the most. Their operations are already marginal in any year when the market is down, while their costs of operation (fuel, gear, crew, supplies) are always higher than in the centers. Largely because of a flood of cheap farmed salmon on the market, the overall Alaska pack value dropped from $781 million in 1988 to $362 million in 1996. It was during this period that Norton Sound fishermen, including my son Frank, were able to buy Bristol Bay gillnetters for half price. Enoch has described how that competition some years closed down buyers completely in Kotzebue.

It doesn't happen every year. Cyclical virus infestations at the farms at times cut their volume on the market and allowed an opportunity for the state-funded Alaska Seafood Marketing Institute (ASMI) and other marketers to partly overcome the dilemma. They concluded that the only way to compete was through better quality fish and developing consumer support for specifically Alaskan fish. They introduced the ideas to net fishermen, with ASMI offering workshops on how to increase quality of their catch for a more competitive product. Bay processors set up ice stations for gillnetters, others installed their own ice-making machines, and now over half the Bay catch is iced. In some regions like Alaska's Copper River, local fishermen's associations charged a tax on their members to help with icing costs and to do their own special brand marketing. Special branding has now spread along the coast.

The FDA says that the majority of imported foreign fish is mislabeled to appear domestic or is a different species than labeled. To counteract such fraud and the unsustainable practices of many fisheries, for several years US fisheries have been able to buy into expensive programs that examine them for sustainable practices, and if the fishery passes, its products can bear labels like the "MSC" (Marine Stewardship Council) stamp of approval that can be a common sight at places like Safeway. To environmentally-conscious customers, and their wholesale and retail sources, some kind of "sustainable" label makes a difference. But are there enough label-reading, eco-conscious, health-conscious consumers to save the "wild" sustainable fisheries? For the average rushed American consumer that won't take time to read labels, probably more important are the cheap, high quality frozen chum and pink fillets that fit right into fast meals. Yet, US consumer purchase of fish is rising very slowly. Most Americans recently surveyed, unlike other countries' citizens, said that their main source of seafood is from restaurants. Apparently they don't feel confident of fish in a store case or in their ability to cook it. Even more education is in order.

In 2008-2011 a major virus epidemic hit many Chile farms, and US fishermen soon saw decent prices again. By 2011 the Chile epidemic was waning, and, as expected, the next year the farmed salmon production was up, and everyone predicted that our domestic prices would drop again. But this time fancy advertising by ASMI and fishermen's coops and the better quality fish being marketed made the difference. Worldwide demand for the popular frozen fillets pushed up the price of every kind of salmon, farmed or wild, such as Enoch reported in 2014 for Kotzebue chum salmon.

A chance exists that a combination of "catch" fishermen, with recreational fishers, ENGOs, and scientists could be strong enough to stop a fish farming onslaught in the US before more catcher fleets and their communities are hurt. The ENGOs have proven that the successful way to close something down when all else fails is through the courts. There is always the possibility that a massive public movement to protect the ocean, such as those to save whales and Pacific salmon, could arise. But it wouldn't stop the global market that now reigns. The international aquaculture industry explosion could be the final chapter for many commercial fleets. Even worse, essential subsistence fisheries worldwide are doomed if industrial fishing is allowed to take over the seas. It will not be just human communities that suffer, of course.

It is confusing to read the NOAA's deep concerns about the ocean ecology and at the same time read that it is promoting fish farms. In 2015 the government announced that the first formal proposal for aquaculture in our federal waters is under review by NEPA. It is planned for 4.5 miles off the California coast, an enclosed pen for yellowtail, striped bass, and white sea bass. I have so far seen no protests reported in the major media. If we let this farm prosper, more will surely follow. The salmon fishing boat of the future might most practically be one converted to a dockside seafood eatery serving farmed fish and chips.

Part I

Photos

Tor Danielson, right, in fish trap, Icy Strait, 1918

Billy Frank Jr. at Frank's Landing, Olympia, WA, 1973 (photo by T. Thompson)

Conrad Danielsen's troller *Frida*, 1920s

George Morford's troller *Rocket*, 1960s

Jack McFarland and Nancy's troller *Deanna Marie*, 1964

Nancy on *Deanna Marie*, 1964

Jack and Nancy's troller *Nohusit* in S.E. Alaska, 1969

Nohusit with Jack and young crew: Dan, Rob, Lesley, Frank, 1966

Rob Thomas cleaning pinks, *Nohusit*, 1969

George's troller *Helen McColl*, 1970s

Jonna Crow, troller *Wanderer*, Cape Alava WA, 1974

Esther Ilutsik's sister Virginia setnetting at Ekuk, 1990s.

Generations of setnetters: Esther Ilutsik, mother, daughter, 1990s

Perry Mendenhall's cabin and camp, Nancy, Cape Nome, early 1990s

Dan Thomas with kids Nicole and Tor, subsistence fish, 1985

Perry Mendenhall at camp, 1997

Lesley Thomas and Nancy, Cape Nome, a meager haul, 1990s

Dan, nice catch, a pike too, Pilgrim River, 2012

Marilyn Thomas, river cohos for smoked fish, 2013

The screened fish rack, Nancy, chums and pinks, 2012

Cherilyn Kavairlook, chums and pinks, finally dry, 2012

George and new troller *Donna*, 2013

Linda Kline, Perry's sister, Copper Valley fishwheel, 2012

Part II

OUR FEDERAL SMALL-BOAT FLEETS
AND THEIR LOST COMMONS

*"Through money, political influence, and the multi-million
dollar horsepower of corporate environmental groups…
fisheries management brought forth the catch share solution.
Somehow, however, the shares all ended up with the big
operations that caused the problem in the first place."*

WEST COAST COMMERCIAL FISHERMAN DAVE HELLIWELL,
AS QUOTED IN *PACIFIC FISHING*, NOVEMBER, 2011.

One source of so much contentiousness in the fisheries today is right in the Magnuson Act, now titled the Magnuson-Stevens Act (MSA). Yet the intent of its creators was to strengthen fisheries. The power of the MSA took decades and many amendments to reach where it is today, and no doubt it will keep on evolving, and along with it the fate of the fisheries. The MSA covers all commercial fishing, but the federal waters fleets are the ones that have seen the most changes through it. These boats fish the waters outside three miles; in the Northwest and Alaska they are chasing pollock, hake, halibut, sablefish, crab, cod, and many species of groundfish. The same is true for British Columbia.

When the US began to industrialize its fisheries, especially following the passage of the MSA, it affected both the fish stocks and the fleets in many ways, and for the small boats in the federal waters in ways especially disastrous. There are still small-boat fleets fishing out there today, but unless we reverse course, or unless they have special protections, such as members of a CDQ have, they are in trouble. Many of the small-boat fishermen, and even their smallest communities, are today clinging to flotsam, or already sunk, while the larger operators and industrial fishing firms have survived the changes in management very well.

Though NOAA has much influence on all commercial fisheries, the states normally manage their own fleets--those fishing inside three miles--as long as they are not in conflict with federal laws. The two systems are of course in complete agreement on many general principles of management: the health of the fish stocks must come first, modern fleets must be managed, technological growth can put too much pressure on fish stocks, commercial fishermen do have to make money at it, and so on. But another "truth" evolved, beginning in the 1970s, one that a great many economists and fishery managers agreed on, that the US (and elsewhere) had too many commercial fishermen. As I described in the first section, during that period many state politicians were also convinced they must stop the flood of newcomers into their fleets and began to pass limited permit laws.

Beginning in the 1990s the two management systems diverged when the federal system chose a path committed to the breaking up of the traditional fishing commons and to seriously reducing the fleets, rather than holding them at frozen numbers. Today, even though many of the problems remain much the same, state and federal fisheries management look very different. But in both fleets the smaller and younger fishermen and the smaller fishing communities are the ones who have lost out the most.

Sixty years ago the US fleets were virtually all small-boat fleets. Through the Magnuson-Stevens Act our federal waters' fleets inherited vast new territory and moved toward industrialization. They inherited industrialization's seamy side: overfishing, waste, market wars, international politics, and as always, fish stock mysteries. We still have many mysteries, and other problems not so mysterious, like bycatch of untargeted fish.

In many US federal waters there are no fishing commons anymore. You can't just buy a commercial license and head out there and start fishing. Since 1995, it has also been difficult to start a new fishery, or even restart a collapsed one, as most stocks are fully exploited, some over-exploited, and almost all of the fleets are now limited entry. In this continent's waters, only in the north--Alaska and Northern Canada--are there a few places where a local fleet could possibly start from scratch with a new fishery. Ironically the new Alaskan ventures that have succeeded have done so largely because of the security they have through their tie to industrial fleets, brought about in a unique deal of the 1990s, the Community Development Quota program. I've already mentioned it a few times and describe it more in the following chapters. All other small boats fishing federal waters, and even larger ones without the right connections, struggle to stay afloat.

At Nome it's different. Here we can see how US fisheries must have appeared 100 years ago, with all of the vigor and promise for ordinary people in a small-boat fishery --the image the public still has of waterfront Alaska if not any other region. Nome and Norton Sound have not, so far, experienced the closing of the commons and traditional rights to the fish stocks, and may escape it due to remoteness. But outside of the CDQ regions, the same North Pacific Fisheries Management Council that put through CDQ has been--with NOAA's encouragement--the leader in developing programs to privatize fleets, first called Individual Fishing Quota (IFQ), now popularly called "catch share". The goal is apparently for every federal fishery to be managed by some form of catch share. Canada has the same goal, in fact often led the way. Catch share programs may stabilize fleets and, in theory not yet proven, fish stocks, but by their nature inevitably hurt the smaller fishermen and the communities dependent on them. Today, however, due to much turmoil on the docks over catch share issues, the MSA requires that each federal plan set aside a small amount of the season's fleet quota for community-managed fishing.

This reorganization of our fisheries should befuddle those who believe our country's strength is based on small business, not big industry, and that our politicians work hard to support it. But before looking at how we have allowed small US fishing businesses to be driven down, we should visit a few of Norton Sound's small fisheries through the fishermen's words and photos. It's glimpse at the life that was, and that now survives as an anomaly on today's seas.

Chapter 19

A FAR NORTH FISHERY: NOME KING CRABBING

The start of commercial crabbing in the Nome area probably has much in common with the way small commercial fisheries traditionally started up everywhere but would be exceptional today. When I arrived in Nome in the early 1970s, you could see a dozen families, all ages, out crabbing for subsistence on the sea ice on a calm spring day. My husband Perry has described in Chapter 2 how as a youth in the 1950s he hand-lined king crab to support himself. Soon after they arrived in Nome my teen children saw the same opportunity and were out hand-lining on their own. We ate crab regularly, gave it away, and bartered for other subsistence food.

Nome's families, typical for rural Alaska in the 1970's, were hard up for jobs in the winter. With commercial salmon fishing off Nome about finished, something needed to replace it. The lower Norton Sound had found a good market for herring roe through Japanese freezer ships, and a few Nome men went down for that fishery. But right outside Nome was our northern king crab. My sons were among the people needing cash-earning work and thinking about those crab. Getting serious, a group of Nome men went to the Alaska Board of Fish in 1977 with a request to open up commercial crabbing for Norton Sound. The Board had a number of big-boat crabber interests in its membership who listened to the pitch, thought ahead of the possibilities for themselves, not for the winter crabbing, but for their big boats in the summer. They went along with the proposal. Nome got their commercial crab fishery, and the next winter a few Nome families had their licenses and were out on the shore ice with sleds, power augers, and pots. I went out a few times with my sons. I wasn't much help with the heavy pots, but it was an excuse to get out on bright days. The new fishery survived, even reviving again and again when windy winters increased and the pots and gear went west. Fortunately, so far, no Nome crabbers have gone unrescued.

Winter crabbing stayed a local fishery. It's hard sweaty work, but some find it a way to chase away cabin fever and catch great meals. For others, the money is important. Either way, you are forced to go out whatever the weather. Even with covers, the holes will refreeze and be more work to chip out, so you must break loose from your indoor life on all but the worst days. You have to be dressed right, not so much on the sea ice where you stay warm working, but on the snow machine ride out and back, especially to the commercial grounds which are farther away from town. When the ice is firm and thick, crabbing is not a very dangerous operation unless there is a whiteout, meaning you could lose track of where the edge of the shore ice is and where new, weak ice might be. Commercial crabbing is definitely a fishery to satisfy one's macho inclination and fills one hole in Nome's seasonal job availability. Some years it's a money maker.

Rob, my second son, started crabbing when he was in high school. Now he is in his fifties and still out there each winter. His recall of his start is a typical small fisherman's story of hope, need, and stubbornness.

"I went crabbing with Marty Howard the first time, hand-lining, so it would have been 1973. Marty was from Diomede, a smart guy that knew about local subsistence and was willing to show me. We tried a lot of things, but the one that actually paid off was crab, later on, when people started talking about using pots. Hardly anyone was doing that then, but I thought I could barter a few crab around town in addition to what we ate. Gilbert--everyone called him Putu--my elder friend who also has passed on now--helped me make up some pots by bolting conduit together, and I got a subsistence crab license. Both my brothers did too. At first we were just running a few pots for friends and ourselves.

"There was no commercial fishery yet, but it was an obvious opportunity. Pretty soon many families were switching over from hand-lining to pots. It was a lot more productive. We hauled all the gear on snow-machine sleds we made: pots, ropes, trail stakes, plywood covers for the holes, gas-powered ice auger, ax, saw, bait, and tubs for water for the crabs to keep them alive. When the commercial crabbing started a couple years later I was right there. My brother Dan also bought a commercial license but before long he went back to subsistence crabbing. Since he was working fulltime, crab to eat was more important than cash."

Before long, Frank, my third son, now in his late-forties, joined in. For years he had counted on summer jobs through the Laborers' Union. Winters could be slow and one winter he ended up at Kodiak commercial fishing. He tells of his indoctrination in the life of winter fishing.

"When I was 24 I went on my first commercial fishing job on a cod boat in winter, out of Kodiak, so I'd found out what work was supposed to be like. Somehow I was the only crewman that never got hurt on that boat that winter, but I went home feeling rich. After that experience, my first year commercial winter crabbing--chopping holes in ice and hauling up pots-- was nothing. My first partner was Eric Osborne. He's a big strong guy with a lot of ambitious ideas. It was a challenge for us to get along as I had my own ideas too, so we practically killed each other several times that winter over that crab operation. But he has some good ideas, and he came up with the suggestion that we could sell live crab to Aquatech, a company in Anchorage that shipped live crab to Japan. When you came in tired from crabbing it would be nice to know exactly where the crab were going instead of having to start making phone calls and run all over town. And by now a lot of people were fishing pots, so we did need another market, not just Nome.

"We had to keep the crab lively for the trip to Aquatech, so we carried them in a tub of sea water on the sled as we went along the ice checking our pots. Then for the jet ride we transferred them to regular fish boxes, wrapped in wet blankets of newspaper to keep them damp. It meant we had to hustle, not have any breakdowns, to get the crab into Nome and into the boxes in time to catch the evening flight to Anchorage. We had to make that flight. But we pulled it off, and Aquatech was happy to buy good "clean" crab. They gave us about $3-4.00/lb. and paid the freight. It was a good arrangement for me, working for the local driller when he had work, and crabbing on the ice in the evenings.

"I did the same thing the next winter, but this time I went out with Fred Topkok from Teller as crew, another good man that has passed on. He was very strong, tough, quiet on the ice. There were no arguments. We did well; caught 10,000 lb. Of course I had to buy all the supplies and equipment and pay Fred something, so it wasn't all profit but it was a living. For the first time I was able to have at least one good snow machine, and I made do with the usual ancient truck. For many years that was my winter income I could count on. There were maybe a dozen of us out there crabbing commercially. With spring thaw the crabbing got more exciting. How long could you leave your pots out and still retrieve them before the ice went out? As it turned out, many years nature won--goodbye pots, covers, lines--the works headed west. The pots by regulation had to have an escape system to cover that event. A few times I've faced the hellish task of having to build a whole new outfit."

Rob recalls this nerve-wracking side of a generally peaceful livelihood: "One time we went out to check our gear and saw the ice breaking up near the shore where an onshore wind had pushed it against the bank. There was a gap of about six-eight feet, too much to fly across with a snow machine and sled, and then maybe be stuck out there on the other side of the gap. We sat there staring at that crack and all at once the wind shifted, the crack filled up, and we rushed out. We had about six hours to get all of our pots up and race back to shore. We all made it before the crack opened up again, hurray! But other times, we have lost the pots. They go up the coast, dragged by the ice, and you rarely see them again. One year a friend did rescue about twenty of Frank's pots as they came drifting by his place. It's just a matter of where the ice breaks off. You can make up some new gear up in a few days, but there goes your profit for the season. A hand-built pot and all the gear to go with it is about $200, and you need to have at least twenty, so there you go. Another time I had just got to shore and twenty minutes later the ice broke off and a guy a ways outside of me had to be rescued by helicopter."

I'm glad I didn't hear these stories from my sons until winter crabbing was over for the year. Frank tells of another time when they were out to catch one last crab:

"One late spring my friend Finn and I were trying to retrieve some of my gear. There was a big crack in the ice parallel to the shore and many more perpendicular cracks, a sure sign our time was limited. We headed out on the ice, pulled the pots, and heaped our sleds with all the crabs, pots, and hole covers--our snow machines groaning with the weight. On the way back the crack along the shore had widened and was now four feet wide. We had to travel along this crack and cross it every now and then, as there was no more snow on the beach for the snow machines. I was in front and always looking back to make sure Finn was still following. Then, one time I looked back, and oh shit, something was wrong! Finn was standing on the ice with a blank expression. No snow machine. His ski had caught on the edge of the crack and the machine and loaded sled were down in the crack, submerged. Luckily the water was just five feet deep. The machine was on the bottom, the back of the sled partly floating.

"After much swearing we managed to unhook the sled and pull it out. Finn wanted to just leave his snow machine, crazy! I told him to jump in and lift it. But this meant he had to climb into the icy water, and hold his breath enough to go under and lift up the front of the machine onto the edge. He tried, but he couldn't do it. So I, cursing, jumped in. I did manage to get the skis high enough for him to drag it out, all very traumatizing and hypothermic. And although it was a warm spring day, we had a ways to go, slogging back in wet clothes, hauling all his stuff. But we made it back anyway, even got the snow machine cleaned up and running again for the next winter."

By 1994 there were probably twenty commercial crabbers out on the ice ready to claim some of the bounty. A local buyer had appeared. Norton Sound Economic Development Corp (NSEDC), our new regional CDQ non-profit, had decided it would try buying crab in Nome and had leased an inactive freezer plant for the winter. Rob recalls, "It happened to be a good winter for the crabbers. I ran 40 pots, and made $8,000 gross in one month, and though that particular operation didn't continue, the idea was planted that Nome needed its own buyer."

Crabs, like fish, have their cycles, and they move around. Weather and ice cycle too, and slumps would come, but NSEDC decided to keep a buying station open, sometimes at a loss, as part of its stated mission to develop local fisheries. Warming currents for a few years thinned the ice, and by 2000 many would-be crabbers were tired of seeing their pot markers moving off toward the North Pole. By 2010 the price of gas was up to around $6 a gallon. Snow machines are prodigious gas eaters and the commercial fleet shrank to the most stubborn, or the biggest gamblers. My sons must be both. In 2013-2014 the ice was late, weak, and the wind was strong and constant. Between them they lost fifty pots and gear one blowy night. But two weeks later they were back out there with new handmade pots. They had no choice, they said, "Bills to pay, and luck could come our way." They were right. The next season saw both higher prices and more crab. A lot of the bills did get paid. But the crabbing was so successful that a record 40 people were on the ice and it was hard to find new spots to make pot holes. A record of almost 100,000 lbs. was harvested. Old-timers wondered if they had to go for limited entry if it kept up.

Fish and Game's local biologists also worried about the risk of more "ghost pots" with more crabbers, noting that the ice was forming later and breaking up earlier. Both Frank and Rob were among those losing pots again. Fish and Game and decided to start the season in January and end it in April. They also set a separate small percent of the quota for winter, apparently to discourage winter crabbing by so many as well as to save more for the summer crabbing. Most of the crabbers were unhappy with that, saying that winter was when they most needed cash. Some of them raised the question of limited permitting, but others opposed it saying that their youth would have nothing to plan for in the future.

Once again the conflict between preserving the resource and preserving the fishermen had risen to the surface. Once again the conflict between limits and an open commons, and between the security of retirement and the outlook for youth seemed to be permanent, inevitable conflicts for managers and fishermen to thrash out.

Crabbing with Skiffs

By the late 1970s Nome crabbing included a summer season. Local men re-rigged 20 to 24 ft. hunting skiffs, all the boats Nome had to offer at the time, and made up bigger pots. But their success at creating a commercial fishery had a bad spinoff. They were horrified one year when they saw a fleet of big crabbers that normally fished Bristol Bay moving in from the horizon, soon to drop their hundreds of pots right off Nome. After three years of this new pressure the local crab stock was far down, and Nome's small crab fishery, including the subsistence fishery, was doomed.

The Nome crabbers were furious at the takeover. They went to the Fish Board in 1981, backed up by local fish biologist Charlie Lean, and convinced the Board to support a "super-exclusive" fishery that would discourage the big crabbers from the long trip north. The Board ruled that if a vessel fished king crab in Norton Sound it could not fish for the same species in another area that same year. It also put a pot limit of 40 to 50 pots per boat, depending on vessel length, nothing to tempt vessels used to handling hundreds. The big boats never showed on the horizon again, and in a few years the local crab stock recovered enough to be seriously fished once more. It was one of the few examples I've found of small winning over big in fisheries. The Nome crabbers had done well. With crabbing back to normal Rob tells how he geared up for the summer fishery:

"About 1990 I'd seen an old fiberglass skiff sitting on the beach, and it seemed to just lie there, not being used at all. I found a fellow, a trader, who thought he owned it, and he traded it to me for $500 in other stuff. It was only sixteen feet, but Putu thought it could be made into a large enough boat for commercial crabbing. He helped design it, and I put an eight-foot fiberglassed plywood stern on it, a splash rail, a covered bow, and a windshield for the driver-- it never did have a cabin. With a gear loan from NSEDC--it would have been impossible without that loan--I ordered two forty-horse Suzuki outboards and a davit with a hydraulic block and tackle. Later on I got another loan for two nice Honda engines and foamed the inside for flotation. It was not a bad idea but the foam kept absorbing water and so the boat got heavier and heavier each year, with more fuel required, and it was hard on an old truck to haul it."

At the time I thought Rob was going back in time. He had started out as a child on a modern troller, and now he was planning to fish as trollers and codders did in 1920 except that he had two engines. Later he had a GPS and a radio, but still, this was an undersized if not primitive set-up, and very limited space for pot hauling. Yet it was not romance so much as necessity. Rob's equipment was simply what was available in Nome at the time. And from Kotzebue south past the Kuskokwim River, basic skiffs were what salmon fishermen all used, so now crabbers thought they should try too.

Rob goes on: "The *Island Girl* turned out to be a good sea boat, and I used her for years. We wanted real boats for crabbing, of course, but had to settle for the gear loans. NSEDC said no to boat loans--too risky. We didn't have to compete with big Kodiak boats that never came back, but it turned out that the forty-pot limit would include the Yukon boats. They had converted herring skiffs, usually larger than ours and better rigged, with bigger, faster engines, more seaworthy perhaps, and had at least basic cabins. So we Nome fishermen still had some serious competition, but it was possible. The *Island Girl* was just barely big enough, some said just big enough to get us in trouble. I could carry twenty pots out and set them, and then run back out with the next bunch.

"Fish and Game ruled we had to stay fifteen miles offshore of Nome to protect the subsistence fishery, quite a ways to go for a bunch of open skiffs hauling pots. Later they moved it in a couple miles. Still, a long ride, and the sea is practically always rough since it's so shallow. And coming into the Nome jetty in rough weather was more thrill than you needed. But that was what we had. My plan was to keep selling to Aquatech, sending the crabs in on the night plane.

"Frank had other work in construction going so he didn't crab in the summer most years. I got used to my boat and knew what it could do, but it was hard on deckhands, they were never totally happy with the challenge, often lasted one trip. I remember one summer the ice just hung around and wouldn't blow away until late July. We couldn't even get out past that boundary line. I never could do very well, being so limited by the weather, the size of the boat, and the deckhand turnover. It wasn't a living, it wasn't even a good idea, but it had the same attraction fishing always has."

I wondered about my sons taking skiffs loaded with pots and gear and crewmen out twelve-fifteen miles in the rough Bering Sea, but I remembered from my youthful adventures with my cousins that even a skiff, if well designed, can ride a lot of chop. I didn't know until later that the crabbers would end up going out farther yet to actually find crab. If it bothered them, they didn't say anything to me.

Crab Buying Experiments

The next winter Frank decided to spend some time in Anchorage. Perhaps he didn't mean to stay all winter, but he was hired at Aquatech, a broker that shipped to Asia, and saw the other side of crabbing--the buying, holding, selling, and shipping.

"The Aquatech people gave me the idea that I could be a more efficient shipper to them if I would install some of my own live tanks to keep the crab in until they were ready to ship. I could buy crab from other Nome fishermen too, and I could also sell locally out of the tanks. I talked my relatives and Aquatech into a loan and ordered the tanks, hoses, pumps, a scale, everything needed to set up the tanks. Back in Nome Audrey was able to get an old warehouse, a historic building that was being surplused, and I hauled it to the property next to Rob's house. I named the operation Chukchi Seafoods and started putting the tanks together. When it was all assembled it looked pretty good.

"I didn't actually go crabbing that summer because the Laborers Union called me, and I could make better money on those jobs, but I still wanted to try being a crab-buying station. That meant I left the management of Chukchi Seafoods to Rob mainly, and I came there when I could. But while

I was investing in all that, I didn't know that NSEDC, our local non-profit, had a plan for starting its own buying station. And here I had all that invested in the tanks and pumps. Rob was still fishing and had his pots out that needed checking, so he hired another fellow, Bill Topsekok, to keep the place open while he was out on the water."

Rob continues the story of the new Chukchi Seafoods:

"We soon saw we had technical problems. The circulating pumps weren't keeping the water clean enough, and the waste from the crabs would build up. We didn't have a big enough clean salt water supply--you can't have fresh water mixed in, like from near the river mouth, and it was hard for us to get clean ocean water. It has to be the right salinity and the right temperature. Aquatech didn't want sluggish crab, they'd really scream, so the crabs had to be shipped right away or sold locally. Then, no sooner did we get the holding tanks running better then we found out that NSEDC had contracted with another guy as buyer who already had contracts with the Yukon-Kuskokwim boats to sell exclusively to him. It was a shock; I don't know why we didn't know about it earlier. So we got only a few Nome boats to sell to us for a better price than that guy could pay. But then, on top of that, the crab didn't show up as strong as we had expected. We worried that stock was still weak from those big crabbers earlier. In a few years the recruits did build back up, but it didn't help us then.

"Between competing with the other buyer and the problems with the tanks, Chukchi Seafoods never made a profit. Some fishermen wanted it to and stayed with us a couple years. But the technical problems were too much too, and after that the tanks just sat there. After all Frank sank in. And that other buyer never did make good on his commitment to build a real plant. I heard he got caught dumping crab wastewater into the city system, got closed down, and left town. The whole deal was disgusting.

"We didn't try any more marketing because before long NSEDC decided to build a real fish processing plant in the harbor. That was the best answer for everyone I decided. It seemed easier to me then to just go crabbing and let them do our marketing. Sometimes I wasn't happy with their marketing, but at least it was out of my hair, and I could spend my extra energy trying to figure out how to get a real crab boat."

The idea of shipping live crab direct to an outside buyer hibernated for a while, but years later they would try it again. For now, finding financing for real boats was now the biggest problem for all the crabbers. What the Norton Sound crabbers were attempting was risky and the profits too small. Banks weren't interested in the loan applications that came their way from Norton Sound. The crabbers would keep on working from skiffs or give up the whole idea. A few did give up; others took their place. Though the tiny Norton Sound fleet was insignificant in the national fisheries picture, it was surviving while hundreds of other little fleets were drying up, hastened along by federal fisheries policy. The Norton Sound fleet faced only the traditional problems of small, remote fishermen.

Rob observed that there were months in the year still un-fished, so why not try to develop another fishery. There was already a successful herring roe fishery in lower Norton Sound, so there should be no reason another one a little farther north couldn't work. Herring roe was the real golden egg for many districts in Alaska, probably paying more per pound than any other fishery during that decade. But it was a very special market based on a holiday tradition in Japan.

Chapter 20

THE GOLDEN EGG: HERRING ROE

In the 1970s the lower Norton Sound herring roe fishery in June was where the real money was for the region's fishermen. But it was crowded with locals and even people that flew in. In the upper Sound herring was in recent years mainly used for crab bait and dog food. But Rob thought there was more potential; herring had a long, rich history. He had a personal need for lots of herring. By the time he was crabbing each summer, he and his wife Audrey, a St. Lawrence Island Yupik, had three children at home---so, fish harder. He needed cheap, good crab bait, but he also a dog team and the dogs needed herring too. His Inupiaq friend, Putu, a retired dog musher, had helped him get the team started. Rob was soon busy netting herring for both for crabs and dogs. Looking at all those beautiful fish, and always infected with entrepreneurial urges, he thought there ought to be a herring roe fishery developed near Nome that would also provide spawned-out herring for dog teams, and maybe even a commercial dog food plant.

My own ancestors valued herring a great deal for food. It is a rich, nutritious fish and to me delicious once you master the de-boning. I'm still working on that. It has also been used for cattle fodder in winter as well, so the milk tasted fishy--I even remember that taste from earlier Alaskan times. But still, it was fresh milk. People have salted herring for many centuries, and today make pickled herring. Though herring reduction for fertilizer and oil is finished, today it is a commercial baitfish for many fisheries. And it is especially important as a forage fish for larger species like salmon and cod. But herring is unpredictable; the runs can be extremely abundant, then simply disappear as their huge schools chase plankton swarms across the seas.

In this region, herring was first a subsistence catch, then, starting after the Gold Rush and through the 1930s, a few families at Grantley Harbor gillnetted the hordes, salted them in barrels, and sent them down the coast by freighter to be sold. It was a good cash economy for a time. From the mid-1960s until the US boundary was moved out to 200 miles in 1976, Japanese fishermen fished Norton Sound herring for their freezer-factory fleet. When a wide border of the Bering Sea became American waters, people living along the coast from Dillingham north to Kotzebue looked at the new possibilities in herring. Perry remembers attending a meeting at Nome in the late 1970s where west coast small fishermen discussed how to increase regional commercial fishing opportunities, and created the Bering Sea Fishermen's Association of Alaska (BSFAAK) with that goal. One of BSFAAK's first efforts was to build up local herring roe fisheries.

Frank Kavairlook, now living in Nome, but until recently from Koyuk, talks about the development of the small-boat herring roe fishery in lower Norton Sound.

"It really got going in the 1980s when a processor and tenders were there to pick up the herring. I was in on the start of that when a bunch of us younger guys agreed to go work on the processor. We started down south and worked processing as it came up the coast with the tenders, picking up herring. Local people saw they could use their salmon boats for herring, and soon were making good money. Some outside fishermen came in, and people learned from them too. I always felt good that we were in on getting that fishery started."

Herring roe was in high demand in Japan for traditional holiday celebrations. With floating Japanese buyers available, in a few years the roe fishery was booming in lower Norton Sound, worth at one time $1000 a ton to the fishermen. It was fished in short, frantic "derby" style openers in different regions at different dates depending on spawning time. In lower Norton Sound there could be one hundred skiffs and ten to fifteen tenders that picked up for one or two Japanese processors offshore and when the run finished ran back down the coast to other fisheries. The fishermen rushed to scoop up their share of the spawn-ready herring in a few days, delivered them, often for big checks. Then they, too, might go on to another fishery, salmon gillnetting. But by the 1980s commercial salmon fishing was no longer a moneymaker near Nome--smaller runs, no buyers--and nothing had taken its place.

Rob recalls, "I thought another fishery could be at Grantley Harbor again, west of Nome only about 75 miles with good road access. Grantley was such a great natural harbor for economic development for those villages, Teller and Brevig. A commercial fishery could be the jump-start. I knew a few people up there to talk to about it. The biggest problem, I thought, would be arranging a tender to come up the extra distance from Norton Sound.

"In 1987, a few of us convinced Fish and Game to give us a permit to fish sac-roe herring up at Grantley as a demonstration project. Sac-roe is tricky, it has to be shipped clean and fresh in salt, but big bucks for everyone involved, for the fishermen, the tender, the broker, the retailer, everyone. I'd hoped to get a grant for the start-up, but I didn't know how to write up those things. But a tender got the word and decided to go up there, along with a processor and a spotter plane, a big power skiff, and about a half dozen boats. My friend Putu and I took my eighteen-foot Lund up there, and we all waited for the herring out by Pt. Spencer. It was foggy, with a lot of ice around, still coming out of the harbor. The spotter plane couldn't see anything. We waited and waited, and then suddenly the fog lifted, and the plane radioed to us that there was 100 ton of herring headed our way.

"We were ready. Putu and I plugged two gillnets and I remember looking down into the water and it seemed like there was a fish in every mesh. But the next thing we knew the fish dragged all the nets under the ice, and what the fuck, they got tangled together under there. There was only the one skiff powerful enough to try to pull them back out, and it was having a time. After Putu and I hauled and picked about a ton by hand we were worn out, so we just cut our snagged net and delivered our fish. We were bushed, pulling that weight one damn mesh at a time, and it had been over 24 hours, so we had to eat and rest. By the time we got going again the herring had disappeared. Gone. I don't know how many tons the fleet got altogether, eighteen I think, mostly from the power skiff. The tender accepted them but then took off to its other obligations. End of harvest. That's what can happen with herring.

"I still believed the Harbor had a chance for a commercial fishery, so Putu and I tried again the next spring. I had the required public meetings at Teller to get the interest up for opening a commercial fishery, got the green light. But we couldn't get a tender to come that time, because at the end of the Norton Sound herring run the tenders all were scheduled by their company to head back down to Bristol Bay, a consistent fishery. We couldn't locate a company that wanted to try the Teller venture again, and that year we had to air freight our roe. Not a big success. Still, I decided I would try herring again the next year. I heard they were getting $18/lb for macrosiscus kelp. That's a very special item: herring eggs laid on a certain kind of kelp. You could find it in the wild, or grow it in pounds, or import it and plant the kelp in a cove at just the right time, and the herring would come in and spawn on it. It could be sold, the eggs right on the kelp, to Asian stores, or shipped to Japan where they were mad for it. I found a diver down at Sitka to gather up the kelp and airfreight it to us.

"My crew that year was Marty and Fred. The kelp arrived, but our timing was off for planting it--there was still too much ice, so the kelp got spoiled, wasted. So we gillnetted a bunch of herring, popped out the eggs, and salted them and the carcasses right in the boat, as there was no local buyer that year. But then on our way back down the coast a storm waylaid us for a couple days on shore, and we couldn't keep the roe cool enough. We sent about ten tubs of eggs to a broker in Anchorage, but he said it got there all spoiled--who knows--and sent me a consolation prize of leftover Halloween candy. The kids were happy. Next spring, what the hell, I ordered more kelp to try it again, but the diver was slow getting it to us and the herring had gone past. I never paid him.

"By then the sac roe herring market was going down. Once $1500 a ton, then it was $900, then $500, then something like $200-250 a ton, still dropping. The culture had changed in Japan, we heard, and the younger generation wasn't so interested in the traditional foods for celebrations. This was before sushi went global. The roe fisheries further south like Sitka that opened earlier could fill the market. Our big lower Norton Sound herring roe fleet faced tough times.

"So I gave up on roe and went for bait herring that I could use and also sell. A bunch of us were still fishing herring up by Teller, but then coming down the coast one year we happened to drop the net out in front of Nome, and here was a great horde of herring right off the Sandspit. We had to laugh at ourselves--from then on we fished bait herring a half-mile from home.

"I had an old WWII Quonset in my yard, so I converted part of it into a homemade walk-in freezer, installed some old discarded compressors. That's where we kept the herring until time for winter crabbing. I was able to get all the bait herring I thought I could use or sell or barter. This system worked well until the old compressors finally broke down for the last time, and I couldn't get anyone to fix them. New ones were way out of my range, so I closed down the freezer bait-selling operation."

Later Rob and a few other fishermen put the pressure on their regional CDQ non-profit, NSEDC, to try to get something going up with herring at Grantley Harbor, and eventually a group from the board and staff flew up to take a look and see if it was feasible. Rob was up there at the time fishing bait herring. He goes on:

"I saw the guys from NSEDC get off the plane, and I knew they were there for the herring, and I thought, *Good!* I can finally relax on trying to establish that fishery, let them do it. But it turned out there was too much bycatch in the hauls, only fifty percent herring, the rest tomcod, char, and whitefish, not a commercially realistic catch, and the local subsistence people protested too. They didn't want their char and whitefish scooped up, so nothing was developed. There's still no commercial fishery at Teller. The red salmon run in June is too up and down, and that's their prime subsistence fish they are getting ready for, right during the herring run. It turned out they didn't want to trade their subsistence red salmon for herring."

NSEDC, however, had an obligation to try to develop more local fisheries, and did not give up, as described in chapters 24-25.

Real Roe Fishing

Rob smiles, "You know there's still more yet to my herring efforts. I thought I should see the real roe fishery, so in 1994 I helped Eric Osborne weld up a boat and went down into Norton Sound with him for the big roe fishery. Then I found out what herring fishing was really like. Timing was everything. The herring had to have at least ten percent roe in weight in order to be acceptable to the buyers, who were buying the whole herring, and the roe had to be ripe, not too far from spawning. They obviously could not be spawned out. The idea was to net them up as fast as possible when they were at the perfect stage; if we were too early or too late, no sale. Spotter planes were hired to fly around watching the herring to see signs they were about to spawn, and Fish and Game took samples to be sure there was ten percent roe. When all looked right, they would officially called an opener and the mad scramble began in the short window. That's derby fishing! You're fishing on adrenalin.

"The crews are shaking or beating the herring out of the nets using mechanical devices. They do have to take care so the roe isn't shaken out. When the skiff is full they run for a tender, unload and rush back for more. My God, it's all over almost before you catch your breath--a huge gamble in an opener of only a few hours. That year the run was poor and so was the market. My timing was off again, I was seeing that famous fishery on a downhill stretch. We only got eleven tons, and I think we were close to high boat. After that, the roe market was so poor those fishermen had to find something else. A few locals were still fishing for bait, for $150 a ton, subsidized by NSEDC just to help out those communities who had depended on it in the lower Sound. So I went back to my search for some kind of damn fishery where I could make a little money--and feed my dogs."

I also asked Frank to tell about his first trip to the lower Sound for herring. He laughed. It was more comedy than anything, he thought, but years later he made the experience pay.

"We were all being towed down by the tender, a whole string of skiffs. When we got down into the Sound, and they turned us loose, we found it was still full of ice, and no herring yet, so we had to wait. And wait. We were eating and some were taking showers at the tenders. I never did get a shower, too many of us. There were about ten tenders and thirty boats with about seventy crew. The tenders tried to put us to work since we weren't doing anything, but that wasn't our job, to do unpaid work for the tenders, so the guys said, hell no, let's just go over to Besboro Island and camp there while we wait. And there we were, over seventy people camped on the beach at Besboro, for three whole damn weeks, waiting on late herring. Unique. What did we do? Ate, drank beer, smoked grass, played cards. I had f—ing jungle rot from no shower for three weeks wearing raingear. And of course I ran out of money."

Ken Waltz, one of the novice skiff captains from Nome, picks up the story:

"Our plan was the Nome boats were going to coordinate and corral the herring. But for days we were only coordinating how to get something to eat. About every third day a couple guys went to the tender to get some booze and food. A couple of the White Mountain guys caught two rabbits. We decided to climb the Island, hunt the rookery for sea bird eggs. You have to climb up the cliffs through the rocks covered with bird slime. That was interesting, actually pretty scary.

"Swarming above us like a bunch of mosquitoes were spotter planes. Every group had their own spotter plane, using code, planning to rush in when they got the signal from Fish and Game. Ha! Finally they called the opening. Off we go. Everything is forgotten between the Nome boats, what the fuck, our big scheme to cooperate no more; it's every boat for itself. We were fishing off Cape Denbigh, and it seemed like you had to get in close to the rocks to do well. Our boat never got any of the roe herring--we couldn't seem to get in close enough. We got a bunch of spawned-out bait herring was all, and got only $50 a ton for them. The boat Frank was on did better, quite well actually. They got roe herring at $800 or so a ton."

Frank adds, "But I didn't get paid and had to borrow money to fly home. I'd lost three weeks of the construction season. Well, I at least had other work to go back to, and Ken finally got me $300 from my skipper. I bought that big skiff and motor, and after that I was always able to net up some bait herring right off Nome."

If these herring stories seem mainly a comedy of errors, fishing can be that way, even for the experienced. When I thought my sons were two generations too late in their ideas about a good living from fishing, I had lots of evidence in the case of herring. My family could put herring ventures in the deep freeze, so I thought. But I was wrong again. In 2010 the salmon fishermen of Norton Sound, hurting from the king salmon collapse, appealed to NSEDC, couldn't it do something about herring? That's what it was supposed to do after all, build up local fisheries. It was a tough call. The roe market was saturated from places like Sitka, still a wild scene with big seiners swarming around each other each spring for seasons of six figures in a few days. The price was by now very poor at $150/ton, with the Japanese younger generation losing interest, but a seiner at Sitka could still make a season at that price on sheer volume. The Norton Sound families meanwhile were hurting because of the Chinook crash. Rob explains how the problem was solved.

"Norton Sound skiffs could barely break even at $150 a ton. But NSEDC decided to do what no private buyer would do--subsidize the dock price for us member fishermen--and talked Icicle Seafoods, a processor out of Petersburg, into sending a buyer up to the Sound for a week before it went for salmon down the coast. NSEDC agreed to bring the price up to at least $250 a ton, and more depending on percentage of roe in the fish. Outsiders could also fish, but they had to bring in their own buyer.

"I had a boat by then that could fish herring, and bought a limited herring permit from a guy who was giving it up, got a junked roller from another guy, and four young strong fellows willing to give it a try. We got down to the grounds, the herring were there and ready, and here was the ice again, lots of it, and about 30 boats. But the percent of roe was great, over 13 percent! It was a record for this region. My crew had to pull the net by hand over the roller and almost rebelled after hours and hours of that, but we ended up one of the high boats and happy. The crew all got about $1000 each for a few days work, about the same for me, so it was a good first season."

The next spring both Rob and Frank were ready to "go for the roe" again. This time everyone went home happy. With help from their CDQ, a sinking small-boat fleet had resurfaced. The next spring, though, the pack ice wouldn't move out and the fishery was cancelled before the fleet untied. The next season was marginal--ice, weather, boats, and herring didn't connect well either. Rob went down, and mostly waited on fish. Some people observed that with today's poor market the local roe fishery couldn't exist without the NSEDC subsidy. But regional fishermen said, so what? They are well aware of all the subsidies that go to big oil and big agriculture--why not them? NSEDC, with royalties from big fishing in the Bering Sea, could and would do that. But the next season NSEDC could not find a buyer to come north. With help from the regional non-profit, a sinking small-boat fleet had for a time resurfaced, but it appeared that Norton Sound families could no longer depend on herring roe. Sitka could supply the shrunken market.

At this point, many might ask, why even talk about such small fisheries, aside from the anecdotal interest? None of them have significance in the larger picture of American fisheries. But they have value because they are modern versions of how a desperate need to make a living on a hard coast, mixed with a good dose of personal challenge, has sent coastal people out on the sea for centuries. Surely that will continue as long as the wind blows in the scent of opportunity, even a whiff. But aside from western Alaska and its CDQs, North America is losing its small boat fleets and their home communities wherever industrial fishing can move in. The Magnuson Act, for economic and political reasons, probably inadvertently, encouraged it. Many social scientists and politicians have spent almost forty years trying to repair the damage.

Chapter 21

SEA CHANGES: THE MAGNUSON ACT

The passage of the Magnuson Act in 1976 changed US commercial fisheries as nothing had since the invention of the steam trawler. This was not all at once, but in a dynamic process as the Act is open for amendments every ten years. Each time the opportunity opens up for amendments there is more stormy political weather. Yet many small fishermen probably have no idea of how the Magnuson Act (MSA) is important to them. Congress itself is probably still trying to solve the dilemma of its dual goal: *"Conservation and management measures shall prevent overfishing while achieving on a continuing basis the optimum yield from each fishery for the US fishing industry."* Finding a balance between optimum yield while avoiding overfishing is an obvious challenge. The National Marine Fisheries Service (NMFS) states in retrospect that for the original 1976 version the primary goals were to install a 200-mile federal limit for our coasts and to begin federal management of the fisheries within it, but that "conservation was a centerpiece".

Natural resource conservation always swirls with conflicting ideas. The 200-mile limit, however, was a goal everyone could understand. Just offshore were foreign freezer ships hundreds of feet long--floating factories like the ones we deplored as we trolled outside Sitka in the 1960s-1970s. They carried double crews of as much as 100 for 24-hour workdays, and were able to stay out on trips of weeks or months. We take such vessels for granted today, but most US fishermen in the 1970s were still operating family-owned wooden beam-trawlers, gillnetters, pot boats, trollers, or dredges that returned to sell every night or every few days. It had worked for the families, but as a country we couldn't begin to compete with the world's industrial vessels. We chased away the foreign fleets from our coasts while we created our own modern industrial fleets. We were not the first to do so; several countries had already had, for the same reasons. It would be 1982 before our 200-mile Exclusive Economic Zone (EEZ) was fully established, and another ten years before all of the foreign vessels were gone from our waters, but then our fleets had vast seas of fish to chase.

The other major goal of the original Act, the conservation and rebuilding of commercial fish stocks, from the beginning was strong on intent. But with each ten-year re-enactment of the MSA, though conservation was addressed more, application floundered. A lack of scientific knowledge meant more funding for research was needed, but industrial expansion, not conservation, was the priority for Congress. Yet although stock conservation is of obvious benefit to all in the long run, an industrialized fleet doesn't benefit all fishermen; it benefits the big producers and the companies that own them. The small-boat fishermen can only hope that some of those benefits will trickle down to them through advancing technology and improved infrastructure like port facilities. Today as fishing fleets shrink, so does support to smaller ports that the large vessels don't use.

The need for serious conservation is obvious: we can't have fishing fleets without fish, so we have to be able to look beyond the next few years' profits, and for a rockfish it could be twenty or more. A halibut isn't a mature breeder for twelve years. We have the choice of the long or short view. Industrialization anywhere has rarely chosen the long view willingly, profits winning over conservation. Conservation in fisheries as everywhere requires strong management and controls that more far-sighted people demand. So, every ten years, congressmen and their staffs put on their armor and limp in to do battle once more through MSA amendments.

Even before the MSA passage the movement toward industrialized fisheries took encouragement at the end of WWII when an explosion of new technology hit the waterfront, made possible from all the upgraded, now underused shipyards and surplused equipment. Efficient, powerful diesel engines, new electronics like radar, and plastics that went into everything:

fish nets, wiring, and rigging all worked fine for fishing boats. By the time I started fishing with the *Deanna Marie* in the early 1960s most of the salmon trollers had installed surplused "jimmie" diesel engines and would buy whatever they could afford in electronics. Fishing became not only more efficient but safer and easier. Many traditional fishing families who had gotten by with basic livings bought into the post-war technology and bigger, safer boats as fast as they could afford them. New people who formerly might have thought fishing was a pitifully hard way to make a living now saw the role of a fish boat skipper as attractive, rewarding work. The number of people in commercial fishing soared.

Generous government loans for larger, modernized vessels meant the owners had to fish harder to pay them off. Before long, small-scale fishermen may have regretted that so many new vessels were pushed off the ways as fast as buyers came forward. The new factory-organized vessels were headed for federal waters that prior to 1976 saw virtually no management. The owners and crews would not especially welcome the new book of rules.

Technological advances were inevitable, of course. Few people are going to refuse inventions that make life better if they can afford them. As a commercial fisherman I always felt a strong need to be smarter and fish harder. But I did want to be safe, and never disputed the purchase of hydraulics that meant a stronger anchor winch, or the cost of a reliable radio, an engine that didn't break down, and finally a radar, installed our fifth year, that helped keep us off the rocks. Few fishermen are like my cousin George, who stubbornly held onto technology of the 1950s--compass, charts and radio--and his sense of where the fish were. But the government wasn't interested in mysterious instinct or traditional knowledge. It was going to have a fleet to compete in the world.

Our government's concern that we were falling seriously behind in the race to harvest fish had been growing for years. Paul Molyneaux, a Maine fisherman turned writer (The Doryman's Reflection, p. 22-23) tells that in the late 1960s two commissions were pondering deeply how to rapidly modernize fisheries. One, the International Commission for the Northwest

Atlantic Fisheries, had jurisdiction over the waters outside twelve miles. Its 33 member countries concluded that allotting quotas of fish to countries based on volume of past harvest would eliminate excessive competition. Yet fisheries were a risky business for investors. Private money would be far more likely to buy in if private rights to a specific amount of fleet quota could be guaranteed to an entity. This concept of individual quotas assigned to vessels ran up against a traditional practice much older than those of modern business: the right of all to fish in an ocean commons for whatever they could catch. The fishing commons was one expression of the public commons that went so far back in time there is no record of its beginning. To allocate individual rights to the fish would abolish the commons much more than limited permitting, the system described in Part I, soon to be adopted by the states.

Nothing in the original MSA stressed privatization, but the concept hovered in the government think tanks, based on a theory, "The Tragedy of the Commons" as expanded to the fisheries. The theory is that as a population grows and competes for resources, what was renewable in a commons is eventually destroyed because no one really owns it. Proponents of this theory talked it up until many politicians were convinced that conservation measures would never, alone, rebuild depleted fish stocks. Only individual rights to a certain percent of an annual fleet quota would make fishermen seriously embrace conservation.

In truth, countless small traditional groups have kept their commons sustainable through the ages using unwritten mutual rules that are passed down through the generations. It is only when too many outsiders begin to use the commons and either don't understand the rules or ignore them that sustainability is lost. Or a human population explosion could put too much pressure on a commons. Both trends happened in the west coast fisheries after WWII. The simple solution was in the opposite direction of the commons. How, then, could rights to fish be given out fairly?

By the 1970s privatization of public-owned sectors was a growing popular political strategy in the US (think railroads, utilities, telecommunications, clinics, even schools), so why not the ocean fisheries? For the free-market economists who advised our federal government, fisheries fit nicely into this trend. Not all public policy economists climbed aboard. Eleanor Ostrom won a Nobel Prize in economics for her proofs of how a commons can be sustainable when a group takes responsibility for it. Poor stewardship of the commons can also be seen in private ventures' overgrazing, or in erosion caused by clear cutting--and in overfishing. But it has not always been the fault of the local population. In a well-known example, the British government and landed gentry were the perpetrators in the infamous land enclosures and loss of the commons that took place in northern England and Scotland in the late eighteenth century that drove thousands of peasants to the cities to factory work and social misery. The new gentry owners then turned the commons into vast sheep ranches that over-grazed the land. No natural resources were saved that time. Breaking up of a commons has often favored a select, politically powerful group, but government economists apparently didn't choose to look at this other side of the "tragedy" theory, history being ignored in favor of increased production and efficiency. This movement has affected every fishery mentioned in Part II.

The second commission, the Stratton Commission, as Molyneaux points out, found that the US had too many small inefficient boats that would clutter a scheme of privatized quotas. Bigger but fewer boats was an essential first move. However, this commission warned that the long tradition of small-boat fishing in a commons could not be undone suddenly without running into resistance. It would have to be introduced gradually, as Molyneux interprets it (p.23), "…rather than rip apart the social fabric of fishing communities the goal of fishery management would be to unravel it slowly, severing fishing production thread by thread from its past, and binding it to a rational model, which would seek the highest net economic return from the resource and nothing else." That says it all.

Such privatizing schemes came to be known as "Limited Access Privilege", or soon, more commonly "Individual Fishing Quota"(IFQ) and "Individual Transferable Quota"(ITQ). Now they are lumped as versions of "catch share". The Commission also recommended formation of the National Marine Fisheries Services under NOAA to oversee the fisheries and to develop these concepts further. The government's official definition of IFQ today is: "A permit to harvest a quantity of fish or processing units representing a portion of the total allowable catch of the fishery that may be held for exclusive use by a person [or entity]." In standard English it means that eligible boat owners gain the right to catch assigned percents of the overall fish quota for the season. A person's eligibility for quota and how much almost always has depended on the boat owner's historic volume of catch.

There were other important goals that had to be addressed at the same time, and to do this the MSA created eight regional fishery management councils. They each included both fishing group and government representatives with the theory that both the industry and the states would accept radical management changes better if such advisory councils played a role. As another bid for cooperation, the MSA would be open to amendments every ten years, an opportunity to fix problems. These councils, with NOAA's encouragement, would be the entities eventually to plan the limited access privilege/IFQ/catch share programs.

A potential conflict in the MSA had to be dealt with however: efficiency versus conservation. With each efficiency gain through vessel size and technology the danger of overfishing grows as well. Support for stronger scientific management would also be needed, a commitment in more funds from Congress, not just talk. But that concern didn't get enough attention. The NOAA website describes the problem that soon developed for management: "Without effective regulatory restraints in place, by the late 1980s Americanization of the fleet and advancements in fishing technologies over-ran the slower growing science and management infrastructures, exploding the rate of domestic-driven overfishing...." No longer could we simply blame foreign fleets for loss of fish stocks, in fact in many cases, such as in the Bering Sea, we did a worse job of protecting them than the foreigners had.

Soon a growing movement of conservationists accused NOAA's National Marine Fisheries Service (NMFS) and the regional councils of not carrying out their responsibility of protecting and rebuilding fish stocks. US scientists at the time didn't know nearly enough about ocean fish stocks to advise the councils perfectly. The lives of many ocean fish stocks are still a mystery:

How many baby crabs are eaten by halibut? How many baby halibut do crab eat? How do we set harvest quotas without knowing numbers? Any form of management needed to cover ecosystems, not just individual stocks. Such research was complicated and expensive, and so was NMFS enforcement of any new regulations created. Adequate congressional appropriations were not forthcoming. Congress had created the MSA, NOAA, and the councils, but it wasn't willing to truly support conservation.

If small-boat fishermen fishing federal waters heard of these developments, a warning bell should have gone off for them: the US could be giving up the fishing commons in favor of support to corporate-owned fleets and their new monster trawlers being turned out in New England and Seattle shipyards. "Bigger" was definitely winning, and "fewer" soon would be. The small boats supporting thousands of families and their communities didn't count for much in international balances of power.

Even with the MSA of 1976, the accompanying 200-mile Exclusive Economic Zone (EEZ), and the surge in boat building, the US was still far behind in industrial-sized vessels. Thus we welcomed certain foreign fleets to continue fishing for years in our EEZ on condition they not target fish species we were interested in. The US also didn't have nearly what was needed in fish processing and transport, so we also arranged joint ventures with foreign floating freezer-processors, and allowed foreign onshore processors to continue. Today many of Alaska's shore processors are still foreign-owned, but by 1992 any off-shore fishing company had to be 75% US owned.

Fixing the MSA

The first ten-year re-enactment of the MSA in 1986 focused on getting control over our federal fishing waters and winning international agreements to stop blatantly bad practices such as miles-long high seas drift nets. US fleets finally had to themselves some of the richest fishing grounds in the world, such as the Bering Sea and Gulf of Alaska. British Columbia had the same riches right across from us in Dixon Entrance. But in taking on full responsibility for protecting the stocks, we still settled for a weak, underfunded system of on-board observers, and our new federal trawl fleets were free to work much of our ocean floors in a way to assure more ecological destruction.

The MSA 1986 moved into much more turbulent waters as the government picked up on the Stratton Commission recommendation to privatize the rights to fish. NOAA asked the regional councils to look at individual fish quotas (IFQ) as a way to organize the fleets. This system would be a real windfall for the larger operators if the US followed the method in use in other countries of allocating quota shares based on a boat's historic catch and then allowing quota to be bought and sold on an open market, or in some cases leased. When owners of small amounts quota sold out it was sure to create an active market and a means of consolidating the fleets to bigger, more efficient vessels. The boats with rights to fish bigger percents of the season quota could then expected to be able to pay more in landing fees and relieve a government of some of a fleet's administration costs. Thus a difference in philosophy of management developed between the federal system pushing for more efficiency and the state managers continuing to manage in a traditional commons using seasonal quotas and restrictions on gear and boat length--intentional reductions of efficiency.

It is not difficult to see why the government-employed economists of the day would find privatization of fisheries attractive, or why the larger fishing entities would see it greatly to their advantage. But individualized quota is clearly not "business as usual" that competitive small businesses usually face in our country. It is government-promoted social engineering. The argument against it, the right of all to fish in the commons as an inherent democratic right, has gotten nowhere in courts because of the legal technicality that the fish are not being given away, only the right to fish for a certain amount of them. IFQ is similar to a transfer of public grasslands to huge ranches for small fees, which, it can be argued, is not a giveaway because it doesn't transfer the land, only the rights to the grass. The courts in one country after another saw no problem with eliminating the ocean commons, as they all were faced with growing fleets, threatened fish stocks, and saw industrialization, and along with it privatization, as a way out.

Twenty years before, states' limited permitting in fisheries had actually been the start of the erosion of the commons through giving the right to fish to established fishermen. But those that had won the permits still had the right to catch as many fish as possible until the season closed or the fleet quota was met. Now even that right was under the knife. The concept behind IFQ troubled many people, not just small fishermen. John, the businessman I talked to at Kodiak, was not alone when he saw it as "robbery of constitutional rights". In truth, it was rights going farther back than that, back to English common law. Though a main IFQ selling point for economists was that it eliminated inefficient vessels, mainly smaller ones, efficiency has always been weak on merit when the stewardship of natural resources is a priority. Strip mines are the most efficient way to mine, miles-long drift nets are an efficient way to catch salmon, and clear-cutting is the efficient way to log. Large numbers of the general public object to them. Few of the general public had any idea of the changes going forward in the federal fisheries.

With NOAA's encouragement, the North Pacific Fisheries Management Council was one of two regional councils that began planning for privatized fisheries. It moved slowly, and its plans weren't active programs until 1995, but through that council and the Mid-Atlantic Council a radical new system of management was launched before the next re-enactment of the MSA came about in 1996.

The MSA 1996 amendments bore down much more on both conservation and fleet issues. Following the general trend in federal program management, the MSA stated that the councils would start using measurable objectives in the plans. Pushed by fisheries biologists, the councils were to identify and conserve "essential fish habitats" and to minimize the adverse impacts on them "to the extent practicable". They were to carry out scientific stock assessments of the most important commercial stocks in order to set maximum sustained yields--the level at which a fish stock could be safely harvested. Deadlines for corrections were included for the first time. By1998 bycatch was to be brought under control, overfishing was to end by 2008, and all fisheries were to be at sustainable level by 2011. None of these deadlines were met. They were unrealistic considering the number of species and the funding Congress provided, and all deadlines were extended.

The MSA 1996 also stated that the regional council plans could not be designed for solely economic purposes. They now had to address three new "Standards", all with the caveat "to the extent practiceable": to minimize by-catch; to improve safety at sea; and to "make provision for sustained participation of traditional fishing communities, minimizing adverse economic impacts on such communities". Any council plan that addressed those standards could not be accused of purely economic motives. Since several IFQ plans had already been activated by 1996, the economic influences were already obvious--a plus for some involved, not so for others. The affected ports on both coasts were full of raging fishermen, jubilant ones, and many simply overwhelmed.

The sponsor of the original MSA, Senator Warren Magnuson of Washington, a Democrat, and later Ted Stevens, a Republican, the Act's strong advocate, were both politically powerful; both were Chairs of Appropriations in their long tenures. Their opponents accused them of 'pork-barreling" as they directed millions to their regions, with fisheries development one of the beneficiaries. After Magnuson died, Stevens became the big voice for the western fisheries. The writers of the MSA amendments of 1996--Stevens undoubtedly among them--had moved the right direction with the addition of the three standards. But they were late, and vague enough that real protections were easy to avoid. The councils and NOAA Fisheries (NMFS) were mandated to come up with detailed policies and regulations.

Federal Conservation Lags

The implementation of the MSA 1996 continued to be weak in fish conservation efforts. The industry's powerful lobby fought any acknowledgement that important commercial stocks were dangerously overfished. There has always been a portion of any fleet that insists there are more fish than the scientists find. The fishermen are out on the water, they argue, not sitting in labs and offices, and see what's going on. They complain that they were not being included enough in the assessments, in determining what stocks are at risk. An example I am familiar with was a count of bowhead whale off Alaska some years back, with the Inuit hunters who'd had their annual quota cut, insisting there were far more whales than the scientists estimated, and in this case turned out to be right. But many fishery stocks were indeed still shrinking.

The industrial fleets with political clout fought the radical action needed in stock and habitat rebuilding, while small fishermen with little political power accused the industrial fleets, especially trawlers, of the major guilt in the depleting of stocks and ruining of habitat. Government funded scientists, state and federal, had to take some blame, as often they hadn't fought hard enough for the fish, spending their energy instead on protection of their agencies and programs from government cuts and environmentalist (ENGO) criticism. But partly due to the growing pressure from the ENGOs and partly due to scientists' better research and program planning, many important stocks did get off the national "overfished" list. Others joined the list, but the net effect was more stocks recovered.

This was not the case for small-boat fishing fleets and their small communities that were also shrinking but with little or no chance of recovering. Yet the MSA intent was clear: Don't abandon the small fishermen; at least think about them. The next MSA re-enactment in 2006 would emphasize that principle more, along with more attention to fish protection. But the opponents of social protections or conservation measures, or even safety measures in some cases--any that could cut into profits--would not give in easily.

Free-marketing Fish

Turning commercial fisheries into private domains was one effect of the "Reaganomics" years' goal of cutting back government functions and costs through privatizing them. Reaganomics can claim the fisheries one of its successes. The philosophy behind a private business take-over for fisheries was popularized at the University of Chicago, the home base of economist Milton Friedman and his followers, known as the "free-marketers", or the less complimentary "trickle-downers". Privatizing government functions is lauded in laissez-faire capitalism. Even during Democratic administrations, many free-marketers are embedded in government positions and think-tanks of influence. Private salmon ranching had already invaded US fisheries by the 1970s as described in Part I. Though that enterprise ran into enough problems that it died in many states, the concept stayed healthy and continued its movement into the MSA as it got amended and into the federal fisheries.

When we look at all the government activities on the waterfront that support commercial fisheries directly or indirectly, and the taxes we pay to support those services, it's clear why the free-marketers have gone after that area of our economy so energetically. Consider the costs of federal, state, and local support for: docks, other port services, navigation aids, fisheries research, federal aid to hatcheries, buy-backs, fishermen's (limited) health insurance, harbor dredging--it's a huge list. Though government still is active in much of this, the privatizing process isn't over.

The fishing fleets were an ideal place to start privatizing as they were already, in the economists' view "overcapitalized", meaning too many boats for the amount of fish available. They couldn't all make a good living, they could only hope to. Yet that isn't any different from other small businesses that only hope to succeed. No one dreams of handing out individual quotas of business to other enterprises, but for fishermen, and to some extent fish processors, somehow it is all right. The justification has been that they are fishing in a public commons, the ocean, which the government assumes it has power over. Small fishermen are, of course, not the only victims of privatization and free markets. All small business can suffer when under the free-marketers' influence the markets go global. Every news story on the economy reports other sectors crying woe over cheap foreign products, many of them produced by European and American firms moved overseas. Cheap foreign fish is just one product on that list.

The MSA would be reviewed every ten years, and each time the amendments would include stronger language to protect fish. Progress continued to be made with some stocks, not all. Progress with reducing bycatch also continued. Though some politicians, fishery managers, and environmentalists actually did believe that the small-boat fleets had value, that they were more than the economic and social anachronisms as viewed by free-market economists, they were not a loud voice. As for the general public, it continued to support popular movements to save spotted owls, whales, wolves, and salmon, as the media continued to lump commercial fishermen together and portray them all as greedy rapers of the seas.

Even though Congress from 1996 on gave more attention to conservation, enforcement of MSA goals and standards continued elusive. The Coast Guard did received funding to implement much improved safety rules and equipment for vessels that saved boats and fishermen's lives. But federal management allowed excessive bycatch and other waste to continue. The social protections of Standard 8 would be set aside until local residents themselves raised a loud enough cry. Congress had passed the MSA and its reenactments and then gone on to other tasks, trusting that NOAA and the regional councils would assure that the law's intent was truly and fairly carried out, even if not adequately funded.

Though NOAA continued to give stronger encouragement to the councils to plan for catch share, from 1996 on there were more amendments addressing the pitfalls of IFQ/catch share. There was no language to stop privatization, just to repair the damage, and prevent at least some of it. Yet none of the amendments would be turned in policies or regulations that could effectively do that except in grants for small projects for local fleets.

Chapter 22

THE REGIONAL FISHERIES COUNCILS
RETOOL THEIR FLEETS

In the 1980s the North Pacific Council had a growing mid-level trawl fleet eager to get better control of the Bering Sea and the great pollock fisheries the MSA had handed to it. There were too many big vessels and processors, too much foreign involvement still, perhaps even too many small operators that needed to be screened out. It was not just the pollock fleet; there were several other US fleets off Alaska seeing too many boats in their domain. IFQ could fix it all, government economists assured the fleets' big operators, and improve conservation at the same time. The Council could work out the process.

The MSA, by mandating that so many interests would sit at the councils' tables, had assigned them a daunting role. Yet the concept behind the councils employed a popular belief that if stakeholders were actively involved in planning a program for themselves they were more likely to be committed to its success. My first experiences attending regional fishery council meetings showed me how difficult it was to get any change that favored small-boat fishermen at the expense of the big-boat fleets. The structure of the councils was supposed to encourage democratic process, and indeed the councils do hear much testimony from fishermen and other citizens during public testimony, but the sleepy faces of the members that I observed over long days of meetings made me wonder how much testimony was carefully listened to. And it is not an easy public process. Few of us are used to speaking at a microphone before scores of people with a timer limiting us to three minutes.

Prior to 1976 the commercial fleets in federal waters worked in a freedom hard to find in modern society. Fleets in state waters were already accustomed to a serious amount of management, but not the new federal waters fishermen. One exception was the Pacific halibut fleet, which earlier had come under the management of the International Pacific Halibut Commission. For new federal regulations to work without heavy, expensive enforcement, they would need to show real economic advantages to the boat owners and skippers. But the diversity in the fleets made program planning a huge challenge. Even among trawlers the boats would soon range from family operations with a 32-foot boat, no bigger than a gillnetter, to the giant US factory-freezers coming on line. There was almost no region where this diversity didn't have to be dealt to some extent. The only exception would be remote areas where there were no big fishing boats, like my own region of Norton Sound.

Aside from fleet diversity, another problem for getting agreement on any council's action plan was the very structure of the councils as mandated by the MSA. One reason the councils accomplished so little in their first twenty years was, as some observers note, partly because there were so many debates leading to so many amendments, and litigation as well. NOAA acknowledges that one council plan has had thirty amendments. The North Pacific Council's years of debate over salmon bycatch, described earlier, is classic. The composition of the Council makes part of the problem clear. Of fifteen members, four are non-voting representatives from concerned federal agencies like the US Fish and Wildlife Dept. and eleven are voting seats. Of those voting, six reside in Alaska, the majority, but it means that all of the Alaskans must be present and vote as a block to overcome a motion that opposes Alaskan interests, usually meaning small-boat interests. The other five usually will support the interests of the fleets from outside. Sometimes, of course, every voter agrees on an issue. Sometimes actions simply get tabled, and more than once. The Council also has a scientific advisory committee and an advisory panel, both to include representatives from various interests, all with their own agendas.

Complaints about financial conflicts of interest on the regional councils have been frequent and at times noisy. Scientists are sometimes even accused of representing special interests. Yet it is impossible to include industry at the table without such conflicts and no point in having councils if industry is not included. We could run the all fisheries from a federal department as Canada does but that is not the normal process in the US, and where most government programs have a board or advisory group. The understanding from NOAA is that members will voluntarily recuse themselves from voting when appropriate, but sometimes they are ordered not to vote. They can debate all they wish. Thus it can take years to get a program passed, and then amendments begin. Another complaint regarding council composition comes from small fishing communities, subsistence fishermen, and conservation groups that don't have their own formal voting representation and believe they ought to. Yet, new voting seats would dilute the power of the original seats, and would require an MSA amendment. Conflict is thus guaranteed by the council structure, yet that is what councils are for--to provide a democratic process that includes debate. Certainly that has been carried out.

By the 1980s American federal fisheries were in a triple-bind: the fish stocks were still shrinking, the fleets were still increasing, in some cases too much, and costs of government management kept growing during the "Reaganomics" era, with its goal of shrinking government involvement in business. Part I describes the criticism that developed over the government costs of managing salmon hatcheries, but every fleet required expensive management. Privatizing them would cut costs, economists assured the councils. But it was inevitably going to be a lengthy process, especially with diverse fleets. If a plan did get approved, and Congress funded its start-up costs, the regional council and NMFS had to create the regulatory detail. More meetings, more public input. Finally, NMFS would carry out enforcement of the new program. Every plan and amended fisheries plan outside three miles needed to go through this long, stressful, and expensive funnel, all intended to ensure democratic process and attention to the government's long-term goals. What came out the end of the funnel obviously depended much on political power at play.

The funnel can be a frustration, partly because it is so slow and cumbersome, and partly because various interest groups believed they aren't being heard or understood, or can't get what they want. Congress had foreseen that introducing new management to federal fleets would never enjoy smooth waters and had included the regional councils in the MSA partly as a way to keep so much of the chaos out of its own halls. Yet it could still work. And on occasion factions determined to get a major fishery change have gone around a slow regional council to propose a plan direct to Congress.

On the broadest level there can often be agreement on a regional council, such as in a belief that fisheries need to be sustainable, but there might be little agreement as to how to make that a reality, or even if a case of overfishing truly exists. Stock assessments to determine if a stock is being fished at a sustainable level are essential but are expensive and time-consuming. Yet, to justify privatization, one main argument put forward was that it would protect fish stocks. Issues of fairness to the members of fishing fleets could be at least as troublesome. How far should a council go to assure fish stock or fishing fleet protection? The council meetings I have attended were calm, at least outwardly, but meetings are just as likely to be full of red-faced testifiers, to say nothing of what goes on in the hall during breaks. Private rights to fish stocks would turn out to be the ultimate in mutual accusations of blindness, greed, and narrow interests, and back hall deals.

Fish processors and corporate fleets soon found their way to work within the council process. But leading conservation groups complained that the "industry" dominated too much, while industry complained that scientists didn't have "on the sea" knowledge to make realistic recommendations, and that conservationists were there protecting their own bureaucracies. The small fishermen complained on their blogs that the councils didn't listen to them, that every idea they came up with was shot down. How could it be otherwise? Aside from wanting to catch many fish every season, how could a family operating a 24-foot halibut skiff have much in common with an 80 ft. high-tech, unionized halibut schooner out of Seattle, or a 150 ft. freezer-processor? Still, the council process requires that concerned citizens will be heard on every new program. And if the North Pacific and the New

England Councils are typical, the opportunity is well-used, with every quarterly meeting jammed with people eager to join the fray.

Each regional fishery council has developed its own personality. The New England Council is known for its diversity in fleets: heavily dominated by large trawlers and the processors that own many of them, yet also the home for many thousand small and very small fishermen. The Mid-Atlantic and Gulf regions are much concerned with the charter fishing industry, many small operators among them too. The Pacific Council has salmon runs in serious trouble and has to deal with a multitude of agencies and ENGOs that have a voice in the solution. The North Pacific Council has earned a reputation for being the most influenced by fishery biologists, never passing fleet quotas that are higher than the "accepted biological catch", the ABC. Even so, fishermen in western Alaska have seen how difficult it is to pass strong salmon bycatch reduction plans with the pollock fleet fighting them. Since this council is the one I am familiar with, my comments apply most to it, but many of the problems I've raised are general to all.

Shrinking the Federal Fleets

The decision to work through the councils to privatize fisheries was not the first effort to stop the influx of more boats into the federal fleets. The councils were already moving many of them under "Limited License Programs"(LLPs) similar to the states' LEP limited permitting. Voluntary buy-backs/ buy-outs of vessels and licenses were a sure way to shrink a fleet, but at a cost, and the government couldn't make loans for all the buy-backs needed to reach its goals of smaller fleets, nor would it find enough boat owners willing to have their boat or license bought out. And, as it turned out, frequently a fisherman who received funds from a buyout of a boat would use them to buy into another boat. The solution was to change to buy-outs of licenses instead of boats. The strategy for fleet-shrinking that government economists promoted, individual fishing quota (IFQ), was sure to shrink a fleet and was predicted to be cheaper for the government to manage.

During the years that our IFQ concepts were being fleshed out, no one in Congress or the administration seemed to anticipate all the social upheaval that could come to regions like Alaska or rural New England (for BC the same), home to so many small family-operated boats. Or if they did expect upheaval, they weren't concerned; the regional councils could thrash it out. The predictable occurred. By 1996, small fishing ports had lost numbers of the local fleets, and Congress had pushed MSA's Standard 8 through. The New England Council had seen enough on that coast and it chose to put IFQ on the back burner for years. But the North Pacific Council went right ahead with privatizing its fleets.

When American fishermen learned what the regional councils were brewing for their future they soon joined two camps with few treading in the middle. It wasn't difficult for boat owners to decipher where they would fall in the allocation of quota shares, on the lean or fat side, based on their sales during certain years that the councils chose. And in each fleet there were people ready to retire, and for them a seller's market in quota was timely. Though many older fishermen had no problem with voting yes on a referendum for IFQ, in the opposite corner would be the future fishermen and the fishermen that moved among several fisheries to make a living. Any small fishing port where other job opportunities were slim would in a few years see the disaster coming for their community as their young people left. That included most of the coastal communities of Alaska, British Columbia, and rural New England. But in almost every case, a fleet referendum on IFQ has favored it.

NOAA, in its own promotion of IFQ, apparently paid no heed to the well-known fact that all of the US coasts had been home for decades to small, inefficient commercial fisheries and the mainly small processors and tiny communities they supported, the kind I grew up in on Puget Sound and the kind I live in now. I think of Nome, population about 4,000, with thirty small boats in the Norton Sound fleet, and what would have taken place if the super-exclusive crab fishery had not been authorized, if the big crabbers had continued to come north, and if IFQ had been introduced. Even though the contributions of social scientists were weak in those first council plans, warnings on the social implications of IFQ/ITQ were there, for those who wanted to look, in what took place in Canada, Iceland, and New Zealand. Yet to Congress, to go down the same track of our failed state management policies, specifically salmon, seemed to guarantee more lost fisheries.

Meanwhile in Alaska a third management alternative by the North Pacific Council was also in motion. It was just as radical, but with a very different intent: to encourage local small fisheries for places like Nome along the whole west coast of Alaska. Alaskans themselves came up with the unique design for "Community Development Quota" first mentioned here in Chapter 1 and described more from Chapter 25 on, and convinced Senator Stevens to support it. Thus federal fishery management off Alaska went in two directions at once.

The fishermen and managers who promoted IFQ/catch share programs refused to accept the whole truth of what was taking place in the fisheries, or else chose a strategy where they thought they could win, even with guaranteed losses. If you triple the number of boats it is true that even inefficient operations like trollers can become a threat to the fish stocks. Yet what is the comparative pressure on the fish stocks of traditional small boats even in large numbers, measured against gigantic, efficient trawlers, the investment in those immense vessels, the pressure to keep them fishing, the profits flowing, and the politics to assure it? So far, no one except certain ENGOs has chosen to take on the trawl industry.

It is also true that the public will always get more excited about what seems to be an obvious problem like disappearing fish than a vague one like fleet structure. The government and its councils, realizing that IFQ was an alien idea that had to be sold, saw that one way was to promote its conservation benefits. A privatized fish would be a protected fish. Several national environmental organizations like the Environmental Defense Council strongly supported IFQ early on. Others, aware of the history of industrialization's effects on small operators, didn't buy it.

Some fish biologists supported the concept of privatized fisheries, believing that they might create more accountability. If boat owners/skippers owned the rights to fish and could fish at their own pace rather than racing other boats, they would take time to move out of an area where they were hauling in an excessive amount of unsellable and wasted bycatch. There was sense in that. A less provable theory was that quota owners would recognize that if the targeted fish stock then increased in numbers and the fleet quota could therefore be increased, then a vessel's share of quota, as a percent of fleet quota, would also increase. Vessel owners would therefore become active conservationists. That argument contained a pile of variables, but could be the case in a tight fleet of inherited vessels like the halibut fleet out of Seattle with a history of being pro-conservation. A dedication to conservation would not be so likely in a fleet where an owner of purchased quota might be a non-fishing investor in Florida leasing to a non-resident skipper from California hired to bring in the most profit possible from the remote Bering Sea where he would never be a resident. And that sort of fleet membership would become more and more common in federal fleets as they privatized.

Fishermen that followed these program developments at their regional council soon could see that the only reason for a big-boat owner or an industrial-sized fishing corporation to be against privatization would be if one had a strong philosophical belief in traditional equal rights to catch the fish in the commons. Or one might have the foresight to worry about what would become of the small-boat owners and the small independent processors and communities relying on them. Many coastal Alaskans, however, did understand the threat evolving at their North Pacific Council, especially at Kodiak, a growing port where, along with boats fishing the state waters, many federal waters boats picked up crewmen. As the Council steamed along with its plans for halibut and sablefish IFQ, many Kodiak residents argued that a radical restructure of the fleets wasn't necessary, and that correction of overfishing was already taking place through reduced fleet quotas and other regulations. But the big boat owners, most of them from "outside", loved that IFQ wind on their backs.

After the first IFQs created so much outcry, Congress ordered a moratorium on new ones until fish scientists and economists carried out more research. Government agencies like the OMB also submitted studies, but these did not change NOAA's recommendations in favor of catch share. The council planning and the debates carried on, especially in Alaska, and in New England where the regional council was under pressure to consider privatization of its trawl fleets. Belatedly a scrutiny of IFQ's social impacts began as anthropologists/sociologists realized that privatized fisheries were interesting experiments in social engineering for study. Today it is not difficult to find independent in-depth criticism of the effects on fishing communities, especially the smallest. But though with each decade the social problems of privatized fisheries rose more to the surface, correcting them turned out to be more difficult or expensive than the government was interested in taking on. By the time the moratorium was lifted in 2002, the North Pacific Council had three privatized programs fishing for years (halibut, sablefish and Bering Sea pollock) and a similar plan for its crab fleets was being thrashed out.

Yet at the same time, in a burst of creativity the Council had also embarked on the very different plan for western Alaska, one designed for the needs of small, economically underdeveloped coastal communities, the Community Development Quota (CDQ).

To this day we have the peculiar situation of traditional local communities happy with their CDQ benefits but furious over the salmon bycatch problem of the pollock trawlers, both managed in programs organized by the same council. The outcomes of its privatization experiments, and its contrary CDQ experiments, and how they each have affected remote communities and small fishermen, are described more in the following chapters.

NOAA would see that the regional councils continued to develop more plans for privatized fishing rights. Since fish stock conservation was what the politicized public cared most about, congressmen would see to it that the next re-enactment of the MSA in 2006 was written with firmer mandates for fish protection, but also some rather feeble ones for small-boat fishermen and their communities. How much any of them would be funded and enforced was the next question.

State Fishery Management: "We Pass"

The states, even though struggling with similar problems of fish stock sustainability and fleet crowding, chose to stay with the fisheries' traditional management that they had invested decades and millions in. It had its problems but it didn't deliberately favor big operators over small. If you had an eligible commercial license and a permit for an area and gear, you would fish with restrictions that applied to everyone until the fleet's harvest quota was reached. States continued to count on harvest quotas, gear limitations, seasons, and area restrictions, and emergency openers and closures as needed to keep fish stocks healthy. Some states added a limiting of "days at sea" and/or trip limits. Some conducted vessel and license buy-backs. Sometimes area closures, however, did have the effect of limiting the commons more if fleets had to fish farther offshore, or move long distance, and thus limited opportunity for smaller boats. But overall in state waters, a commons under limited permitting, though more restricted, was still an effort to maintain a commons.

Whether this traditional system works for a healthy fishery or not depends on many factors. It hasn't done well for Pacific salmon, outside Alaska. Lately, even the Alaska success story is showing clouds. Some state-managed fleets, like the Oregon salmon trollers, have been practically eliminated through restrictions; Washington has a very reduced troll fleet still fishing and is recently doing better. The gillnet fleets of Washington and Oregon, as described earlier, are about finished through legal actions, not management, while recently those on the Yukon and Kuskokwim have mainly been closed for several years, and it doesn't yet look brighter. Even though there are plenty of hatchery-raised seine fish in Alaska to go around, seiners have used buyback deals at a surprising rate--a way to finance their old age. And another attack on market price by farmed fish competition is always lurking.

State management is far from perfect the managers would be the first to admit. But except in rare cases they have a chance to fix their mistakes. State-managed fleets can at least hope for more intelligent policies and regulations and better future seasons. With privatized systems the gate shuts, the fish locked in preserves for a privileged few, and it is almost impossible to unlock the gate. I know governments have tried to improve on the serious problems of privatization for the future, but I know of very few remediations of the errors already locked in.

The issue of privatizing is just one of the paradoxes contained in the MSA. They will come up again and again in this history, and the councils will have to deal with them. Increasingly power politics drowns out the human story of the small fishermen who start out with basic equipment for a life of challenge, self-respect, and hard work close to nature. But with each MSA re-enactment there is also new opportunity for our regional fishery councils to do a better job for fish and fishermen.

Since so many of the problems of US fisheries first surfaced in New England, and continue to be an example of the worst in fisheries management, it's valuable to take a side-trip into a history that began back in the misty 1500s when the first commercial fishing fleets from Europe reached our shores.

Chapter 23

New England Fishing Lessons

A video-documentary made by Port Clyde, Maine small trawl fishermen a few years ago tells in their words how the industrial fleets and the regulations that favor them have ruined the region's centuries-old small-fishing economy. Watching those fishermen describe their struggles was the next best thing to being on the scene. New England fishermen are carrying a lot of freight for all of our fishermen.

The European fleets, too, have history lessons for us. But though they began the decimation of the bountiful North Atlantic fish stocks, we on the American side followed right along. The overfishing and destruction of the huge cod schools on this side of the ocean, incredibly to be repeated in this decade, could be the most famous fishery scandal to date since it was the main protein for much of Western Europe, and later east coast Canada and New England. Dried and salted-dried cod was a storable protein that made the Vikings' long voyages possible. Cod unfortunately is one of the species that has a difficult time to recover from overfishing and habitat destruction.

When explorers Cabot and Cartier arrived in the Gulf of St. Lawrence, about forty years after Columbus made his landing, they each found as many as a 1,000 Basque vessels hauling in cod as big as 50 pounds to salt and sell in England. Their secret got out, and the Basques were soon joined by fleets home-ported all along the eastern Atlantic from Spain north.

My husband Perry has his own Inupiaq forbearers' history with cod--the one we call tomcod. When the sea-mammal hunting season failed, people counted on salmon and other fish. But when all else failed, people jigged through the ice for tomcod that migrated into the estuaries. They were also good bait for crab, another lifesaver in old famine days. I have watched flocks of elementary school children jigging for tomcod in the spring--a perfect inefficient fishery, one not yet overfished.

The Euro-American commercial small-boat fishing culture (really many ethnic subcultures) that evolved in New England tells a graphic story of what can happen to fisheries through industrialization, a story available to our regional fishery councils as they started to look at radical new ways to manage our commercial fish stocks. The New England fisheries began with traditional handlining from dories, then evolved to sailing schooners carrying dories with two-man crews that were dropped off each day to handline, just as they were for hundreds of years in the old country. The cod were so huge and abundant that no one dreamed they could be overfished. No matter how many thousands of hand-liners, and no matter how managed, the fleets were blessedly inefficient and the stocks stayed healthy. Soon pots, small-scale traps, and gill nets were gears in use, and the stocks stayed healthy.

More efficient gear was slow to reach our shore, but it was coming, and with it the problems we fight today. The conflict between efficiency and healthy stocks is not a new one. Mark Kurlansky's *Cod* tells how angry handlining fishermen burned trawl vessels in Europe in the 1850s. But they were scoffed at by scientists of the day who believed fish stocks were "inexhaustible", that if grounds were depleted the fishermen would move on, the stocks would rebuild, and the fishermen could return. In New England the fleets fished as if they still believed that and began to use more efficient small beam trawlers. But the fish stocks still stayed healthy. Iceland's rich cod fishery, their only real industry and still harvested by an oar-and-sail fleet, had an early warning for us when it was practically wiped out by steampowered beam trawlers, mainly from Britain, beginning in the 1930s, and so threatened by the 1970s that Iceland was one of the first nations to install a 200-mile federal limit.

European steel stern trawlers first dumped their giant nets off our own coast in the 1950s. They could be four times the length of New England wooden beam trawlers then in use, nets to match, with their factory crew living aboard to fillet and freeze the fish non-stop. Our own groundfish draggers were still small beam trawlers, family affairs. Yet our east coast fishermen had meanwhile reduced the Atlantic salmon and halibut to where they were no longer profitable to fish, and didn't draw the conclusions they should have. Where were the managers, one has to ask. By the time the MSA was enacted in 1976, many scientists on both sides of the Atlantic warned that the new trawl fleets were too efficient for cod and other groundfish stocks. We have to accept that either the government didn't have the power to control those fleets, or didn't believe in its own scientists. Though halibut and salmon had been decimated, we still had haddock, redfish, flounder, sole, herring, shellfish, even cod. Inshore, families were still making a living with the traditional artisanal gear: pots, gillnets, jigs, and rakes, small seines, and small dredges and trawls.

When the Georges Bank cod harvest started down, every US boat fishing the Bank blamed the shrinkage on foreign fleets. But when the 200-mile Exclusive Economic Zone (EEZ) forced foreign fleets to leave, modernized, efficient Canadian and American trawl fleets moved right in as fast as we could build them and with the same attitude. There were also international waters beyond anyone's 200-mile limit, like parts of the Grand Banks, where huge new international factory freezer vessels could operate without restriction. Trawlers dragged the bottom of our state waters.

The tradition among New England's small fishermen had been to move between the stocks and areas, counting on many, not leaning too heavily on any one stock. The US now was promoting large industrial-sized vessels. To compete, New England fishing families had only two choices: to go for government-subsidized construction loans for the larger vessels built to government specifications, or to take the leavings. Paul Molyneaux, the small trawler crewman turned journalist, writes how the man he fished with for many years, Bernard Raynes, a ninth generation Maine fisherman, bucked the tide and chose to build a new traditional small trawler with limited electronics that didn't qualify for a government subsidy. Most people probably thought he was mad.

Since my cousin George had visited Maine looking for a new vessel, I sent him Molyneaux's book "The Doryman's Reflection". He commented:

"That book you sent--very interesting! He shows how the small fishermen were the scapegoats for poor management. And the guy he crewed for so many years, Bernard, was an important character in the book. Well, I actually I met him back in the 1970s. He brought his boat, a nice small trawler, into the dock when I was there looking for a boat. He was not in a very good mood--things were not going well. 'I don't want to fish anymore,' he said. There must have been quite a few that agreed with him on that coast. But they kept on, well, most felt they had no choice. And here I was, trying to buy one of those boats to take back to the west coast where we all believed there would be plenty of fish."

The New Englanders weren't blinder than fishermen in other regions. The rush for modern technology and its efficiencies in the fisheries has been told around the world--different stocks, different markets, same story. The New Englanders moved into trawling in large numbers, and the fifty-ton subsidized vessels outfitted with all the available technology dwarfed the old boats hanging on. As predicted, bigger loan payments forced the skippers to fish even harder on the stocks. Governments on both sides of the Atlantic that encouraged this kind of "overcapitalization" and "overfishing" invariably pointed fingers at the fleets' greed, shrugging off their own role. The effects of Northeast industrialized fishing on not just cod but all commercial fish stocks were obvious. The reason my cousin George was able to buy a well-kept 60-foot herring tender in Maine for $15,000 was because local herring stocks had been wiped out for the time.

As the trawler fleets grew they also evolved more and more to be a part of vertical organizations of large processors. Such groups could afford to hire lobbyists and build influence in national politics in both Canada and the US. New England did not at the time have limited entry laws to prevent the region's trawl fleet growing from 570 in 1970 to almost double that by the passage of the MSA six years later. NOAA and two of our regional fishery councils probably took heed, but industrial lobbies and government bureaucracies blockaded any change in our fishery management except that which favored industrialized fishing. By the mid-1980s Mark Kurlansky says that New England was taking an unsustainable 60-80% of the Gulf of Maine groundfish stocks each year. The North Pacific Council was embarking on its planning for the solution: privatizing of fishing rights.

The same destruction to the commercial stocks, amid warnings from independent fish scientists, was taking place in Canada. By the late 1980s the Canadian cod biomass was estimated at one percent of what it had been in the 1960s, yet it cut its cod quota only ten percent. As both Canada and the US began to commit more funds to research it was late in the game. Economic and political realities ruled with the northeast shores full of communities completely dependent on their fisheries. The social costs for any harvest cuts would be huge, costs to make the governments cringe.

Although bottom trawling is the worst way to fish ecologically, it makes short-sighted sense economically with large numbers of low-valued stocks. It makes no economic sense with badly diminished stocks, which was by the 1990s the case with cod and many New England groundfish species. Our Pacific cod stocks are healthy still, but not because west coast fishermen and processors are necessarily more virtuous. With so much higher priced halibut and salmon available, US fleets did not pursue the Pacific cod in such earnestness, leaving it to foreign vessels, and we were able to get sustainability regulations in place before a similar tragedy took place.

Trial and Error

It was not that NMFS and the New England Council didn't try various strategies to save fish stocks. In the 1980s they introduced harvest quotas for the trawl fleets. Immediate protests erupted as the fishermen challenged the stock assessments used to set the quotas. Were they guesswork or solid science? It was a fair question, yet the stock depletions were obvious. Harvest quotas didn't work well for a mixed stock fishery like groundfish trawling as they encouraged a race for fish between boats, dumping of unsellable bycatch, and also dumping of sellable but damaged fish when the quota was reached on one particular sellable stock--the "choke" stock--in order that a vessel could keep fishing other sellable stocks mixed in. Before long New England cancelled harvest quotas for a time. Some other management was needed, but when the concept of IFQ was first introduced it was so unpopular that the Council dropped it and instituted other ways to limit catches more: limits on catch per trip, limiting days at sea per week, or per season, per vessel. It tried closing of the more overfished inshore grounds, while offshore grounds and winter grounds stayed open. This favored larger boats that could safely fish them, and so did "days at sea" as smaller boats like Raynes' were slower and had to count their travel time as part of their days at sea. NMFS apparently had neither the will nor the political power to make regulations fairer to smaller boats. Nor could it force the fleet away from trawling and back to more conservative gear like pots or hook-and-line. It could only manage the death of a thousand cuts: more areas cut, more days cut, more pounds cut, and soon more species cut.

The New England Council and Canada's Dept. of Fisheries and Oceans both finally decided to cut hundreds, eventually thousands of east coast vessels from the fleets through buy-backs or buy-outs. But even buy-backs were not free of controversy. In some cases boat owners were then free to take the funds and buy into a newer, more efficient boat. This error was corrected, but when fifteen years had gone by since the enactment of the MSA and ground stocks were still falling, a collection of conservation groups decided it was time enough and in 1991 sued NOAA Fisheries (NMFS). They charged it was not carrying out the intent of the MSA regarding to protection and rebuilding of fish stocks. They won their case, one of many to be entered against NMFS over the next two decades.

The New England Council meetings became a cauldron of growing anger and frustration as no regulations, even with smaller fleets, seem to bring back the stocks. The percentage of families and firms involved in fishing and fish processing is much higher in New England and Eastern Canada than in the Pacific Northwest and British Columbia, and the fishermen historically have been more outspoken in their own defense than our west coast fleets. Their extended family ties and communication lines are strong; only Alaska is comparable. Even in Alaska half of the people working in fish processing are not residents and not part of the state's politically active public.

Despite the design of the regional councils to assure a balance of interests, as Paul Molyneaux and others observed it the processor/fishing corporations largely controlled the New England Council. Small fishermen who had been warning for years that we were overfishing had little influence in the meetings or behind doors. Mainly they weren't present; they were desperately trying to find fish. Meanwhile the environmental movement and recreational fishing sector were more and more present with their agendas. Legislators, state and federal, were pulled into debates to defend their constituents' interests.

Canada's Cod Fishery Crash

Canada's Dept. of Fisheries and Oceans finally felt forced to act in 1992 and closed the Atlantic cod fishery altogether. It was a death sentence for an industrial network employing over 30,000 people involving 1500 communities. About one-third of the Newfoundland population went on semi-permanent dole. The cod crash, most writers agree, was due to both overfishing and habitat destruction by a fleet with powerful connections, but also by a government too petrified by a looming economic dilemma to take timely action. It was as if everyone that studied the situation knew what was coming, but sealed their lips rather than be politically pilloried and unwilling to take responsibility until the cod were almost gone and their sea-bottom habitat scraped clean. The scandal of the destroyed cod stocks focused on Canadian mismanagement, but the public humiliation it caused was probably related to the percent of the population thrown out of work and the amount of long-term government assistance required.

The New England Council had to face the fact that its own cod biomass was also dropping rapidly, and finally in 1998 it introduced limited entry and closed the Gulf of Maine to multi-species groundfishing. But three-fourths of the Gulf fishery had been boats under 50 feet, so once again small fishermen took the beating. It was a very tardy action, as one study of boat captains' logbooks found that the stocks off Massachusetts were one-third of one percent of what they had been in the 1850s.

By the time of its cod crash, Canada, following the lead of the Netherlands and Iceland, had pilot programs going with privatized fisheries like IFQ, but New Englanders still wouldn't touch it. The besieged Council struggled to find harmonious solutions where none were likely. Regulations were evolving into a maze hard for anyone to understand, let alone to fish. The trawlers most adaptable to every new change would survive, meaning those most able to move more offshore, even to international waters, or switch to other gear would win what was left to win. When Molyneaux (p. 226) asked a spokesman for a national ENGO, "What about the wealth of knowledge we're losing when small-boat fishermen disappear in a consolidated fleet?" The answer was, "That can be replaced with technology." Later more conservation groups would begin to see the smaller commercial fishermen in a better light.

Molyneaux (p. 206) writes he didn't like what he saw happening to the families with small boats, mainly trawlers that he had fished for. He gave up trawling and went hunting periwinkles with a rowboat, a bag, and a wetsuit, refusing even to buy an outboard motor. Periwinkles were the perfect inefficient fishery, but in a few years they too were overfished by the desperate small fishermen. He dove for sea urchins, but too many saw the opportunity and soon that breathing hole closed. Molyneaux put away his fishing boots and in 1997 completed one of the re-training programs offered for fishermen: a degree in writing and literature. He used it well, attending New England Council meetings, reading stacks of reports, interviewing families, and writing articles for *The Fisherman's Voice*. His worries for survival of local traditional fishing livelihoods turned out to be on target. Up to nine generations of New England small fishermen like his old boss Raynes had been forced from a sustainable fishery using skiffs and hook-and-line or handmade traps, then to small trawlers, now to find the inshore grounds being fished out by big trawlers--and closing. However, he had some hope for the small fishermen remaining, as he observed the emergence of more respect for the eco-scientists.

Meanwhile eastern recreational fishing boomed, having the advantage in fewer restrictions, or none at all. Their skippers argued that the charter fleets now brought more money into most local economies than commercial fishing. Some small communities were able to transform themselves into charming destinations for vacationers with charters for rent. In Maine small boats turned to lobster that had moved in when the cod left. Farther south, dogfish and skate, always considered trash, became "money fish".

Finally a council believed the economists, took the risk, and introduced IFQ. The Mid-Atlantic Council decided to pilot it with two big clam dredge fisheries and a small one for wrack fish. Thus began a shift in fisheries management. The New England Council resisted the temptation for another fourteen years. Then, under heavy pressure from NOAA (described more in chapter 33) it began a major restructuring for its trawl fleets in the form of "catch share", and set off another uproar on the waterfront. It was privatization under a new name.

Chapter 24

MODELS FOR FISHERIES:
ICELAND, NEW ZEALAND, CANADA

When the North Pacific Council took on the task of restructuring its fleets it wasn't starting with a blank slate, as working models of privatized fisheries already existed. The Individualized Transferable Quota (ITQ), a fisheries structure like IFQ except that it assures transferability of quota, was already operating at least in Iceland, Netherlands, New Zealand, and parts of Canada. ITQ hadn't saved the first British Columbia ITQ fishery, abalone, but perhaps abalone was anomaly.

The governments in those countries promoted the belief that owners of private resources naturally took better care of them, and so it would be with fisheries that went to ITQ. But they should have known that wasn't necessarily the lesson from history; many private groups that seized public commons had been even guiltier of exploiting them to exhaustion. The European fishing industry spokesmen, however, apparently saw nothing but bright days in ITQ, and especially in the popular version that the larger the vessel owner's prior production, the more original free quota shares he/she should be awarded. Mid-range boat owners were a mix in their understanding of the issue, but if they saw a chance for their own survival they would typically vote yes on an ITQ fleet referendum.

The smaller fishermen, if they followed what was taking place, took a darker view. In the ocean commons they fished they had a least a chance at a living; ITQ would be a worse risk for them. But their protests usually built strength enough to be taken seriously only after the quota had been allocated. Then, restoring fishing rights to a group after they had been awarded to another group, and especially after the quota had been sold and bought on the open market, was a muddled task few governments would attempt.

New Zealand IFQ

New Zealand started its IFQ fisheries in 1986 by first eliminating all part-time coastal fishermen. Then, with the rest under IFQ, the number of vessels dropped from 2331 to 1277, while the number fishing offshore--larger vessels--increased. The Maoris, however, used their indigenous rights, and as part of their larger settlement with the government regained 20% of the total fleet quota that had been allocated. In the end, seventy-five percent of the quota ended up owned by ten companies, meaning that there was almost none available for independent non-Maori small fishermen. The government reports that it paid off in efficiency, with fish landings more than doubling, while quality of fish and stock conservation is much improved. Not everyone is happy with IFQ, and it has been challenged in court.

A Sami group who traditionally fished in northern Norway is fighting a similar legal case to get their rights restored after ITQ virtually eliminated them from the fishing grounds. Everywhere that IFQ/ITQ has been installed, the "T" ("transfer") part creates a social problem when the originally allocated free quota becomes available on an open market. Unless controlled by regulation, speculators have quickly moved in, and the quota price soon rises to a level no ordinary fishermen can afford. Leasing of quota doesn't help if leasing also goes on an open market and it, too, is soon out of reach for many. This history was available for study by the time NOAA promoted the concept of individual quota to our regional councils.

Iceland's ITQ Saga

Iceland's history with ITQ, beginning in the early 1980s, revealed problems early on. It had many lessons, especially for Alaskan fisheries that the

North Pacific Council had responsibility for. In the 1960s Great Britain's diesel-powered trawlers were scooping up the cod stocks just offshore. Iceland finally saw the only way to save its only economy was to force out all others and was one of the first countries to declare a radical 200-mile EEZ. Britain fought a fish war to keep its industry, complete with military action, and finally agreed to honor the new boundary only when Iceland threatened to close its strategic NATO base if Britain and others would not accept the 200-mile limit.

The small country was then in a hurry to catch up, and its own freezer-trawler factory vessels began to slide off the ways in the 1980s. Better conservation of fish stocks had been a selling point for ITQ for the Icelandic public, just as it would be for the BC and American publics. But efficiency was at least as big a goal for the government, meaning fleet consolidation. In 1984 Iceland introduced ITQ as a closed system for its large-vessel cod fishery, and soon others. The new industrial boats were too efficient and soon other stocks joined cod on Iceland's at-risk list. The government then bought out almost two hundred of the small-boat fleets and organized the remainder of all fleets by ITQ in 1993.

Consolidation was major; the herring fleet, for example, dropped from 200 vessels to 30. It continued, with the number of fishing vessels, overall, dropping from 1173 in 1993 to 762 by 2007. Meanwhile the factory fleet predictably upgraded to bigger, more efficient, more ecologically destructive vessels. According to a report by Michael Clark submitted to NOAA, unlike some countries, in Iceland the number of fish landed has remained about the same; they are just caught by fewer people.

The social problems created were soon in the public eye. Iceland, except for the capital, is a country of very small rural towns and villages like my Norton Sound region is today. Economic catastrophe can happen quickly. Iceland's rural people protested as they saw their local boats disappear when corporations, or the most successful fishermen, bought out the smaller operators and their quotas. The government then tried to correct this by not allowing any one entity more than 12% of the total fleet quota, and no more than 35% of any one species. It was a good step they should have ruled at the outset.

Rural fish processing plants also were affected when the new freezer ships that intended to take their loads to Europe for sale saw it was more profitable to install processing right on board. The number of municipally-owned processors dropped from 79 to 45, dooming the little towns where those plants were the main employer, and during a period when technology upgrades already had called for worker lay-offs. A mass migration of unemployed to Reykjavik, the one city, was inevitable. About twenty villages were all or virtually abandoned. The government at last stopped quota from being sold out of region after so many places had no fleet left and/or no processor. Again, it was a good idea but tardy action.

Alaskans on the North Pacific Council should have been keeping notes. Maybe they were. Iceland is a good example of a small economy being able to show graphically the effects of socio-economic change. The government's promotion of privatized fish stocks had created a new class of people called "sea barons" by the locals, like "armchair fishermen" in America. They were former fishermen and their children, or not even fishermen but investors, who now lived off fishing leases. Their children came back from college with advanced degrees, looking for ventures. Outside of government jobs there were two industries in Iceland: fisheries and aluminum production. Being a trawler captain didn't interest enough of them anymore. By 1999 the Iceland government tried to stop the rural flight by awarding small communities 10% of the overall fish quota that they could lease out to local boats to work for them. But the community quotas were offered as an alternative only after the local economies were desperate.

Meanwhile the offshore freezer-trawler giants, using the gear most guilty of destroying habitat, grew from a fleet of 31 vessels in 1998 to 63 in 2006. Ironically, some of them discovered it was most profitable to lease out their quota to shore boats and move their own operations entirely to international waters outside 200 miles where the open "race for fish" was still in effect--supposedly what everyone had wanted to eliminate.

Although by the 1990s many Icelandic social scientists were reporting serious social disruption to rural communities, our regional councils might have missed this. US writers that interviewed Icelandic businessmen found they were pleased with the outcomes of ITQ, one explaining that the fishermen remaining after consolidation of the fleets made more profit. Thus they had more money to spend, and could hire the people laid off in new jobs in services or domestic work. Unfortunately this well-known trickle-down theory has not proved out well in the US, and it is unlikely that it did any better in Iceland. By 2004 there was so much public objection to the ITQ giveaway and the still shrinking fish stocks that the Icelandic government introduced a 6% dockside "catch fee" for the trawlers, later raised to 9.5%, in addition to the usual administrative fees already paid, but to be used for research and stock rebuilding.

Meanwhile, some of the displaced young fishermen/sea barons who chose not to fish or run aluminum plants became investment experts. Due to enthusiasm coupled to lack of experience, they took part of the blame for the crash of Iceland's banks in 2008 and the banks' takeover by the government. Iceland has succeeded in crawling out of its financial hole but not its fishery dilemma. Today on the web one can find much commentary about ITQ by Icelandic social critics, some calling it a reintroduction of feudalism they had thought long in their past. Fishermen have sued the government several times, claiming ITQ is illegal. But their appeals court, like the US courts, has so far found ITQ legal for the same reason--it's not a gift, only a "privilege of use".

In 2007 Icelandic fishermen took their complaints to the UN Human Rights Commission, which sided with them, stating that it was illegal to turn over the "rights" to a public commons to private enterprise, but the Commission has no enforcement powers. (It also agreed with the Norwegian Sami complaint.) The Iceland government, smarting from its own public's criticism, considered that perhaps it should buy back all of the quota, to own it and lease it out by the season, with the earnings to go to research. But by then the state bank was mired in global economic turmoil, and as of 2012 Iceland had not moved on the idea of a quota buyback.

BC Casts Off With "ITQ"

Before IFQ came to the waters off Alaska, the Canadian government had already installed ITQ programs in British Columbia for herring roe, clams, abalone, and sablefish before turning to halibut. Like the US, Canada did use advisory groups to get ideas and reactions from the fishing industry and public, but its Dept. of Fisheries and Oceans (DFO) would make the final decisions. Over the years DFO often jumped into experiments before NOAA did, giving us the opportunity of learning through trials and errors close to home. The benefits of BC's first ITQs were consistent and easy to measure. Vessel owners that survived the process with a healthy quota allocation were always happy, especially if they were able to buy or lease even more shares. Quota prices rapidly shot up and that market was seen as a fine place to invest. And though boat owners that received very little quota usually did sell it and drop out of the fleet, economists saw this as positive, though the labor department and welfare offices that had to dole out assistance may not have been so pleased.

Eric Wickham, a BC small fisherman who had turned from salmon to black cod (sablefish), provides an alternate view, that ITQ or catch share can work well in particular cases. Despite being a heavy critic of DFO, as the black cod fishermen's association president Wickham (p. 148-149) tells how he worked with the agency in 1990 to design a ITQ program for a fleet of about 45 boats. First, about a quarter of the fleet--15 vessels--agreed to and received government buy-backs. That probably removed many or even most of the fishermen that thought they wouldn't do well with quota allocations. The remaining boat owners created a coop, decided how to distribute quota among the members, and set rules for conduct. After a year's trial fishery, the fleet voted to accept it. Wickham recalls that the association had been meeting every three months, with a sulky low attendance. Now everyone turned out. Eventually annual fees for research and enforcement were $5,000 per boat, willingly paid he says.

The timing was good. The small fleet made healthy profits with sablefish, the priciest fish on the west coast market. DFO, too, was happy since it now had little cost for management. The program apparently worked out well even for smaller boat owners who, like Eric with his converted 42 ft. troller, were not the real elite. His message is that a privatizing program carefully planned by a fleet, not excluding anyone, keeping the quota in the hands of active fishermen, and fishing a healthy stock, could work out.

With the sablefish success behind it, DFO initiated a halibut ITQ only a year later involving a much larger fleet, about 435 vessels after vessel buy-backs. In the one year evaluation, once again DFO gave ITQ high marks, listing significant fleet consolidation, elimination of the "race for fish" meaning a safer fishery, and improved conservation, as boats had stayed within the fleet quota and had less waste or abandoned gear. The dock price was better as fresh fish came in over a longer season. A big plus for the government was the consolidated fleet's higher revenues per boat, allowing it to raise license fees from $10 to $250, as well as increased landing fees. The fishermen remaining after the pilot year were apparently satisfied, as about ninety percent voted to continue with ITQ. Soon this fleet was paying almost all of the costs of its own administration.

Shortcomings of BC's ITQ programs also came to light but were given little coverage in government reports. Small traditional fishing communities gradually saw their fleets disappear with hundreds of crewmen left on the beach. With no race for fish, the skippers remaining could operate with fewer hands, meaning more layoffs. Crewmen complained that with fewer crew the boats were less safe. Worst of all, as with every privatizing program, the open market value of quota shares soon surged, often far beyond the value of the boats themselves. Those who were not allocated enough original quota would try to lease more, or lease their own quota out. But an open market in leasing soon meant many skippers couldn't afford to lease more unless they could borrow for it. Crewmen earned less since cost of leasing came off the top with other expenses, usually 10 to 15 percent, before their shares were calculated.

Alan Haig-Brown (p. 102) quotes BC fisherman Bill Wilson, " 'I was disgusted by the way that each year I was offered so much money to lease it [his quota] by the halibut buyers. They would offer me $3.00/lb. They would then lease it to fishermen and at first guarantee them only .50/lb. when they delivered their catch. Later this went up to $1.50/lb for the boat. One company managed to lease 600,000 pounds this way."

And again, Haig-Brown (p. 122) reporting another fisherman's fairly recent experience: "In 2008 Wick explained that fishermen were paid [by the broker] between $4.20 and $4.70/lb, but many had leased quota at $3.70/lb [and] were earning only $1/lb to be split amongst the three or four men and the boat."

Obviously this kind of leasing arrangement did nothing for small fishermen who earlier would have been fishing in a commons. Was this all Canada could come up with for the fleet? Yet before long NOAA and our councils would be allowing similar arrangements, not with halibut or sablefish, whose fleets were aware of the negative trends, but when the North Pacific Council turned to the federal crab fleets off Alaska.

For aging fishermen, however, leasing out their quota, or both boat and quota, was an ideal retirement system, something they'd never had, better than selling. They owned floating goldmines. A fleet of BC armchair fishermen flourished rather soon, people who never fished again, or never had. Today close to 100% of BC halibut quota is leased out. Thus the quota owner's economic state was indeed improved, but those forced to lease in order to fish were soon paying up to 80% of their harvest revenue in leasing costs. To DFO, however, the effects of leasing were apparently not a big concern since the ITQ system continued to be introduced to more Canadian fleets and the BC fleet majorities consistently voted in favor. Those referendums, however, were confined to the boat owners that had survived the first "pilot" year, so the opinion from those not happy with their allocation that had then exited a fleet couldn't be known.

DFO continued privatizing its fleets using ITQ wherever it would work to consolidate fleets. By 2000 the BC trawl fleet was reduced to 75 vessels through a combination of buy-backs and IFQ consolidation. Next on its list were 200 large longliners and thousands of smaller hook-and-liners fishing rockfish.

Some positive policy changes had also taken place that were not dependent on ITQ: DFO ordered that BC fleets be fully monitored either electronically or by government observers, with the fleets paying for most of this through landing fees, clearly an easier task for bigger producers. Bycatch thus had been greatly reduced. Social assistance had been introduced after the negative effects from the halibut ITQ came to light. Fishermen could sign up for retraining programs and crewmen were eligible for unemployment compensation and other government assistance, but these obviously didn't take the place of genuine livelihoods.

Other corrections were made in the groundfish IFQ, so that today DFO reports that unlike its halibut IFQ, 90% of the groundfish fleet quota remains in the hands of active fishermen, and vessels must have active owners on board. Recently DFO also began to include opportunities for community-based quota, especially in the case of First Nations communities. These changes were good to see, but it is strange that every country, so far, must repeat the same learning process with privatization, leaving the same environmental and social destruction in its wake, then spend much time and energy trying to fix the mess it has created. And the corrections didn't alter the general course of fisheries management. DFO claimed it used consensus-based decisions involving the fleets, but in April, 2015 the president of the fishermen's union UFAWU-CAW commented regarding the Herring Fishery Advisory Board representation: "The government thinks it is getting consensus advice from a broad range of stakeholders [but]....When one company (Canfisco) owns 226 herring licenses and controls a good deal more and independent fishermen own an average of one or two each, who is electing the seine and gillnet representatives?"

BC's Catch Share for Salmon

Canada's DFO jumped ahead of us in its experiments again and did what I assumed was impossible when in 2007 it introduced catch share to its BC salmon fisheries. (Thus salmon get a place in this section of the book.) DFO had earlier set its priorities for fishery quotas thus: Conservation is first priority, then aboriginal fisheries, then sport, and commercial last. But the DFO budget allowed for BC conservation projects, just as in the Northwest, wasn't nearly enough to cover the stock rebuilding needed, and Ottawa told DFO it could expect more cuts to fishery management. Meanwhile, the BC commercial salmon fleet had already been cut in half, mainly through buy-backs and the Mifflin Plan, and the landed value of its BC commercial salmon catch was only one-fourth that of a decade earlier. The cost of limited entry troll and gillnet licenses had doubled between 1994 and 2002 and were still rising. DFO apparently felt it had gone the limit with all other strategies and decided to introduce pilot projects in catch share for parts of its BC troll and seine fleets.

A quota of 450 Chinook was allocated to each troller owner, rather than allocated by history of catch, but the trollers had always hoped for more harvest than that. Predictably much selling and leasing of quota took place, as they questioned if they could actually make a living from their assigned allocation. Before long half of the northern troll Chinook catch share quota was leased out, and a new fleet of shore-side fishermen was sitting in Vancouver enjoying TV fishing channels. DFO had expected this would take place, since it had for other catch share fleets. After two years a vote was taken among the trollers still active as to whether the program should become permanent. The younger, rural, and northern-based fishermen voted no, but the needed 2/3rds of the trollers voted to continue the program.

Fortunately, a survey asked for their reasons for voting pro and con the program, and it shed some light on the popularity of privatized programs. DFO found that the majority who voted yes for catch share expressed worries about their future. They were looking at retirement security in quota shares, and believed that they also offered some protection against what they saw as the growing power of competitive groups: conservationists, First Nations, and the recreational fishing industry, with their fleet relegated to lowest priority. They concluded that no matter how low the fleet quota dropped, at least by owning marketable shares they would have some security. Older members of a fleet therefore had voted to give up their fishing commons and, in effect, let the younger generation find their own way.

I try to imagine how I would have reacted if offered a chance to go for a salmon IFQ for a set amount of fish when I was 30 years old, knowing as little as I did then about fishery management. I don't know if I would have gone for the security of individual quota, or the gamble in the commons. How would I have understood the trade-offs since I never went to meetings--I don't even know if there were any meetings then. The gamble has its eternal attraction, but poor harvest years would make a set quota attractive too. I picture younger fishermen trying to decide where they stood, and I believe Canadians knew a lot more than I did. But in the end DFO's decision would prevail.

Meanwhile the restrictions for all fishermen--First Nations fisheries, commercial, and recreational--were each year tougher. No wild Chinook could be kept, and almost no coho, wild or hatchery. Sockeye were so fluctuating that they could be closed early or never opened some years. By 2010 it was clear that history had repeated itself with the BC trollers. Seventy percent of the salmon quota owners were retired former fishermen or investors leasing out quota to people that wanted to fish. Smaller fishing communities would once more pay the social costs as they continually lost more of their local fleets. The same trend had taken place in the salmon seine fleets that had gone to catch share, with majority of the boats now owned by investors from three Vancouver-based companies that mainly leased out the vessels and quota and were paying more of management's administrative costs. DFO justified this, arguing that the public couldn't be expected to support the fleets' administration when joining the fleet wasn't really a public option anymore. To DFO's critics this was a self-fulfilling prophecy, since the agency had itself removed the option of public access in a commons.

DFO decided it should expand the salmon catch share further, and the West Coast Trollers Association took DFO to court in 2012, 86% of the declaring they didn't want catch share and wanted instead to be paid for the fleet quota they had lost to sport fishing and the closures they had suffered. It is doubtful that they won, as the next year DFO proposed to go entirely to catch share for all commercial salmon fleets. But DFO also gave a positive response to political pressure when it is strong enough, as in 2008 it bought back 48 salmon licenses, and gave--not sold--them to First Nation communities to use as commercial communal licenses managed by bands.

I have trepidations of what catch share would mean for US salmon fleets and hope it will never be proposed. Every government that embraced privatized quota has argued that it would aid conservation, but in few cases has this proven out. (New Zealand claims it has there.) Much good, indeed, has come from waste reduction that was introduced along with catch share but was not dependent on it. The privatized fisheries saga continues, so there are clearly big rewards for important groups for social engineering like this to be promoted and to be embraced by so many.

Chapter 25

American Fisheries Act and the Community Development Quota

Commercial fishermen in western Alaska probably stayed largely unaware of fisheries developments elsewhere until the 1990s, but in places like Nome people wanted and needed to fish for any species that they could hope make a living from. Except for occasional surges in gold mining at Nome and in government construction jobs, there weren't a lot of choices. Rob, with his herring ventures, is an example of the limits people faced. Support services for fisheries were rudimentary or nothing at all, and no bank wished to sponsor boat loans for such marginal ventures.

A new very different opportunity was born when Norton Sound Economic Development Corp (NSEDC, or as known locally, "Norton Sound") registered with the State of Alaska as a non-profit fisheries organization for the region. It was one of six regional groups formed through the unique Community Development Quota (CDQ) program for western Alaska. The CDQs began to change the lives of families and communities all along the coast from the Aleutians north to Wales on the Seward Peninsula. Nome's harbor today looks very different than it did 25 years ago.

The CDQ program was a new concept for fisheries, created specifically for the economically struggling west coast communities. It was also a new opportunity for the industrial fleets in the Bering Sea offshore of those communities. Though the six regional CDQs are non-profits, they are not traditional government aid organizations. They have similarities with a producer/marketer's coop but the profits go to regional development. Once the CDQs were rolling well they were expected to support themselves, make profits, and use those to support and expand local fisheries as well as other worthy community projects. Twenty years after start-up all six CDQs have survived, are self-sustaining, and at various levels of profitability, most doing very well. How they are doing with community development, in particular local fishery development, is a more complex question, as each potential fishery has its own promise and problems. Aside from the corporations formed from the Alaska Native Claims Settlement Act, the CDQs are the only model of government involvement that has given Alaska's west coast communities a chance to develop their natural resources to become economically viable. Most aren't there yet.

The CDQs' tie to the corporate fishing industry, especially through the American Fisheries Act (AFA), is their means of receiving regular influxes of money instead of from government grants. Today their harvest royalties from the Bering Sea industrial fleets are just part of the CDQs' value. Almost all of them soon decided to actually buy into those corporate fleets and processors off their shores, and into other investments. Thus the CDQs have the Janus face: they have indirectly become a partner in the loss of small-boat fishing to industrial fleets, (halibut, crab) yet new small-boat fleets for some CDQs have been a positive spinoff.

Some Alaskan observers worry that the focus on profits from industrial fishing might lead to a loss of support for the traditional subsistence and commercial salmon fishing. Others on the scene say that the CDQ profits have enabled investment in infrastructure the salmon fleets had lacked, such as easily accessible buyers/processors. Certainly that is true in Norton Sound. In many places that did have local processors, they were too small to handle the peak of the runs, the fishermen couldn't get boat and gear loans, there were no repair services, and so on. CDQ investments have helped correct that. The CDQ communities have the leg on each deck, industrial fishing and small-boat fishing, a condition not likely to change, and an odd fish for some critics.

The history of two major changes for fisheries in the Bering Sea--the CDQs and the pollock fleet's AFA--and how they are so important to each other is a fascinating example of how things get done in fisheries. The CDQs were conceived by Alaskans who had the political clout to talk fisheries development where it counted--people like Senator Ted Stevens, Governor Wally Hickel, and Alaska's "fisheries czar" Clem Tillion, a member of the legislature and the North Pacific Fisheries Management Council--all of them used to dealing with corporate leaders in fisheries.

The 1980-1990s had already seen a major change in the Bering Sea since the passage of the MSA. As the last of about 3,500 foreign vessels were finally evicted from our federal Exclusive Economic Zone (EEZ), American trawl fishery/processor companies saw the opportunity to take over pollock, a cheap but bountiful fish they had until then ignored.

At the same time Henry Mitchell, the head of the Bering Sea Fishermen's Association, Tillion, Harold Sparck of Bethel, and other Alaskans were brainstorming how to reap benefits from the EEZ for the economically struggling Alaska Native communities bordering the Bering Sea. The coastal residents knew the Bering Sea was rich in fish stocks. But though most fished salmon, they had no way to go after the vast schools of groundfish and pollock that required larger boats and expensive equipment to be profitable. They had little way to market such bountiful fish as cod or flounder. Pollock, rapidly growing to soon be the biggest fishery by volume in the country, is profitable only when caught by boats that can handle huge quantities. Not only did the west coast skiff fishermen have neither the boats nor gear, most had no real ports, no place to keep or repair large boats, and, with few exceptions, no way to process or market anything but salmon. Many didn't even have salmon buyers. They had no money to start such enterprises, and no credit.

The Bering Sea Fishermen's Association (BSFAAK) had been trying for years to get more fishery opportunities rolling. It saw the chance for an arrangement that would benefit the economy of west coast villages and at the same time would interest US trawlers and processors. Earlier the North Pacific Council had allowed a portion of the halibut fleet quota to the western small-boat fleet. Why couldn't that idea that be expanded to other fisheries? Phil Smith, a staff person for years in fisheries, recalled to me how things changed with a brilliant idea. Harold Sparck is the person that people credit for sparking the idea for the CDQ program. He had long pondered how to solve the economic problems of the western villages. Here were all those foreign vessels just off shore the Bering Sea villages, scooping up the wealth of the oceans, but no benefits were being returned to the local residents, or even to the state. Yet the North Slope villages had gotten a return from the oil production on their lands and were now doing far better. Why couldn't something similar be done for western Alaska? The ocean out there had once been totally the domain of Yupik and Inupiaq people until Russia, and then the US, decided the waters belonged to them. The BSFAAK and other supporters fleshed Sparck's idea out and the North Pacific Council passed it in 1991.

The plan was that qualified US trawlers and processors would organize a new kind of fishery with the rights to the fish--the structure called catch share now--to include the trawlers Alaskans favored, those that sold to onshore companies that aided the Alaskan economy. It would also cover offshore catcher-processors based in Seattle, as any proposal would need to deal with all of the US fleets. In return for getting better control of the pollock, the trawlers involved would give up a percent of their allocated quota to the coastal communities on the Bering Sea.

Those communities, over 90% Alaska Native residents, would form regional non-profits (CDQs) that could fish the quota or take it as royalties. (Later that quota was raised from seven to ten percent as other Bering Sea fisheries were included in the arrangement. The CDQ royalties would support small fisheries and other development for the coastal communities. It was an ingenious concept. Without our CDQ, Norton Sound Economic Development Corporation, our small crab and halibut fishermen would no doubt still be working in open skiffs far out in shallow, rough seas.

At the time, only a few of the regional Alaska Native corporations had yet seen fisheries as a business they wanted to risk their funds in. Nome, Kotzebue, and Bethel were larger towns on the coast but little different in their lack of support for fishing fleets. Unalaska/Dutch Harbor was a large processing sector, but almost all the profits went to Japanese owners.

Dillingham alone had reasonable support services for a large salmon fleet that included local boats.

My homeport of Nome and farther northwest Kotzebue were the worst regional centers for commercial fishing support. After the mid-1980s there was rarely a local fish buyer in Nome. There was no dock for small boats, so you fought your way through a scary jetty, often risked swamping, and pulled your skiff up on the beach or threw out an anchor in the river. You could order a skiff and equipment shipped in, but there was no diesel marine mechanic, no electronic technician. Loan applications for commercial boats went into a deep hole somewhere. Lower Norton Sound was better off due to the herring roe fishery that had developed with BSFAAK help, and it also had a salmon fishery, both with buyer arrangements. In some years people at Kotzebue, the US's farthest north commercial fishing port, could actually make a good part of their living gillnetting chum salmon. Nome had neither herring nor salmon commercial fisheries active. Far more important economically for all west coast communities were the subsistence fisheries. The brainstormers of CDQ were not unusual in observing the "third world" condition of Western Alaska. I'd seen it ever since I arrived in 1970.

Other parts of coastal Alaska could be quite a different story. In Southeast in the 1960s-1970s, we had no trouble finding all the marine services and supplies we needed at Juneau, Sitka, or Ketchikan. Kodiak, and later Unalaska, also had small versions of Seattle marine services. There was only one other northern commercial fishing area that was as poorly serviced as northwestern Alaska, and that was Nunavut, in northeast Canada, where boats had to unload at one port, Iqalluit, or you took your fish to Greenland. But in the 1980s Nome fishermen didn't even have that option, Russia being off limits.

Fishing coops had been tried in western Alaska, but if they succeeded in other ways, they usually fell down on the difficulties of marketing from a remote area. The CDQs would be a different story. They wouldn't be dependent for long on government money, as they would have guaranteed revenue from their industrial fleet royalties. Although the intent was to aid Alaska Native villages, and the CDQ plan was modeled somewhat on

Alaskan Native Corporations, the regional CDQs would be governed by boards elected by all voting residents of the communities and with oversight from the state. Since the CDQ quotas would be community quotas, no individual would be able to accumulate a personal treasure, nor could the quota be sold. Profits would go back to development of the region after administrative costs were covered. The regional CDQ boards would choose the fisheries and community programs to support. There would also be jobs reserved for local people as crew on the Bering Sea floating factories. Over sixty coastal villages from Bristol Bay north were potential members for the regional CDQs.

The BSFAAK wanted Senator Stevens to introduce the CDQ program to Congress as a bill, but he believed it could never get past the Washington delegation, which would oppose it as taking away profits from the Seattle-based industrial trawl fleet. He recommended it go through the North Pacific Council as a proposal, then to the Sec. of Commerce, where it would have a better chance for getting approval and an initial appropriation from Congress, especially with Stevens pushing it. Like so many of Steven's maneuvers, the strategy succeeded. Norton Sound Economic Development Corp (NSEDC), based in Nome, and representing sixteen communities, was one of the six CDQs born. The benefits that went to the pollock trawlers and processors are described further on.

From the beginning the CDQ program had its adversaries as well as supporters. For some it was philosophically a bad idea, for they saw it as not only a raid on the Bering Sea pollock fleet's profits, but a raid on private enterprise in general. But private enterprise hadn't done much to raise up the communities in western Alaska since the gold rush. Other objectors were not opposed to coops but feared or hated the CDQ reliance on industrial fishing royalties. They feared the potential to undermine support to the traditional salmon fishing and the cultural traditions that went with it. Yet it was a unique opportunity for Western Alaska, as the federal or state government would never have funded such wide and long-range community development as promised through CDQ. Washington DC had made an effort during the "war on poverty" years through construction of schools, water projects and so on, and the Indian Health Service had built hospitals and clinics, but there never had been investment that could create permanent economic stability. The CDQs went forward and soon showed their potential. More about the changes they brought to western Alaska, and the lively discourse over them as they continue to prosper is in the next chapters.

Seizing the Pollock: the AFA

The birth of CDQ was soon tied to a huge gain for our Bering Sea pollock fleet through the passage of the American Fisheries Act (AFA), a short time later. That history is a more classic example of how things work in the world of fish politics. The young people from our region who today work on the pollock ships most likely know nothing about the AFA, and all the changes it wrought, but they can describe their job. It's a far different fishery from the ones I've been close to. Comments from my grandson Scott, who crewed aboard a pollock ship for two trips, explain to me why so many people from this region go for one or two seasons and then decide to find other work. But there are always people to take their places.

The pollock catcher-freezer vessels are giant floating factories. The largest one, now partly owned by our NSEDC, is over 360 feet long. To build such a ship today would cost $70-80 million. There are also smaller catcher-processors that market direct to retailers, and there are catcher boats that sell to shore-based processors. Some catcher-processors have 100 workers aboard, recruited from all over the world, the vast majority non-resident to Alaska. The crews experience pollock as a massive product they haul in from giant mid-water trawls--along with it salmon bycatch--then send to other crews who work in assembly fashion on a "slime-line": washing, cutting, cleaning, wrapping, on sixteen-hour shifts, seven days a week, to prepare the fish for market. They work, eat, watch TV, do their laundry, sleep, and work some more until the trip is over. They don't drink or do drugs. The dining area is like a modern hotel, the food fine and plentiful. The rest is work. Like the crews of thousands of cannery workers that come to places like Bristol Bay for the season, they get a low wage and depend on the ample overtime they build up for their enriched life ashore. They might get a promotion to the deck. It's endurable for young people who can stand on their feet all those hours with brief breaks, who like to eat, and don't mind the confinement. Aside from being on the water and handling a lot of fish, it is the opposite of our traditional small-boat fishery, the opposite of how my sons fish.

A few determined people from this region take a short training course paid for by the CDQs and go aboard a factory ship each year as part of the CDQ opportunity. They take their earnings and buy things or do things they would have a hard time paying for any other way. When it was time for my grandson's third trip, he had a cut on his hand that needed doctoring and they left port without him. He didn't try to catch up. Yet that short career paid off for him in sea legs. He is a cook ashore now in winter, but summers he may go as local crab or herring crew--hard work but shorter trips.

The AFA would end up, like CDQ, a new model for fisheries reorganization. But it would get much more attention from would-be imitators. I first heard the term "AFA" when my friend John from Kodiak stopped by on a business trip to Nome. We sat in the kitchen over coffee and talked about the fishing industry, a natural topic for us since we were both ex-commercial fishermen. He was in close touch with the fleets through his work and was on the scene when the AFA and other individualized quota programs began. The purpose of the AFA, as John and others explain it, followed the intent of the MSA to strengthen a promising American fishery and complete the elimination of foreign competitors. But its promoters also intended to consolidate an already "overcapitalized" US fleet. Too many US boats, in the opinion of the more powerful firms, had already moved in to take the place of the foreign fleets driven out of the Bering Sea. AFA promoters also presented it as a move toward better fish conservation.

For our federal government, the chance to reorganize the pollock fleet was a huge catch, and not just for that fishery but as a role model for others. Not long before, the US considered pollock a "scrap fish" that the Japanese could have. But in the 1980s American as well as foreign entrepreneurs figured out that the humble fish could be the basis of popular fast food like fish-and-chips and surimi. Scrap could turn to yet another golden egg, bigger than herring roe. For some boat owners it could also take the place of the crashed king crab fishery. But at the time our young trawl fleet was made up of bottom draggers who had to learn how to mid-water trawl for pollock. We also didn't have the at-sea processors to handle pollock catches. Some of our new pollock fleet won permission to enter into joint ventures with Russians, trading our pollock for Russian crab, while others sold to the onshore Japanese processors, such as at Dutch Harbor. "The Billion Dollar Fish: The Untold Story of Alaska Pollock", by Kevin Bailey, gives a scientist's perspective on the rise of the US pollock industry, worth reading even if you never intend to fish pollock or eat them.

.As soon as we had our own catchers and catcher-processors ready to fish, the pressure began to halt the joint ventures and expel any foreign companies still fishing our waters. Congress passed an Anti-Reflagging Act in 1987 that was intended to put a stop to foreign fishing vessels that had re-flagged as American, but it had loopholes. Among the pollock fleets were trawlers registered as US businesses, US flagged, and paying US taxes, and legal to fish, but with shareholders that were largely or all foreign nationals. Of special annoyance to American companies like Trident Seafoods was a large pollock fleet berthed in Seattle that was Norwegian-owned, subsidized by the Norwegian government and a Norwegian bank, but legally American vessels. That fleet invested millions into Seattle shipyards and other support businesses, and the Seattle businesses and politicians did not want it expelled from the US waters. The largest of these companies, American Seafoods, had grown to include sixteen catcher-processors. Those floating factories, described above, were too large to be serviced, except for fueling, in any Alaskan port. Seattle loved them, Alaska did not, and would be happy to see American Seafoods and other "foreign" vessels out of the Bering Sea.

Our government at the same time wanted to shrink the US pollock catcher-boat fleet, due to a "race for fish" that had developed. As the number of vessels and their technical powers grew, the fleet quota was caught in an ever-shortening period. The results were not good for the image of the industry, especially the masses of "over-quota" fish being brought aboard and the dumping of bycatch--untargeted fish, including prohibited fish like salmon. Commercial fishing also had the status of the most deadly occupation in the country, blamed on poorly equipped and maintained vessels in the race for fish. After an exceptionally bad year for lives lost in the Bering Sea, the public had come to view that fishing grounds as a widow-maker as well as a treasure horde.

The North Pacific Council was under pressure from both NOAA and the industry to restructure pollock fishing, but it had more than one plan on its table at the time. For several years it been working on plans that would change the Pacific halibut and sablefish fleets to individualized quota structures (IFQ). American pollock corporations proposed the Council should do the same for pollock, and through that eliminate open access and competitive fleets like American Seafoods. It sounded right, but some observers saw the Alaska-Seattle controversy as simply an industrial competitors' dispute with nationalism and conservationist arguments thrown in to win support.

The Council, influenced by the strong voice of pollock executives, but also eager to get a plan approved for an IFQ program, needed agreement among the trawlers as to how to allocate the fleet quota among quarreling in-shore and off-shore delivering boats. With the pollock plan stalled, the Council went ahead instead with its halibut and sablefish fisheries restructuring. But meanwhile CDQ promoters saw an opening for a proposal that would benefit both the CDQs and the pollock trawlers. To speed things, Clem Tillion and other Alaskans, along with the US trawler groups and on-shore processor companies like Trident Seafoods, decided they would bypass the stalled Council and write their own plan for pollock fleet restructuring.

Their proposal included proofs, not difficult, that the on-shore processors, mainly Japanese but including Trident and others US owned, were better for the Alaskan economy than the at-sea catcher-processors out of

Seattle. They provided accident statistics that could be blamed on the "race for fish", which they said their plan would stop. With the great numbers of vessels and processors working the Bering Sea a likely overfishing threat, they lined up ENGO groups like the Environment Defense Council to favor their plan as environmentally friendly. (Others, like Greenpeace, saw it as the opposite.) The final draft of their plan divided up the Bering Sea pollock pie between thirteen US entities.

Senator Stevens, with his typical talent, took the plan, turned it into the American Fisheries Act, and moved it through his appropriations committee as an earmark, later to be formalized and enacted as law in 1998. It was another brilliant, expensive Stevens maneuver observed by the rest of the industrial fleets. Through the AFA, the federal government bought out nine of American Seafood's sixteen catcher-processor boats, with $20 million paid direct, and $70 million more to come from loans by the government to the AFA fleet, to be paid back from dockside landing fees. American Seafoods (and others) had to convert their remaining pollock industry holdings each to no less than 75% US owned. A number of small processors and small US trawlers were also eliminated, another benefit for the powerful.

The situation in the Being Sea was now even more unique. Since the CDQ program was already embarked on its own transformation of western Alaska, the birth of the AFA meant there were two new and very different new fisheries operating, both following the guide of the MSA for modernized fleets. One, the AFA, a cooped trawl fleet with privatized fishing rights, and the other the CDQ model for non-profit community development including new local fisheries, but with funding through royalties from the pollock fleet. The AFA would quickly become the envy of other industrial fleets and the model for more federal plans. The CDQ model stayed a Bering Sea special case, not copied elsewhere on the Alaskan coast, or in any other US region. Either the win-win situation wasn't available or the energy to push it was missing.

The federal waters open-access pollock fishery was ended. No longer could anyone one just steam out there and drop a net. As John puts it, the members of the AFA coops, like IFQ, have all the "rights" to fish the pollock. The AFA also included regulations on fishing practices. It put a limit on the size of vessels, with those larger grandfathered in. It included controls on monopolistic practices, limiting the share of the fleet quota for any one fishing company to 17.5%. Processors could own quota, but no company could own more than 30% of the catch.

The AFA pollock fleet as of 2010 was composed of three groups. According to government figures, there were two catcher processor coops, including about 26 vessels, some of which lease out their quota to other AFA vessels; about 80 catcher vessels organized into coops with seven shore-based processors; and 17 catcher vessels that sell to three "mother" processor ships. This is a sizable fleet of big boats. All of the catcher boats are over 60 feet, and 24 of them over 125 feet. Almost all of the larger trawlers are home-ported in Washington or Oregon and all of the catcher-processors.

The Japanese shore-based processors that serviced catcher boats were not driven out of Alaska as the fleets needed them, and they have held onto the majority of processor quota. The Japanese at-sea processors were also allowed to continue for a time due to American need, such as at Bristol Bay, Norton Sound, and Kotzebue Sound. But a few years later Alaska passed a law that foreign at-sea processors could come into state waters only when the governor declared a shortage of American process-ing available. No permissions have been granted since 2000, and a few Alaskan fisheries have suffered some years because of it. Trident Seafoods built a huge on-shore plant at Akutan, a small village that by 2013 would become the second fish landing/processing port by volume in Alaska, that year beating out Kodiak.

This American Fisheries Act story, much simplified, is a classic example of high-level fishery politics and had great influence on government plans as it seemed to work well for all the vessels/companies that survived the restructuring. It obviously had little to do with the lives and fortunes of the crews that work the pollock boats. It was all about getting generous support from the federal government to seize hold of a new opportunity for im-mense profits for a select group, the AFA members, who would pay seven percent to the CDQs for their unique privilege. It was also about feeding the world, the pollock companies remind us. But the pollock rationalization was different in an important way from other huge corporate deals because, due to the existence of the regional council's powers, the fishery had the opportunity to bring overfishing and waste completely under control. Even more unusual, the new CDQs received the assurance of guaranteed ongoing source of support from their share of the harvest--as long as the pollock stocks stayed healthy.

The AFA didn't mean, however, that scientific management of the fish stocks was now a given, even though that argument was partly what sold it to the ENGOs. The pollock fleets (there is another in the Gulf of Alaska that was not included in the AFA) went on to produce the largest volume harvest in the country, but bycatch remained high, not in rate but in volume, and many suspect that it affects fluctuating salmon runs. By 2010 salmon bycatch reduction was a hot agenda topic at the Council, with hearings on both Chinook and chum bycatch held throughout Alaska, but more effective bycatch regulations have been harder to win than passage of the AFA.

It is also not clear just how healthy the pollock stock is today. Scientists don't agree on what pollock maximum sustained yield is. The stock could be fine or it could be depleting; more research is needed for this stock and many others.

Though the AFA was a major change for US fisheries, left to the Council something similar would have eventually come about. My friend John at Kodiak has an ambivalent take on it: "It was good to exclude the foreign fleets, but what we did was privatize a public commons. The corporations that received individual quota were actually named in the AFA bill! It may have worked out all right for the pollock fleet with its problems. I can't say as I don't know much about that fleet, like how many pollock fishermen were driven out when they handed out the quota. But I'm sure privatizing was the wrong way to go with other fleets I'm more familiar with, like halibut and crab. They included many small locally owned boats that lost out. But at least one good thing did come out of the AFA, royalties for the western villages with the CDQs. I do give credit for that."

Pollock was just one fishery being reorganized by the Council. The rational harpoon, already having struck the halibut and sablefish commons, would soon be aimed at others, creating even more commotion than the AFA on Alaska's waterfronts. These dramas are reported, as seen from various perspectives, beginning in chapter 28. But the results from CDQ in one region, as an example of the six regional experiments, need to be covered next.

Chapter 26

Launching Small Fisheries: Norton Sound

Community Development Quota, or even more, "CDQ" is a term probably everyone in Western Alaska over age eighteen recognizes. Although CDQ blew in as a bright new brainstorm for overcoming the economic malaise of the coastal villages, reliance on the pollock fleet's CDQ royalties did carry some risk. Pollock, like herring, are known to come and go. But the risk was lessened when the North Pacific Council soon added CDQ royalties from other Bering Sea fisheries: crab, sablefish, halibut, and groundfish, and raised the royalty rate to ten percent. From then on, assuming that the fish stocks stayed healthy, most of the problems for the CDQs would be in developing their own management and markets and deciding how to best service their regions. In Nome, enduring our usual economic doldrums, we watched with great interest.

At first there was much required in-house business. Each CDQ needed to form by-laws, dedicated boards, and skilled managers to run it profitably and harmoniously for the public good. The boards brainstormed anxiously over what projects to pursue with their royalties: fish plants and harbors, new fisheries, and a host of other economic development that would develop the regions, bring cash to families, and even could cut the excessive costs of living in the remote regions. Though fish was the top resource along the west coast, in the past the regions that had won government grants for fisheries projects had always run into difficulties. The marketing logistics were too complicated, or they couldn't get experienced managers, and the old fishing coops rarely were profitable enough to stand long on their own. The CDQs could be different.

As reported in the *Nome Nugget,* the establishment of Norton Sound Economic Development Corporation (NSEDC) as one of the CDQs was a challenge for the region's residents as many elements were not clear. It was a tax-exempt non-profit, yet it was a corporation that needed to follow Alaska corporate law. It was like a regional Native corporation in some ways, but it was for all residents in the eligible communities that wished to be included. The start-up activities were the main local news stories for a while. At Nome there were a few vocal fishermen with ideas about local development that they wanted to explore, but once the board was elected, a process that took considerable debate, it was not clear just how the board should tackle such proposals.

Rob was at the meetings and observed that some members were not used to the idea of speaking up publically, as they were not politicians but small fishermen, or very small town politicians, and suddenly they were jumping into the corporate world. Perry commented that smaller places were historically suspicious of ideas coming from Nome as the regional center. Some board members were unsure about anything so new. It was easiest to go with what the hired managers with their business experience had to say.

Rob recalls the start-up from a fisherman's point of view: "I thought now with NSEDC we had a fisherman's organization, and they could help us using those pollock royalties, like get us decent gear for summer crabbing and an easier way to sell our crab. We needed to develop more local fisheries that I was always thinking about, like the herring in the Northern Sound, and now here was a chance. Right away the CDQs began recruiting people for jobs on the pollock ships, and one of my sons went for a couple trips. Then they offered $1,000 scholarships for every region resident that wanted training that could relate to fisheries. And we did get our gear loans for crabbing pretty soon. But at first our board was mostly trying to get organized.

"I was one of the few always speaking up at the quarterly meetings, usually on my ideas of promoting new local fisheries. But the Board hired financial and program managers from outside that they trusted more than locals. It was a big responsibility, and none of the Board were businessmen. The program managers NSEDC hired at first, except for a couple, weren't from the region, and I believed they weren't really thinking about what we needed for the long term. The managers recommended investing the royalties back into the Bering Sea fleets as the safest way to go. I was questioning that as I wanted NSEDC to concentrate on developing fisheries for local stocks that were under-utilized. But the Board decided to go ahead and buy shares in the big Bering Sea boats, a safer bet, and Norton Sound [NSEDC] did start making money there.

"Pretty soon the Board agreed to move the main office to Anchorage, a wrong move I thought, and I complained about that too. Meanwhile it went ahead with things it felt confident of like scholarships and other training opportunities, and grants to villages for obviously useful projects like clinic renovation and youth programs. And I kept trying to get fishermen to speak up more about what they wanted for our region's fisheries.

"The tug between local fishery development and profit-making in outside fisheries went on over the years. We fishermen were invited to express our opinions, but in the end the Board mainly went along with the managers' advice. They favored supporting local developments that were relatively risk-free such as scholarships and public services that no one expects to make money. They were agreeable to helping out already developed fisheries like salmon. That naturally frustrated us crabbers still trying to fish out of skiffs.

"Norton Sound put out an annual report that was somewhat understandable to the residents, and after a while it also put out a regular newsletter that went to every region resident. They also created the part-time position of "Community Liaison" for each of our communities to help improve communications. I was surprised when they offered me the post in Nome. I was such a thorn in the management's side, maybe that's why they chose me. I did that a couple years and accomplished a few things in that position, but not near what I wanted to. It didn't shut me up of course, but for a long time I couldn't get much support in meetings or on the Internet when I brought up my ideas about local fisheries."

The complaint about board and staff listening too much to outsiders was bound to be a problem for the CDQs. The locals wanted to make the decisions for their communities' benefit, but the professionals they hired were competing in a corporate world, dealing daily with the industrial fleets that were obviously doing very well. Ignore them, or join them? The CDQs in the main decided to join them. Rob acknowledges it paid off:

"Norton Sound started getting quite a lot of second tier money from our investments in the Bering Sea fleets. Then they began to give out more grants for support services for the region and the fleets, like boat ramps in villages, boat trailers, and other things the people asked for. Nome badly needed a small-boat harbor, and a NSEDC grant to the city about 1998 helped get one built. We finally had an actual float to tie up to. What an improvement that was! And finally about that time our CDQ was doing so well it was able to build a modern fish buying and processing plant in the harbor, Norton Sound Seafoods, and also a plant in Unalakleet. And a lot of money also went into fisheries outside the region and into stock market investments.

"I kept asking at the meetings when we could get loans for real boats. The *Island Girl* couldn't compete with those bigger, faster converted herring skiffs from the Yukon. While I was bringing in 500 lb. loads that fleet was bringing in 5,000. I couldn't make my gear loan payments to Norton Sound, and a lot of others couldn't either. The Board forgave us on the interest that year. But local boats? No, it was intent on expanding into the Bering Sea fleet where the more sure profits were."

Rob's remarks give a good picture of the balancing act that all CDQs had to manage, exacerbated by the unusual relationship with industrial fisheries, like pollock, that many local small fishermen felt alien to their interests but that CDQs depended on. Yet the CDQs survived politically and did grow in their willingness to support local fisheries.

Another New Fishery: Small Boat Halibut

When halibut was added to the CDQ fisheries, some communities chose to develop their own halibut fishery or further expand one. A few Nome men had been fishing open access halibut out of large skiffs and selling them to a local buyer, and subsistence fishermen at Savoonga and Diomede caught subsistence halibut. The Nome crabbers' quota was usually filled by mid-August, meaning they had time to go for halibut before the fall storms took over entirely. But the crabbers soon found that their skiffs were just large enough to get them in trouble.

Halibuting is never an easy fishery, and it wouldn't be in the northern Bering Sea. I could understand why my father chose to abandon that life of rugged, cold, dangerous work. Now, to hear that Nome fishermen, including my sons, would attempt to fish halibut from skiffs? And with no "mother ship" standing by for them? It was going back to my grandfather's times. Where would you put the halibut? Find a crewman? Would you survive? Yet this is what fishermen in Nome and several other communities decided they would do, and the local skiff fishermen talked NSEDC into backing them again by providing buying stations with freezers, and gear loans for floats, buoys, line, hooks, tubs; and motors, and power blocks if they didn't already have them from crabbing. But no boat loans. I was tempted to lecture my sons on how this kind of fishing was no way to make a living, but they weren't interested in history lessons. Rob recalls how he jumped into the new venture:

"I got into halibut with the *Island Girl* right after they opened it. I figured a twenty-four foot skiff was good enough to manage a couple skates if I didn't have to go too far out. You did have to find the school, but once you did, you could load up. We soon found out the fish moved around a lot, and it was hard to move gear fast enough with a skiff. But that first year I had good luck. The first two weeks of August we had unusual fine weather, and fishing four miles of line and 1000 hooks we caught 500 lbs. right off Nome and got $2.00/lb. so it seemed worthwhile. Of course late August weather always got stormy, and September was worse.

"Before long we realized we didn't have proper boats for it. For summer crabbing, skiffs were possible, this really wasn't. Fishing halibut, especially in the fall, it was tough to get a crew that would stick. I went out every few days, taking whoever I could find, even had to go alone sometimes. They'd get sea sick, scared, or pull a muscle, or get a hook in the hand, or decide they didn't like all the blood and slime, so I was always training new guys, a couple times girls. I tried to get my sons to go, and they would, especially Scott, for crab, but they hated halibut. There was no tradition for it here in Nome. Some of the crew's work is ashore, baiting hooks with smelly herring, and no one wanted to do that part, of course. They figured they got their percent out fishing, that's it. They'd end up AWOL and the skipper ends up baiting. Getting the crew located and out of town was sometimes the hardest part.

"Then, out in the skiff, rocking and rolling, if you were lucky and got a few big halibut aboard, the skiff got crowded and you really had to be careful with the hooks coming in and going back out on the power block. You had to have alert crew, not daydreamers, or hung over, so I had to yell at them, and they didn't like that either. When the wind came up it could be pretty miserable, next to impossible. And we hauled a lot of empty skates, pretty boring. It takes a certain kind of stubbornness to stay at it, a wish to be out in a boat, which I guess I have, from somewhere.

"We found out the major body of the halibut were farther off-shore than we expected, so we really had to watch the weather, and it was more of a risk than crab. You can let a crab pot soak for days if the weather is too bad, but with halibut on a skate you have to go retrieve them before too long or you lose them to the sea lice. So usually we did go out and took our beating. And our beating again, coming into the river through the jetty. We didn't even have a boat harbor yet."

Years later I asked Rob to give me more a feeling for fishing offshore out of a skiff. I was glad I was spared such tales at the time.

"I can tell you about our last crabbing trip with the *Island Girl* one summer. Simon Jack from Teller was my deckhand and we were straight out from Nome. I think Eric was out there with his skiff too. Most of the rest of the fleet was down Golovin way, but that was just too far for me. Then one day, the end of July, here they all came back and the tender with them. They headed out to about 30 miles west from Nome. I thought I'd put my pots out there with the rest of them and then in a few days pick up my pots, sell and leave my pots with the tender. Then I could go check some halibut gear

Nancy Danielson Mendenhall

I had out--you aren't allowed to have two kinds of gear on the boat at once. So four or five days later, it was the same deal, time to go pick up those pots for the year, take them to the tender. And we found ourselves, thirty miles or so northwest of Nome in the *Island Girl*, hauling gear in six-foot seas. And it was getting dark.

"I was damn glad at least the tender was out there and I had a radio to talk to them! It took two long hauls to get all the pots onto the tender. Then we got to have dinner and sleep there, which was pretty nice considering the choices. The next morning when the tender dropped us off we found out it had drifted out to forty-mile during the night. That was how they did it, but it was a little strange for a skiff way out there. And now we had to go find the halibut gear--we did have a GPS by then. Fortunately the seas had dropped to six feet. I got 1500 lbs. of crab from that last trip. And I knew the *Island Girl* was a good sea boat, but forty miles? I knew I had taken her to her limit."

The Norton Sound fishermen had their choice it seemed. Fish out of skiffs or stay ashore. Along the Bering Sea other fishing villages, too, continued to see their fishermen head out in skiffs. Rob's brother Frank was skeptical of fishing halibut from a skiff, and the first summer he stayed away. He knew what long-lining entailed with his memory of his winter off Kodiak on the cod boat and the conditions fishermen from Kodiak to Unalaska accepted in order to earn a living. If you couldn't take it, you went on to another career. Now Frank was looking at Rob's *Island Girl*, all rigged for halibut, a fishery that would have looked pretty risky to the Kodiak fleet. Yet he soon was rigging his own skiff.

"About the second year of our halibut fishery, I thought I should try it for something to do in the fall. I had been working construction, getting Nome's small boat harbor built, but that job was finished. So I bought a large open skiff and an old outboard, total about $5,000, and rigged up with the rest of the Nome fleet. I figured out how to operate the whole set-up alone because I suspected I wasn't going to make a whole lot. When I went farther offshore, which we more and more had to, I took Fred along, the

man who fished crab with me winters. We'd come into port every night, but
even so, we really were too far out for only having one old motor. I remember once we barely got it to start out there--jeez, it had one more cough left
in it. We were figuring out how to rig a sail out of some trash bags and an
oar when it finally kicked in. Yeah, it was a marginal operation. But I went
every fall after that."

Rob picks up the halibut story: "We both did it for a couple more years,
all the time trying to get NSEDC to sponsor us for loans for real boats. But
no, they would give us gear loans, but boats? Too risky financially I guess.
So there were only about six local boats fishing halibut out of Nome--a small
fleet, no wonder! Meanwhile NSEDC continued to invest more in the big
Bering Sea fleets. The Board did approve some regional fishery projects too,
like research on how to build up the local salmon runs."

NSEDC continued to bring in revenues from the Bering Sea and use
them for community-based development, awarding each of the small cities
in the region at least $100,000 per year--later double that--to use as they de-
termined. It organized a bulk fuel order to cut fuel oil costs for householders
and later began to help with all of the households' excessive power costs. Yet
it continued to be cautious about more local fisheries development. Norton
Sound was not Bristol Bay; salmon fishing in the lower Sound was up and
down, after 2000 mostly down. The halibut schools moved in rather unpredictable ways, and were hard to catch up with some years. Herring roe price
was no longer what it was. There could be groundfish out there, but the last
thing local subsistence fishing families wanted to see was bottom trawlers in
their waters and their own ecosystem torn up for the sake of industrial profits.

One day the tide turned. The Norton Sound Board said okay, they'd go
for boat loans. Rob tells how continuing profits were part of it, but public
pressure helped.

"It was pretty obvious to everyone that our skiffs were not really safe
for fall halibut fishing in the northern Bering Sea. We'd so far been lucky.
I remember one local man, not a fisherman, just a citizen with sons that
cared about what the young guys were getting into, speaking up at a public
meeting. He said if Nome wanted a fishery they better make it a safer one.

The decision was a huge breakthrough. Right away I was filling out the loan forms. I found a local boat for sale that had been a small herring tender in the Lower Sound and figured I could make it work for both crab and halibut."

Frank tells how he reacted. "I was working up on the North Slope that winter. I was due for a rotation, and I didn't plan to go back, as I'd made enough to put together a dredge for beach gold mining in the summer and into the fall. I had the mining plan drawn up, I had all the permits. That was what I was thinking about, coming back from the North Slope. I got off the plane at Nome and the first person I saw was Rob. He was excited--NSEDC had finally opened up boat loans, and he had the application forms. Overnight, all my plans changed from gold mining to fishing."

Chapter 27

FROM SKIFFS TO SMALL BOATS

The Nome fleet scrambled to find real boats and gear them up to fish crab. Frank tells how he was able to get a boat:

"It turned out there were a lot of Bristol Bay gillnetters for sale down at Naknek. Eric found them on the Internet. They'd been getting such poor prices for their red salmon--probably because of the foreign fish farm competition--they were dumping their boats for half-price.

A bunch of us flew down to Naknek--Eric and Farley and a couple others. I picked out a boat that needed some things; it seemed well cared for. So I flew back home and came back down with Fred and another friend, and we spent a couple weeks working on my new boat, the *Mithril*. I put in new watertight hatch covers and a new radar to replace the loran, a new transducer, rudder bearing, a life raft, some other stuff. I was able to get another NSEDC loan for all this. The other guys all got their boats too. We had luck on weather coming up the coast, a beautiful trip all the way, typical for June.

"I got together enough crab pots to start with, found out the crab were down by Golovin. The Yukon-Kuskokwim fleet, that other CDQ group we shared quota with, was doing well, so we went down there. Right away we got into a big storm, wouldn't you know it, and had to anchor behind Caroline Island, which is just a hump, not a real island, and it was so shallow we were afraid we'd hit bottom in the swells. Rob was down there with his new boat, the *Stephanie Sue*, with an engine problem already. He couldn't get in and was anchored out in it, stuck for three days, and we didn't know just where he was. What a great way to start! But we had real boats at last and we all caught crab that year.

"The arrangement with NSEDC was that we had to sell to them until we got our annual loan payment to them made, a portion from each fish ticket, so that took the risk mostly out of it for them. Ideally you would have your yearly loan payment paid off by the end of the crab season, and I always have, and all or most of my other fishing expenses. Then you get your real profit from halibut. But halibut fishing can turn out bad. It's partly the weather, partly getting onto the fish--sometimes they just disappear. Rob has a way of finding them, though; he almost always does.

"The next year we fished up by Sledge Island and did okay again. My only problem was keeping my crew. Fred was a real workhorse, always good on the boat, but I could lose him in town. The idea was to keep him out of town all season but of course that wasn't possible. Rob had his crew problems too. The Nome guys weren't used to the demands of commercial fishing. Finally I did find one steady crewman for crab, Conrad, and what a difference that made. It's harder to hold them for halibut, tougher weather then, and end-of-season fever."

The new small fisheries were a great help to regional families. Along the Bering Sea coast in 2012 there were over two hundred CDQ permits just for halibut, most for boats under 30 feet. These small fishermen also had an important benefit in that they were allowed to keep undersized dead halibut and cod on their lines as subsistence catch. But it's a truism that once a family buys into a boat its whole economy changes, with most spare cash going into that hole in the water. Rob tells about his hole in the water:

"The *Stephanie Sue* caught my eye because it was in Nome and had twin outboard gas engines. Supposedly it would be faster getting to and from the grounds, which would make up for the extra fuel it burned. After all the beatings I'd taken trying to get into port, speed sounded great. I worked like mad to get the boat ready for the season, but she had been on the beach six years, and on the first run down to Golovin, I had out half the pots and one of the old engines seized up. Frank towed me back to Nome, and on the way we threw out the rest of the pots to try to rescue something of the trip. So, more loans, new engines, and back out fishing again."

In Norton Sound there were soon about twenty to thirty crab boats, some of them from the Yukon, almost all of them converted herring skiffs or gillnet boats. At first they were averaging over $30,000 gross for a season, not counting halibut, but much more later. Thirty thousand would be scraps to big crabbers out of Kodiak, but as a comparison, in 2004, commercial coho fishing in lower Norton Sound was averaging only $3,300 per boat, and it was the highest in six years, with Chinook fishing soon closed due to the poor runs.

Rob continues the story of boat ownership, one I can identify with:

"The next few years I spent my earnings on boat upgrades, and had to get more loans yet from NSEDC to get it powered right. But I was out fishing, and not in a skiff. In the winter I did crafts for tourists, and some real art at times. Eventually when fuel prices went sky-high I put in a Volvo diesel to be more efficient. As usual I had to do the installation myself, had to learn everything from scratch except for what I knew from old truck gas engines. There was something wrong with the new engine and no local marine diesel mechanic. No mechanic would take the time to fly out to Nome for any damn thing I could offer. Marine mechanics, you can threaten them, beg them, they are always swamped with work in their own ports, and Nome is the end of the world to them with our small fleet.

"When I finally threatened enough, the place I bought the engine from sent a mechanic out, but he didn't bring the spare parts recommended. I seriously considered suing someone. Eventually they admitted to me I'd bought the wrong engine--the one they had recommended. After six years it finally gave up, not fit to rebuild they said. Lucky for me, after long discussions my insurance company saw my side of it and said they'd buy me a replacement engine. I only had to pay the deductible. But they said it had to be the same model that had failed me. Whatever. So far, it's running."

Frank comments, "He wasn't entirely alone. In 2004, the *Mithril's* engine gave out before halibut season was over, and he had to tow me in. It's sure nice that we can take turns towing each other. I was lucky to find a friend, a driller, that knew diesel engines enough and would rebuild it for me, and I'm still nursing it along. NSEDC really needs to have its own mechanic, and we'd pay for the work. Where else do you have a buying-processing plant on the Pacific coast with no marine mechanic available? I bring it up to the Board every year."

Aside from such frustrations, the two fisheries, crab and halibut, worked well together. By the time the region's crabbers were finishing that quota, the halibut had arrived, and before the halibut quota was reached, the harbor was freezing over, the harbormaster pulled out the float, and everyone was glad. Later the revitalized spring herring roe fishery was added. The fishermen's problems were now the clear-cut traditional problems: weather, equipment, crew, safety, finding repairmen, available fish, and market. Market price is still a frustration especially for remote fleets. The tremendous effort that the state put into marketing salmon paid off. As Enoch pointed out in Part I, competitive buyers then came into Kotzebue in 2014 and the fishermen's chum price went from .25/lb to .70/lb. The price paid to Norton Sound crabbers was typically quite a bit below what was paid in Unalaska or Kodiak. The locals argued that their CDQ needed to market more aggressively to get the prices its unique high quality summer king crab deserved.

Other problems emerged as the fishermen learned that converted gill-netters were not ideal boats for crab or halibut. Frank was one that decided he had to make a change.

"After a few years I realized I had to add some length to the *Mithril*. A thirty-two foot gillnetter is not made for hauling crab pots or halibut gear, and you take a risk overloading when you have to move pots. And for the safety and comfort of the crew they needed more deck space too. But there was no one to hire for the job, so in the winter of '07 instead of trying to find work, I got another loan and ordered everything for an almost eight-foot extension. I built a huge tent to keep off the weather and welded it myself. I'd been welding aluminum for years so that was okay. The surveyor said he'd surveyed more than a hundred extended Bristol Bay boats for the same reason, switching salmon gillnetters into another fishery. My insurance came through, what a relief, and I was in the water. Fishing's been much better since then."

I wanted to hear a little more about a day in the life of a Nome fisherman. Rob volunteered another untypical day:

"Something a little more dramatic? I can tell you about the day I almost lost a finger. I don't need to tell you accidents happen on fishing boats, and the kind of crew you get--how important that is. For years I had a revolving door with crew. One halibut trip my young crew had jumped ship, and here I was, all baited up. As a last resort, I took along an older guy with health problems who could drive the boat while I worked the deck instead. We were four miles out of town, hauling, on a nice day. But Joe was up front and just learning how to drive up on the line and keep enough slack in it without getting it into the prop, and here came a fish. The line got tight, so I yelled at him to run up on it more, but then he got too much slack, so I was pulling it in, hollering, 'Haul it, haul it! And for some stupid reason I had the coils in my hand, and they tightened up fast and mangled it. I yelled at Joe, 'Stop hauling!' And he's fumbling with the controls, and, whew, he did figure out how to stop! I yelled for a knife and he called, 'You have one!' I must've been in f--king shock to forget that knife at my waist. I got the line cut off my gloved hand, and one finger was broken. Thanks to Joe I didn't lose all my fingers. Of course I should never've had the line coiled in my hand--just what I tell my crew not to do!

"We went into the hospital, and I got stitched up and a cast put on. And here's a funny thing, the nurse, a man, was so intrigued with my story that instead of being horrified he said he wanted to go out on the boat, so the next day we took him along. This time I drove, he worked the deck with Joe, and we three caught 2500 lbs. From that one day's fishing the nurse probably thought there was nothing to it."

These stories are entertaining, funny at times, scary at times, yet a row on a lake compared to the daily risks fishermen take in winter fishing in the southern Bering Sea or the Gulf of Alaska. But Nome boats are small boats, and the Savoonga halibut fleet out at St. Lawrence Island is still fishing out of skiffs, having no harbor for anything else. We are almost at the edge of fishable for small commercial boats; the Kotzebue Sound gillnet fleet is at the exact edge.

The most welcome improvement for the Nome fleet, after real fishing boats, was the creation of a real harbor and services like tie-up floats, fish buyer, fuel, water, bait--everything fishermen take as normal that they'd never had. The real purpose was not so much to aid the new fleet as to make docking more possible for other marine business Nome hoped to attract, such as cruise ships, but the fishermen certainly benefited. Earlier the city had built a long causeway out into the ocean using a NOAA loan. Now the city won another loan to create a calm passage into an actual harbor. The horrors of surfing through the old jetty into the river were over.

Frank in 2012 told how the Nome fleet continued to evolve: "Of that group of us that brought the boats up from Naknek, not many of the originals are left. My crewman Fred is gone and I miss him; it's the long winters that killed him. I lost another good crewman, an older guy, Conrad, to a house fire. But Howard's kids are fishing, Bill's are, and some younger local guys like Johnnie Noyakuk got interested in running their own boats, including some from down the coast. So for years now our fleet is a permanent part of the regional economy."

The Norton Sound fleets, and several others funded by other CDQs, now were made up of usable small boats in several fisheries and were what the local families needed. They could make a part or a whole living each season. As for the skiffs they had started out with a few years before, Frank sold his, but Rob's *Island Girl* was no more.

"I had no more use for her. She was too big to run around in, and too heavy to hunt in--they all use aluminum skiffs now for that--but too small for the fishery. I thought some beginner in the fishery might want her, but no, now there are loans available for real boats. She was just sitting there. One day a guy took her to the dump for me. I couldn't do it."

The history of the crab/halibut fleet of Norton Sound is important beyond itself because it is such a contrast to what is taking place in fleets outside Alaska. One of the smallest fleets of any kind in the US, it is doing well, and so are the stocks it fishes. The skippers are people of modest means, and among them are younger men who have learned to run boats, catch fish, find financing, and become skippers themselves. The number of boats from regional villages outside Nome increases. These small-boat fleets are the exception to the "graying of the fleets" that social scientists and the fishermen comment on, and that privatization like IFQ has pushed along. The next chapters focus on how the North Pacific Council undertook massive restructuring in its fleets and what the results were.

Chapter 28

Privatizing Halibut:
The North Pacific Council's Showpiece

Pacific halibut and sablefish would be our first west coast fisheries to change to an Individual Fishing Quota (IFQ) system. British Columbia was ahead of us, providing a model as it would time and again, and other countries would as well, with all the results of their experiments laid out for viewing. The west coast halibut fleet was already a good model for profitable, conservation-conscious fleets, having introduced many changes since the dory-hooking days. But it was also a very diverse fleet, ranging from open skiffs like my sons had started with to the old wooden seaworthy schooners still in use, to today's modern steel-hulled industrialized operations. That diversity would prove to be a challenge for instituting an IFQ that would be fair to the smaller fishermen.

Halibut stocks, like salmon, had a history of being over-exploited on the east coast, with few lessons learned as the fisheries developed in the west. By 1880 Atlantic halibut had been fished until it was no longer viable commercially. Pacific longliners and their processors jumped to take up the eastern market. They successfully shipped the first fresh Pacific halibut east that year and found hungry, grateful consumers. They were so successful that by the time my uncles joined Seattle's Ballard fleet, sometime soon after 1910, the grounds off Washington and BC were also producing smaller harvests, and the hardy fleet was moving always north.

The Pacific halibut fishery is a classic story of too much success, but it also gets credit for the first large fishery conservation efforts in the Northwest. The International Pacific Halibut Convention signed by the US and Canada in 1923 had among its goals: to keep foreign vessels off the grounds, to have a closed winter season during spawning time, and to promote conservation. Except for salmon trollers who were allowed to keep and sell "incidental" halibut--a real benefit for small boats that I later experienced--all other fleets had to throw back their halibut bycatch. Today the International Pacific Halibut Commission (IPHC) sets harvest limits and conservation measures, the regional fishery councils develop programs, and NMFS creates the regulatory detail and enforcement.

Halibut fishing is still a tough fishery. I have told how my sons have a hard time finding a crew for stormy fall waters after their crab season closes. In the early days when it was pursued in dories by handliners, the human mortality was high. It is easy to understand why the transplanted Norwegian cod fishermen, so recently from far north Lofoten waters, dominated the fleet. In the next decade the halibut fleet abandoned dory fishing from the mother-schooner for running the gear directly off the schooner deck, and it did get a little easier. The fleet now uses block and tackle or a drum to haul longlines baited with hundreds of hooks, and the big boats may even have automatic baiters, one more example of the fisheries' constant technological advance. The open access halibut fishery was close to maximum sustained yield by the 1960s.

At the same time concern was building over the halibut bycatch of swelling foreign trawl fleets, then only three miles off the coast. The US and Canada felt forced to reduce their own quotas, and that eventually became part of the push to move the US and Canada EEZs out to twelve, then 200 miles.

I have personal reasons to admire the halibut fleet and what they accomplished. Halibut boats are one of my early memories. One of my father's favorite excursions for me, at about seven years, was to Fishermen's Wharf in Seattle where we would climb aboard one of his brothers' boats. These men fished off Alaska, were all married with families and had good houses by then, not just boats, so they were doing well enough. They sat in the boat cabin with glasses of whiskey, chewing snoose and talking in Norwegian with my dad, while I took in the details of the big workboat.

This was during WWII, and he always insisted that I sing the Norwegian national anthem for them, in a terrible accent, I'm sure, and my only parlor trick. They were all very proud of the heroic Norwegian Underground, so I was too. Stories of the North got embedded, guaranteeing I would end up in Alaska.

Today's halibut fishermen brag that their grandfathers' generation pioneered the grounds off Alaska. Norwegian-Americans also played a major part in the Bering Sea pollock and king crab fisheries development and were in on the start of the Fairweather salmon grounds and even a winter salmon grounds off California. They did a share of pioneering in fish conservation too. In the 1930s the halibut fleet, wishing to improve its price by offering fresh fish, rather than mainly frozen, decided voluntarily to self-manage the amount of fish being unloaded at any one time through a system of "lay-up" that a fleet committee oversaw. Boats would choose to fish for a certain number of days, then lay-up for another period, about a week, then fish again until the fleet's harvest quota was reached. Fishing stayed closed completely during the spawning season.

The system seemed to work for decades but broke down in the 1970s when a swell of new people entered the fishery, many of them fishermen squeezed out of the new limited permit salmon fishery. They didn't feel the obligation to do voluntary lay-ups, and the fleet gave up on the system. IPHC said the lay-up system was too hard to enforce and instead instituted four scheduled openers for everyone together, with closures of two weeks or so in between. Many skippers fished other stocks during the closures. But this did not stop the flood of new boats into the fishery. Enter the free-market economists in the late 1980s with a solution already being tried in Canada, New Zealand, Holland, and Iceland with Individual Fishing Quotas and promoted by NOAA for American fleets. The North Pacific Council (Council) was one that took on the task starting with its halibut and sablefish fleets.

The arguments in favor of IFQ were convincing enough that the Mid-Atlantic Council had already voted to start three programs. But the clam fleet consolidation that took place very soon was a shocker, with the clam dredgers dropping from 125 to just 33 vessels owned by a small handful of corporations. Was this what the government had in mind for all of the federal fleets? The New England Fishermen's Federation declared that such extreme consolidation was a violation of the anti-trust law and took the issue to the US Dept. of Justice (DOJ). But that agency found that there was no violation of anti-trust because all species of clams fished had to be considered, not just two. The New England fishermen didn't like the IFQ results and the idea of private rights to the fish soured in the east and was set aside for some years. But the North Pacific Council' IFQ plans for halibut and sable fish did go forward, even though amid much controversy, so much that Congress soon called the moratorium for years on new IFQs. The experience of the North Pacific halibut fleet is what I will summarize, with the smaller sablefish (blackcod) history being so similar.

The Pacific halibut fleet had reached the point of extreme "derby fishing" by the 1980s, due to the increase in boats and improved technology. In 1984 the catch went over the season quota by more than 80% in only five days. The fleet, worried over so many boats racing for the fish, proposed a federal limited permit program as a solution, but the government said no; it had its sights on more radical change through IFQ.

The North Pacific Council (Council) formed a planning committee. The Alaskans on the committee were familiar with the operating Canadian ITQs, and the problems they had caused and knew they must design a better version. IFQ was a strange concept for western fleets used to fishing in an ocean commons open to anyone with a license, but it soon won favor with the big-boat Seattle fleet. After the rush into halibut by salmon fishermen squeezed out by limited permitting had come a new surge of former crabbers following the 1981 crash of Gulf of Alaska king crab. The crowded halibut derby was out of hand in the opinion of many of boat owners, large and small. Other fleet issues were also like BC's: dock price, stock conservation, and crew safety. The government had another issue of its own: the cost of administration. Small boats with marginal profits--open skiffs rolling around out in the federal waters just as my sons would for years--couldn't be expected to pay much more than they already were in dockside fees.

Dr. Evelyn Pinkerton of Simon Frazer University, a Canadian who has written extensively on ITQ results in BC, comments that much social disruption to come for smaller fishermen and their communities could have been avoided. Limited permitting could have been instituted, and the IPHC could have agreed to formalize its old lay-up program with a realistic enforcement strategy. This would have allowed smaller fishermen to survive on an equitable basis with the big operators. She is not alone among scientists in believing that DFO was under pressure to privatize and that the major concern was not stock conservation so much as the Canadian government's determination to move toward fleets of larger, more productive vessels that could be ordered to pay higher administrative fees on their catch.

The Council's halibut IFQ plan, developed over several years of heavy debate, did reflect lessons learned from BC, where the original allocated free quota was rapidly bought up by larger firms and within a decade was mostly leased out, the quota owners enjoying the view from the beach. The US fleet didn't want to see that outcome and demanded many provisions to keep the quota in the hands of working fishermen. The boat owners to receive the original free quota allocations, many by now older men, would be allowed to have a hired skipper on board while they stayed ashore. but any new purchasers of quota from them would have to be aboard while fishing went on. Within the years of fishing history used to determine eligibility for free quota, someone who had a poor season could average it in with better ones, and a medical excuse could be used to set aside an "owner on board" rule for a year. Furthermore, much of the quota would be kept in blocks of shares to prevent too many smaller portions of quotas from being sold off piecemeal as need for funds arose. All of these, and more, were genuine social protections for the many smaller operators as firms with heavy pockets showed up on the docks.

Alan Haig-Brown, ("*Still Fishin*" P. 134) writes of how his friend, Mark Lundsten, skipper of the longliner *Masonic,* worked hard on the halibut planning group to make the program fair for all. But that even after the program was signed by the Secretary of Commerce that "…many in the Alaskan fishing industry fought on, even with the support of some NGOs…

that claimed it would hurt the resource and small fishermen." They entered and lost court cases through the same argument as in other countries--that IFQ did not award ownership of fish, only "rights" to fish for a certain number.

The Council gave the green light to its one-year pilot program and the fleet went fishing. Buying and selling of the original quota shares soon took place, just as the economists hoped, the share price quickly rose, and any serious observer could see that NOAA's goal of consolidation to a smaller fleet of bigger vessels was guaranteed. How had a fleet containing so many Alaskans considered such an elitist program? Council planning required opportunity for public testimony, and Council staff did travel to some of the remote ports with small-boat fleets to get their reaction. What the locals had to say may have generated some note-taking but didn't stop the rising IFQ tide. John, my Kodiak contact who had fished halibut for a time and still regularly talked with the fleet through his business, described for me how he saw firsthand the federal pressure to push IFQ through. He recalls a hearing he attended in Washington D.C. where the true purposes of the Department of Commerce, if not the fishery councils, were exposed:

"A group of us were in DC for a different meeting, and we heard there was a discussion of IFQ going on so we went there. After listening for a while, I moved that we take a straw vote on how many there were in favor of IFQ and someone seconded it. The chair protested, but I knew my Roberts Rules and said you have to recognize a motion from the floor. So they took the straw vote and it was overwhelmingly against IFQ. You could see the lobbyists running to make phone calls. The next day we went up to visit the Alaska delegation and they said, 'What's going on, why did you do that?' We told them, because that's where we are, we are against it.

"To give you an example of their thinking at that meeting, one of the young government staffers gave an amazing presentation. He probably never caught a fish in his life. He put up a chart that showed an empty rectangle, that's all. He said, 'What do you see in that rectangle? Nothing, right? That's what we want the fishing fleet to look like.' He spoke for them, right out in the open, for their long-range goal. They wanted the fish to become just another commodity, totally managed by huge corporations. I really blew up; I yelled about how they were deliberately cutting families out, and that this was not what America was about. One of my bunch had to drag me out of the room. Of course we let everyone back home know what was coming. But maybe our protests simply made them push it along faster."

To get majority support for such a radical change in fishery management, the Council's plan included several non-economic arguments in IFQ's favor. One was that it would promote safety by slowing down the fishing season. Many blamed "derby fishing" (the race for fish) for the high death and injury rates in the fishing industry nationally. Certainly if skippers had their own personal quota to plan around rather than racing with others, they would make better decisions, such as not feeling forced to go out and compete if bad weather threatened. There was no guarantee that a skipper would slow down, but there was no doubt that, overall, the fishery would be slowed and thus probably be safer.

There was little scientific evidence that IFQ by itself would promote better conservation, but it was another important argument to get congressional and public support. Nor could IFQ guarantee better dock prices with fresh fish coming in over a longer season, rather than a derby flood of fish that had to be frozen, but the argument had some sense.

Critics observed that the Council had also neglected to set up many normal administrative requirements for programs: general standards, a data collection process, and an evaluation system to assure that the program followed the goals of the MSA. But there was no way for a council to manage all this for a pilot program without much technical assistance, and this was not forthcoming from NOAA, the logical source. The federal government wanted the fleets privatized but apparently chose not to commit the resources to do it properly.

When the halibut and sablefish plans were signed off by the Secretary of Commerce, Alaska's Senator Ted Stevens used his earmarking talent to rush an appropriation through Congress for startups in 1995, and they were formally included in the MSA re-enactment the next year. Before long it was clear the halibut/sablefish IFQs were a Pandora's box afloat.

Weighing in on IFQ

When after a year it came time for a program review, the fleets (and the communities they hailed from) chose sides. As it turned out, neither the halibut nor sablefish plan was the immediate slaughter that some Alaskans had feared. Of over 8,000 applications for IFQ submitted by fishermen, all but 1,100 were declared eligible for receiving quota. The plan was intended for working fishermen, not speculators, and in that respect it was a better plan than BC's. But as expected by NOAA and the Council, many small-boat owners, mainly Alaskans, had already sold their shares and left the fleet. They had either not gotten enough shares to fish, or were willing to sell for other reasons. But all of the socio-economic effects of fleet consolidation wouldn't show up for some time, and even when they did the implications didn't seem to be clear to the Council. At Kodiak, however, at the time the biggest west coast halibut port, the economic effects of IFQ were understood very soon. Yet, a survey by researcher Gunnar Knapp of the University of Alaska found that of the halibut boat owners still active and selling at Kodiak (many berthed in Seattle), the majority approved of IFQ. They voted for it (and the sablefish IFQ) to be permanent. Knapp was not required to survey the boat owners who had already sold out, and one could assume they would have been less positive, but they apparently no longer counted to the Council.

Faced with much dockside grumbling at Kodiak about lost crew jobs through vessels leaving the fleet, the Council asked Knapp to discover the actual rate overall of transfer of quota shares for halibut that first year. The rate was 13 and 16 %, depending on district, but this selling and buying of shares would continue, and the transfer would be much higher over time. The Council's figures show that by 2011 the quota loss in 42 smaller Alaskan ports would range from 55 to 66%, and all had lost boats but one.

But that was in the future. The remaining fleet saw many positives in the new program: a smaller fleet, the end of the "derby" race for fish, less abandoned gear, and fleet deaths were zero that year. Fresh fish had brought a slightly better price than frozen. The skippers could work with a smaller crew. Negative findings, many from non-vessel owners, and very similar to those from the east coast clam dredges and the BC halibut ITQ, included: excessive consolidation in vessels, resulting layoffs of crew, and negative economic and social effects on small fishing communities as they lost parts or all of their fleets. Hired skippers and crews were bitter that their own investment in fishing--their rigorous, high-risk work, which was such an issue when the safety factor of IFQ was promoted--had counted for nothing when it came to awarding quota. Later, after more debate and a Council amendment, hired skippers and other crew who could prove 150 hours work at sea were allowed to purchase quota. But by then the market price of quota was in the clouds.

Though the Council felt it had done what it could to produce a fair IFQ, and believed it had, as required, "preserved the traditional nature of the fleet," some planners admitted they were shocked at the speed of consolidation. Yet, calling a halt to the program, even if a majority had wished it, would be almost impossible due to all the investment in quota buying and selling by the vessel owners. It wasn't just the Council majority and NMFS that lacked concern for the small-boat fishermen and the larger boat crewmen. It would be about fifteen years before the state of Alaska funded a loan program making it possible for crew to borrow for quota purchase. The federal government delayed even longer.

Up in Nome our CDQ small halibut fleet heard little of the Council's IFQ experiments, but elsewhere in Alaska opposition to IFQ kept growing, including from the processors. They tended to prefer the fast race for fish where they could better control the price, and had invested in extra equipment to handle short, crammed seasons. They wanted a share of the quota. Other objectors on both coasts continued to believe that IFQ was an illegal grab of the traditional public commons. There was so much strife over it that Senator Stevens decided to push through the idea of a moratorium on new IFQs so that the opposition could calm down and more study could be done. The Washington State fleet and politicians objected strongly but Stevens won, and a five-year moratorium was included in the MSA amendments in 1996. The Supreme Court refused to hear new cases against IFQ. However, the processors who didn't like the design of IFQ had power, and the next privatizing program would answer to their demands.

During the moratorium both the National Resource Council and the Government Accounting Office looked at IFQ issues of initial allocations, fleet consolidation, effects on fishing communities and processors, and the quota market. It was too early for them to observe the longer-term effects, but both agencies recommended that processors not be included in quota distribution as they wished, and instead to simply be paid off for their revenue losses from stranded capital. But they did not recommend a halt to IFQ programs, and the Council's planning for the next one went ahead.

The idea of privatizing the fisheries (and many other industries) had become a groundswell, but protests had too. Ten years later I asked for a retrospect from a few Alaskans who had been involved in the halibut planning. Though their sympathies were with the Alaskan small fishermen, they pointed out that the planners had been dealing with a publicly-owned resource, not just fishermen's concerns, and they stuck by the value of the program. They'd felt they were forced to deal with the problems of the derby fishery: the excess fish that flooded the market all at once, most of which had to be frozen with the resulting lower price, the abandoned gear when a closure was called, and the terrible safety record of the federal fleets off Alaska. Something had to be done. Yes, the IFQ had eliminated many boats, they knew, but "those small craft had no business being out there." One said, "The IFQ saved lives; I'm sure of it."

I'd enjoyed a taste of fishing halibut from a skiff when I went out in Rob's *Island Girl* one time. He had done everything he could to make his large skiff safe: he had installed flotation, a radio, and two motors, but it was still an open skiff on the fall Bering Sea. The planners were right, people shouldn't be out in skiffs on that rough, cold sea, and I'm glad that the Nome fleet before long graduated to real boats. But at several CDQ ports they still fish halibut out of skiffs. People have to make a living there.

The planners had more justifications for the final halibut design. They had struggled with how to distribute the quota fairly, and believed that going by a boat-owner's sales history was the fairest way, that it wouldn't be right to give all boat owners equal shares, as many of the smaller boats were part-time fishermen. They pointed out, further, that the fishery had been pioneered by the Norwegians with their big schooners out of Seattle, and they needed to be recognized for that. I saw the point. I thought of how my uncles had made their livings: the bitter cold, the rheumatism, cramped hands, boredom, exhaustion, whiskey, months away from home, all to get to where they were: small comfortable homes, a car, kids finishing high school, wives holding it all together on land. However, I also saw that the "pioneer" argument was used as one more way to sell a controversial program. I suppose all people have a myth about their glorious past, or it would be good if they did. But the halibut and sable fish IFQ/ITQ in both the US and BC, even with all the protections included, produced the closing of a public resource and the promotion of a profitable business for a favored few. They had indeed put their lives into it and had the clout to demand the program, but if the IFQ was about recognizing "pioneers", the descendants of ancient halibuters from around the Gulf, still fishing in small boats, would have been better "recognized".

The dilemma for the smaller operators was that there would always be pressure to sell your quota shares when you needed cash and didn't receive many shares to start with. But in the small Alaska ports, unless you were also licensed for some other fishery, there was little else to invest that money in, and it would be soon gone for household needs, or to send a kid to school, or to pay off an IRS bill dragging on. What, then, happened to the home communities built around the small fisheries?

The halibut fishery, despite all the quota trading, stayed important to Alaska. Today over half the total fleet are home ported in Southeast, especially at Petersburg-Wrangell, with the others mainly at Kodiak, Homer, and

Washington State, and with a scattering from smaller Gulf of Alaska ports. In 2012 the total fleet numbered 1060, with 211 of them CDQ (non-IFQ) boats from western Alaska like my sons' tiny fleet. Almost all of the active 740 Alaska-based boats were less than 60 feet, a small-boat fleet. But the real winners in IFQ were the big boats from Seattle that bring in huge loads and also build their nest eggs in quota.

The difference in culture between the large and small fishermen is a large one throughout the fleets. Few coastal people familiar with the local boat owners would agree that those who sold their quota and dropped out were "part-time fishermen" and therefore were better off out of the water. The argument for eliminating part-time fishermen didn't make sense. My sons fish every species and every license available to them but they still have to find other work in the winter. Most small fishermen do, or wish they could. Small fishermen everywhere in the world tend to move from one opportunistic fishery to another, based on the harvest forecast, the weather, the condition of their boat, the available crew, and so on. It is as viable a way to fish as another. The season landings aren't so impressive but sustain families. Crewmen were another discounted group. They hadn't invested in vessels, just contributed their own lives and health, which didn't count much, as it was not a monetary investment. Some crewmen eventually would save up or borrow and buy quota. But available quota soon was hard to find.

Even though much consolidation and much controversy had taken place, NOAA considered the halibut IFQ a model program, and certainly the Council, its advisory groups, and NOAA/NMFS had slaved to create it. Where they fell short the most was in not anticipating how fast the value of quota shares would rise and how the sales of those shares out of community would affect the small Alaskan fishing ports. And this effect continued with new IFQ programs and with larger fishing communities.

The Gulf of Alaska Revolt

Far to the north, my sons and about ten-fifteen other small-boat halibut skippers at Nome and Savoonga had no worries over IFQ. They were part of a CDQ, the stock was healthy, and there weren't enough people interested to create a mad race for fish off Nome. But for small coastal communities in the Gulf of Alaska and Southeast, left out of CDQ, within two years it was obvious that IFQ threatened their economic and social health. The Gulf communities saw population drops tied directly to loss of local boat ownership or activity through IFQ.

The forty-two small ports that had suffered most organized the Gulf of Alaska Coastal Communities Coalition (GOAC3) to protest and to ask the Council for help. The level of losses to these communities from IFQ can be seen in the final report of the Council's "Review of the CQE program"; the figures surely caused the Council chagrin. By the end of 2008 the number of IFQ quota holders lost to the 21 eligible Southeast communities averaged 55% and the amount of quota shares lost to them averaged 49%. In South Central Gulf the 21 towns/villages averaged a 50% loss of quota holders and 26% of their quota. Another official report states that the figure was "25% of the [quota] issue to residents of small coastal communities had been transferred away." Either way it was a serious economic blow. These communities were actually thrice burned: they had suffered through limited permitting and its local fleet losses, then were not included in the CDQ program, and when they asked to have it extended to their region, got a "Nay." Now, more losses with IFQ. Towns with economies based on local processors were hurt again when local fishermen sold off their quota and boats stopped unloading there. Pelican is an example. Its small processor, very busy when we sold salmon there in the 1960-1970s, would open and close several times in the next decades.

An award of community quota could help put those little towns back on their feet. But one observer told me the big-boat owners strongly objected, as community quota set-asides would drain off quota from the fleet. Where was the tradeoff for the big-boats such as the Bering Sea pollock trawlers had won with the CDQ deal? The Council dropped the GOAC3's request for community quota for a time. But GOAC3 didn't give up and eventually the Council did recognize their well-documented grievance, developed a new program to help such places, and in 2002, nine years after its first IFQ began, passed the Community Quota Entity (CQE) program. It began to purchase quota from boat owners interested in selling, up to ten percent of the fleet quota, to be held for community sponsored quota.

CQE would be available to coastal communities with less than 1,500 population that had traditionally harvested halibut or sablefish and had lost local fleet. On paper it looked like a good program--except for the cost of quota. The regulations developed had many of the social protections of the IFQ, such as owner-on-board, and limitation on blocks of quota, and also had elements like CDQ in that profits would go to community development, but with the big difference that there would be no awards of free quota; the communities would have to buy it. How would they do this? The federal government didn't provide a means. By 2005 the Alaska Legislature did authorize a loan program, but it was more years before it appropriated the funds for it. Meanwhile the cost of quota shares was shooting up. One of the staff commentaries at a technical support workshop for the communities states, "The success of the program will depend on the imagination and hard work of the community and the organization they form to represent them." If I had been one of those community members I would have been offended by the patronizing tone of that remark, but there were bigger issues to come.

The GOAC3 communities had reason to feel poorly treated by the Council, NOAA, the Secretary of Commerce, and the Alaska congressional delegation. Senator Stevens apparently didn't view a community quota program as important enough to push through federal funding for it, or else felt it had a slim chance with the "outside" fleet opposing it. Buck-passing was one reason for the stalling. One argument that circulated was that if the Gulf communities wanted quota, let their regional Native corporations grant or loan them the money to buy it. But it turned out that there were legal as well as economic reasons why the regional corporations couldn't or wouldn't do so. There may have been legal reasons why the state couldn't help. But legal arguments have never stopped an IFQ plan from going forward.

The community quota solution seemed dead in the water. A social scientist at the University of Alaska, working on a technical assistance grant to get CQE moving, found out that there was ample local expertise for working up grant applications, administration, and reporting. The real problem was financing the purchase and administration of halibut quota. "Hard work and imagination" weren't the real issue. Seventeen years after IFQ had begun its destruction of small fishermen, the State did at last set aside $10 million for the loan program, $1 million per village, for purchasing of quota and expenses of administration. Seventeen years--CQE must not have been very high on its agenda. By then the cost for a 10,000 lb. block of halibut quota--the poundage for a "reasonable trip", says the Council report--could be almost $300,000. The loan asked for a 35% down payment, there would be interest owed on the balance, and a deadline for payment--the usual loan requirements. There were also restrictions on the size of the vessel that could be leased and on the community full-time residence of the leaser.

Thirty CQE villages of the 42 had done their paperwork and were eligible to apply, but almost all of them realized that the total costs of a down payment, annual payments on the loan, interest, the administrative costs to manage the program, plus the vessel and skipper restrictions made CQE unrealistic for a small town to sponsor. One might wonder why the Council staff that developed CQE didn't realize this? Why didn't the State? As of 2012, two towns, Old Harbor and Ouzinkie, had found ways to make the program work for them, and two or three others were possibly going to reach that point. So far, the rest apparently had given up.

One community that was able to get CQE working for it, Old Harbor, was out fishing its quota by 2006 and by 2008 had five boats producing. It was able to buy community quota with funds from its Native village corporation and a grant from Ecotrust, an environmental organization that sees small fishing communities as part of local ecosystems needing protection. But Sven Haakanson, a fisherman through his youth at Old Harbor and for many years the Alutiiq Museum director at Kodiak (now curator at Burke Museum at University of Washington), was disgusted with what the village had endured to get community quota operating. "If you want to kill off the Native villages, IFQ is the way to do it...It cost us two million to get that community quota working. Two million." (In addition to "hard work and imagination".)

By then another problem was emerging--the TAC (total allowable catch) set for halibut was going down each year. Could a CQE vessel's catch even cover its loan and other costs?

The GOAC3 still didn't stop its pressure, and recently the Council decided it could solve the problem by renting available quota to the CQE villages. They would still have to raise other money for start-up and administrative costs, but it was a possible route. The Council also agreed to provide free federal groundfish permits with cod endorsements for the eligible western Gulf CQE villages that had a history of cod fishing. Both belated actions seemed to show a shift at the Council.

All of the regional councils could have learned from the experiences of the Gulf of Alaska and Southeast small ports: how not to structure a diverse fishery, and then if having done so, how difficult it was to undo. The GOAC3 determination to right a huge wrong helped make the truth of IFQ visible to the world and was a warning to regional fishery councils that they needed to be more alert to the potential social effects, long term, especially on small fishing communities. But the majority on the Council didn't take the lessons to heart, nor did NOAA.

More IFQ Issues Surface

Fish stock conservation through IFQ continued as an important strategy for the federal government, and the arguments over this drew in several ENGOs. Ecotrust, Friends of the Earth and Greenpeace were among those that opposed IFQ, saying it was not a conservation strategy at all and hurt small communities. But some of the most powerful ENGOs like Environmental Defense Council (EDC) and Pew Environmental Trust, defended it. At least one national conservation group promoted the quota market as a way to make money through "green" speculation. Most ENGOs stayed out of the debate, as it was difficult to prove the conservation value of privatized fishing rights. In most cases IFQ had been introduced along with other reforms, such as waste reduction. As for the lauded improvements that could be specifically attributed to IFQ, many of them, too, turned out to be pipedreams, for example the promise of IFQ to rebuild halibut stocks. Despite IPHC and Council efforts, the size of the individual fish harvested would continue to drop. Each year the halibut fleet quota on which the individual quotas were based was cut more. Possible causes of fish size shrinking were suggested: a warming ocean, predator increase, trawler bycatch, weak assessments allowing overfishing, all of the above--no one was sure, but IFQ could not deal with it. Conservation had been a useful selling point for Congress and for a general public, which, even in Alaska, mainly believed simply that we needed to "save the fish".

As for saving fleets, before long a number of independent and university social scientists found IFQ rich material. Today, hundreds if not thousands of research articles on the social and other effects of privatizing programs like IFQ can be found on the web and have been published worldwide. As the problems with IFQ emerged, the Council went through its ponderous amendment process to take care of several issues. While efficiencies like safety records, prices, sales, quality of product, and so on, had originally sold IFQ, values like social fairness, diversity, and community health would gradually get more attention in the ongoing debates.

The federal government continued to see the halibut and sablefish programs, despite their snags and snarls, as role models. All of the regional councils watched closely to see how those first programs evolved as the IFQ moratorium ran its time. But even with so much to be learned from those first efforts, the North Pacific Council's next program, for Bering Sea crab, tossed out many of the best parts of those first IFQs and added new design to disrupt even more the small-boat fleets and their communities.

Chapter 29

Small Crabber Outrage

The North Pacific Council turned its spotlight next on crab fleet restructure, but the little Norton Sound fleet, even though it fished federal waters, was bypassed and remained a fleet of about thirty boats, just one of them over forty feet, still fishing a commons. Because of it being a "super-exclusive" fishery, with the pots allowed per boat only forty, or fifty for a longer vessel, the big crabbers from farther south stayed away, and the local fleet kept its niche spot as a summer king crab fishery. And it was a healthy fishery, ranked high for sustainability. It didn't have to pay attention to crab fleet happenings outside the region, beyond "What price are those guys getting?" As of 2012, the small Norton Sound crabbers had never had a "bad season". Then, in 2013, the good luck streak ended, at least in part due to unusual summer weather conditions. Rob, Frank and the rest of the fleet found out more what the gamble in fishing fishermen can mean. They were surviving, just arguing to the NSEDC fish plant managers that they should be getting a better price for their special summer crab. By 2014 they did, and the next winter it was better yet. Such a thriving small-boat fishery as Norton Sound's was an unusual phenomenon on any US coast. Then in 2015 the winter king crab price, heavily counting on Korean sales, crashed due to Russian crab competition. The locals said it was illegal crab, another hard lesson for them on international marketing.

Northern Crabbing–the Big Boats

The big-boat crab fleets fishing Bristol Bay/Bering Sea, the ones we see on a TV entertainment series, endure a very different challenge from the fleets in the Norton Sound: the worst possible winter weather to fish in, with incredibly rough waters and icing conditions. The boat owners have hundreds of thousands, maybe millions, invested in boat, equipment, licenses, quota, crew, and often leasing costs. A single pot can cost $1,000, and a boat may carry more than a hundred. Crab biomass tends to fluctuate, and management of them and the fleets that fish them is always challenging. These were the fleets the North Pacific Council proposed for the next IFQ.

Though people had been commercial fishing for crab in the Bering Sea since the 1950s, the big rush began in the late 1960s with Kodiak as the main processing port. According to people who joined the rush, all through the 1970s anyone who put out pots made money, experience unnecessary. Technology played a part as usual as boats upgraded from wood to steel and aluminum, and to all electronics available. Seattle boat builders had a line of would-be crabbers waiting for their vessels, sure they could pay them off in a single season. Even crewmen had a chance to make incredible gross. One purchase of a processing plant was paid off in its first year. As more boats entered the fishery, most of them with big loans pressuring them, it became another race for fish in a dangerous winter fishery.

Then, in 1981 the Bering Sea/Aleutian red king and tanner crab fisheries crashed. A typical red king harvest of 130 million dropped to less than ten million. Some scientists said the cause was faulty stock assessments that had allowed overfishing. Others said disease may have been a factor, or temperature changes, while others blamed predators such as halibut on juvenile crab. Later scientists produced detailed studies, including photo evidence, of the destruction that the US had allowed of an important, well-known crab nursery off the Alaska Peninsula, and by joint-venture bottom trawl fleets we had organized. Commercial red king crabbing near Kodiak was closed to allow stocks to recover; they never have. The Bristol Bay red king crab fishery remains open, but is tenuous and closely monitored. Why hasn't the other stock recovered? Another "fish mystery". As a comparison, our Norton Sound kings were able to rebound from several years of overfishing by the 1990s invasion of big crabbers. Perhaps the difference is that though the Norton Sound crabs were overfished for a few years, they never had to endure draggers across their nurseries. But that's just one possibility.

After the king crab crash innovative crabbers converted to other fisheries--halibut, sablefish, groundfish, or state waters Dungeness crab--but others stayed with federal waters crab and were able to do well many years by fishing snow crab and golden king crab, or by moving out to the west and north. Then, with the advent of halibut IFQ, boats left out of that were converting back to crab again. And again the Council heard complaints of "too many boats". In 1995 the Council introduced a moratorium on new federal crab permits and in 2000 introduced limited permits (LLP) for its crab fleets that required landings to have taken place during 1988-92, thus excluding the newer entries. It also eliminated unused licenses over time.

More changes came in 2002 when Congress directed the North Pacific Council to consider "crab rationalization", IFQ by a new name, and many of the big crabbers, observing how well IFQ had worked for the big halibut boats, eagerly lined up. Alaska's Senator Stevens pushed through a buy-back program for 29 crab licenses and told the Council to have a crab plan submitted by 2005, to be ready when the moratorium on new IFQs would be lifted.

The Council's experience with halibut and sable fish warned that crab rationalization would not be smooth sailing with local small crabbers. Nor would the locals that crewed on the larger crab boats be enthusiastic, having seen how many halibut crewmen had been left ashore. And none of the crabbers were likely to be happy about passing some of the fleet quota to the processors, as was being proposed. By now debates about privatized fleets were rumbling across the fisheries on all coasts. A 2004 Government Accounting Office (GAO) review by Anu Mittal, looked at two issues: the effect of the halibut IFQ on processors and the need for protection of fishing communities. The review questioned the reliability of a state study that gave statistics showing that more processors were harmed by the IFQ than had gained from it. The review expressed concern about "the potential for IFQ programs to harm the economic viability of fishing communities," and urged that the crab program follow closely the 1996 MSA's Standard 8 to protect those communities.

The GAO report listed many ways that an IFQ could be structured to protect communities from flight of quota (and boats) out of a community. An example is the "owner on board" rule, or a geographic limit on where quota shares could be sold. It also pointed out in each case that these social protections had the risk of reducing efficiency of the fishery. It did not give a recommendation on any of them, but said a choice had to be made of efficiency versus social protection. However, this report, coming out in 2004, was late to have any influence on the crab proposal, ready to go forward. It did have a chance to be considered for the 2006 amendments to the MSA under the pen. When the quota shares started to be sliced up, some would have to be set aside for the small communities, and for crew and "new entrants" to purchase. GAO also had earlier emphasized a structured program evaluation process practical enough that any council could make use of it in planning a program.

Alaskans by now were more skeptical of federal fleet restructuring. In 2005 the Alaska Marine Conservation Council prepared recommendations it believed essential to make a more fair, stable program, including regulations to discourage armchair fishing, encourage conservation practices, and keep options open for new entrants. Unfortunately, its reasonable recommendations also came too late for the crab proposal. This sort of delay seemed too often to be the case. It was as if people weren't giving the issue enough attention, still weren't knowledgeable enough about what privatization entailed, or didn't understand the Council's process. Perhaps the more long-range social effects of the halibut and sablefish IFQs weren't yet visible outside the communities affected, but the OMB understood and the Council itself should have. The crab crewmen certainly understand the effects of IFQ and would demand more rights than the halibut crewmen had won. However, as crewmen they had no vote.

The States' Dungeness Crabbing

Alaska has another important crab fishery that was not involved in the crab plan--Dungeness crab. "Dungies" were another subsistence fishery in which I experienced very early success. Managed by the states, Dungeness crabbing has survived well with traditional management. The biomass has held up along the entire Pacific coast using regulations that restrict by season, size, and sex. Boats are restricted to the number of pots they can carry. This management system obviously is only possible for a species like crab where you can bring them up unharmed, quickly distinguish sex, measure each individual, and throw illegals back alive. But the same ability to select is true for federal crab.

Dungeness crab stocks, like others, cycle up and down based much on ocean conditions, and the fleets tend to cycle with them. The Dungie fleets of Washington and Oregon were among the fleets flooded with new boats following the 1970s, in part caused by limited permit programs pushing boats out of other fisheries. Like the rest of the fleets, the Dungie boats soon grew in size and improved technology, but due to fleet resistance it was the mid-1990s before limited permitting went through for the Washington and Oregon fleets. The Northwest fleet also had to give up 50% of the quota during the period when the treaty tribes won 50% of the salmon.

Though clearly red kings and Dungies have experienced many similar impacts, the Dungies are still healthy stocks, while the other crab stocks that were rationalized are mainly still having problems getting past the edge of sustainability. State crab fisheries are apparently successfully managed, causing one to wonder why NOAA was so determined to restructure another fishery. Was the answer simply that NOAA was riding the worldwide wave of privatization that the Milton Friedman's free-marketers had promoted?

Selling "Crab Ratz"

The North Pacific Council's plan for federal crab rationalization included all federal waters crab except Norton Sound red kings and Pribilof golden kings, which had special designations. The state also retained an important role for all crab as it would set seasons, crab size limits, and gear restrictions. I will focus on red king crab, since the privatizing plans and results were similar and involved much the same boats. The arguments put forward in favor of "crab ratz", as it came to be known, were the ones already heard for the halibut and sablefish IFQs: overbuilt fleets, a race for fish causing waste and increased danger, and abandoned gear during closures with the result termed "ghost fishing". But there were also significant differences in the crab plans.

In one large change, non-owner crew were allowed to buy three percent of the total quota shares allocation if they could verify 150 hours of on-board experience over three years, as proven through signed fish tickets, i.e. if they had been in the role of skipper who signed tickets. Later, they could also buy more quota shares through a federal loan program, but this did not come about until 2011. The opportunity for communities to form non-profits and purchase quota shares to lease out to boats was also added and nine communities were declared eligible, similar to the halibut CQE, but based on history of crabbing. But in another change, leasing of quota would be allowed, and owner-on-board was not required, so the prospect of speculators as well as owners living off leases was guaranteed, something the halibut plan had worked to avoid.

The change arousing the most controversy across the fleet was that despite the GAO and the NRC recommendations against it, the processors shared in the windfall. They won a percent of the original free quota distribution, and had boats specifically assigned to sell to them. The processors had argued that unless they were included in the quota distribution, big crabbers would create their own processing network and squeeze them out--or so they claimed. All independent crabbers, big and small, opposed processor inclusion but it stayed in the plan.

Fish conservation and fleet safety issues became main political means of winning public support for a program growing more controversial with each season. In the late 1990s there had been an alarming loss of life in the winter fisheries off Alaska. Most infamous, in 2001 the catcher-processor *Arctic Rose* sank with fifteen lives lost. Politicians used that tragedy and others as part of the argument for continuing to introduce privatized programs that would stop the race for fish. Due to the terrible publicity over fishermen's lives lost, Congress passed legislation and the Coast Guard introduced many new mandatory safety requirements for the boats over a certain size. They included things that are commonplace now even for smaller boats: immersion suits, gas-inflatable rafts, limitations on pot loads, safety training, and stability tests for larger vessels.

Though I had been active in one of the safest fisheries on the coast, shore trolling--I was safer than riding on the freeway--others in my extended family have been active in fisheries that definitely have higher risks and need strong safety regulations. One of them, maybe two, are alive because of mandated immersion suits. I've been in favor of every Coast Guard regulation passed except where they are impossible to install on the smallest boats. But these improvements don't require privatized fisheries, and the Kodiak crabbers included smaller ones who argued that there were other ways to make the fishery safer besides eliminating them through fleet consolidation. My friend John at Kodiak agreed, "The safety improvements were already happening. I believe the government let the derby fishing build up for years to where it was unmanageable. They wanted that argument for rationalization, so they deliberately let the derby go on till it could be called unmanageable."

There is no way to know if it was deliberate, but derbies did go on until they were a total frustration to manage, and were probably in part to blame for the accidents off Alaska. But as one anonymous Coast Guard officer commented, "The downturn in deaths [on crabbers] began long before 2005 when crab ratz went in…The stabilization reviews required, with a maximum number of pots allowed, were the most important part in avoiding capsizing and saving lives." The use of older boats with aging problems, or converted boats not built for carrying pots on deck, or for fishing the Bering Sea winters was the kind of problem the Coast Guard had responsibility to take care of and did. The survival rate also climbed fast after the Coast Guard began to require immersion suits aboard. Still, all the improvements for safety prior to crab ratz were included in the public arguments in favor of the program.

But another large safety problem for the crab fleets all along the coast was injuries due to fatigue. Though IFQ/rationalization would give skippers a looser schedule and the option of allowing their crew more sleep, that didn't guarantee they would. In my own experience, we were always short of sleep trolling and we had no race for fish, just our own pressure. We would go without sleep or risk missing fish that might be gone tomorrow. I remember that Jack fell off the boat once at sea purely through fatigue and irritation. It's of no benefit to have personal quota if the fish had gone down the pike. Privatizing the crab fleets couldn't eliminate fishing's risks: skippers' choices, human error, equipment failure, and the special demands of winter fishing.

A Nome man with experience in Norton Sound's small-boat fleet, who has since left the scene, described to me his first experience winter crabbing in the Bering Sea in 2010. This was after crab was rationalized, and the picture is not as different as you would expect:

"It's an incredible job. I probably won't go back next winter. I'm too old for it, 33, too old to start it as a career. The rest of the crew were in their twenties. Winter crabbing out of Dutch is a whole different thing than crabbing off Nome. It was a very nice boat, 110 ft. But you never get more than four hours sleep, never, so you're totally spaced out. You get a break for meals. And then when you're running between the pots, it's about a fifteen minutes break, and the crew just goes into the mudroom and lies down in their raingear and sleeps until the skipper yells "One mile", meaning time to get out there for the next pot. And dangerous? Yes! I was hit twice by pots when they had not been secured right, 800 lb pots. One came swinging and knocked all of us over. Then, the icing up and chopping ice off the rigging and everything. No, I don't think I'll do it again. I'll let my buddy do it, he's younger. I'm already too old for that."

I gather that what he described are the normal demands of winter fishing in the Bering Sea--after rationalization. I would never discount safety measures; I just wish to show how fishery safety was another politicized argument for programs that needed help to sell them. The fleets did get safer, just not especially from privatizing the fleets. Total deaths in federal waters fleets off Alaska dropped from an average of 37 each year in the 1980s to an average of about thirteen a year between 2000 and 2009. The Bering Sea/Aleutian crab fleet death rate in recent years has been almost zero. The highest rates of death on the North Pacific today are in the salmon and Dungeness fisheries, especially off the Washington and Oregon coasts' river bars. There too, improvements have come through ever-stricter regulations from the Coast Guard.

With all the inevitable risks, some men and women love the grueling work, the pace, the money, and the risks of winter crabbing and similar fisheries, and go back year after year. Though I would never miss the derbies, they say they do. Their complaint is that they deserve proper compensation for their work and not to have to pay tens of thousands for quota shares. As one crab crewman put it plainly, "All those boat owners ended up with this quota, and it was built by guys like myself that were out on deck all those years; they didn't get anything out of it."

Better fish conservation continued as the other argument used to sell Congress and the public on fleet privatizing programs. Obscured for the much of the public was the fact that modern stock conservation is mainly under the control of managers, not individual fishermen, that the managers must have the resources needed to set fleet quotas, seasons, areas, monitoring requirements, and other regulations, and to enforce them. Today it's not easy for a skipper to evade regulations for long, and fishermen with any foresight will follow conservation rules based on good science. Others will choose to believe that any effort that way is foolhardy, that the vagaries of nature and fishery management will dilute any of their individual efforts to conserve stocks. Still others say they would feel better about the regulations if they were sure the stock assessments were accurate. Other skippers have no argument and just succumb to simple greed. How much does privatized fishing improve on this? So far, according to the independent scientists, not much, in some cases it makes things worse, but saving fish continues to be a selling point for privatizing commercial fisheries. There is one clear gain: privatizing bycatch quota makes bycatch reduction easier for a fleet; it works like carbon trading--sharing the bad effects.

Crab Ratz Protests

Crab ratz was soon swimming in noisier disputes than halibut IFQ ever did, and Kodiak wasn't the only noisy Alaskan port. As far back as 2003 the crab fishermen at Homer had submitted a resolution to the Council showing that they understood well the issues and wanted to be sure the Council considered their position. They wanted the crab program to include: no processor restriction for boats, support for the small processors, original quota allocations to include hired (non-owner) skippers, boat owners to be aboard actively fishing, no leasing out of quota, fishing community protection, bycatch caps, and entry-level opportunities available for youth and new fishers. They had done their homework. Other groups also asked for a limit on the duration that any owner could hold quota shares without re-application being required--that is, a sunset clause. Homer fishermen and others hit all the key concerns of the smaller operators.

The president of the Alaska Longline Fishermen's Assn., Linda Behnken, knew that a proposal for Bering Sea groundfish rationalization, which was already being discussed, could include longliners. She wrote Senator Stevens strong recommendations that included many of the goals and standards already in the MSA, her point being that a concrete enforceable process for carrying out the goals was, so far, not in motion. She added that a processor quota would create "a processor cartel that will economically strangle independent fishermen and the coastal communities they support." I don't know if Senator Stevens answered her.

The big crabbers--like the halibut fleet they were largely out of Seattle-- were happy with the one-year pilot program, but they often carried crew that lived at Kodiak who were not so pleased. The local Kodiak fleet consisted of boats of all sizes and descriptions, many falling under the economists' category of "inefficient" boats with little to gain from rationalization and its inevitable consolidation. Their noise would get attention but not enough to stop the tide of privatization sweeping the country.

All of the independent crabbers, large and small, did agree on one thing, that allocating a percent of the crab quota to the processors was heading a bad direction. Both the Government Accounting Office and the National

Resource Council had advised against this, so why was the Council so taken with this idea? I asked my friend John, and he gave me the answer I would hear a lot from fishermen: "Money speaks". But there was another side to it. Small Alaskan communities had suffered when their local processors closed, or threatened to, due to halibut fleet consolidation. Processors in places like King Cove, Larsen Bay, and Adak were the local economic base. As many as fifteen small or more remote processors were at risk of closure if fleet consolidation went forward with crab as it had with halibut and sablefish.

The Council's plan to assign specific boats to processors for 90% of the quota addressed the MSA Standard 8 as a means to help keep the more remote processors operating. A good example was a Trident-owned plant at remote St. Paul. No boat except one close by would pick St. Paul as its first choice for a port to steam to in the middle of a winter ice storm, but St. Paul needed to hold onto that base of its economy. Crab ratz also included mandated arbitration in any price dispute with the processors. That took away vessels' negotiating power, and they assumed their dock price would no doubt be lower.

Because of the processor inclusion and the uproar it generated, NOAA submitted crab ratz to the Dept. of Justice (DOJ) to review for anti-trust issues. DOJ responded that it, like GOA and NRC earlier, didn't approve of including processors. It acknowledged that the companies could suffer from "stranded capital" but suggested that it would be better to find a way to compensate processors for this and to move forward, rather than including them in a quota system. It also recommended against the binding arbitration proposed by the Council. But NOAA and the Council took none of the recommendations regarding processors.

The Outcomes

Crab rationalization turned out to be even more socially and economically disruptive than the halibut program for some communities. More inclusion of social scientists as Council advisors early on could have revealed that likelihood. It was only when the Kodiak and /Aleutian East Boroughs contracted social scientists like Gunnar Knapp to interview community members that the Council majority seemed to recognize, too late, the extent of the socio-economic problems it was creating. Yet the Council knew by then the effects from the more socially protective halibut IFQ, now ten years old. The pilot program covered about 275 catcher vessels organized into coops with 26 assigned processors that received about ten percent of the overall quota. The other major change from the halibut program was that leasing of quota by anyone was legal. An informal government review of the program after one year found that it was creating all the same issues as the halibut and sable fish IFQs, but with the processors' new power added. A few of the small or remote ones possibly avoided closure due to their inclusion. It was too early to see any long-term improvements in safety, but some positive changes could be measured in conservation. Though wasteful "high-grading" (throwing back less valuable crab for higher priced ones) had actually increased, "ghost fishing" (hasty abandonment of pots in the water) had decreased. Still, even under rationalization fishermen could be tempted to fish in an unauthorized area, and a few did.

Aside from processor quota inclusion, a big effect of crab ratz was excessive consolidation in boats. The government economists must have been pleased, as after just one year the Council found only 89 crab boats still active of the original fleet of 275, and 66 of those were the bigger boats from outside Alaska, a consolidation far more extreme than from the halibut IFQ. A new wrinkle was that leasing of quota was approved. Kodiak was the port hit the worst in numbers. The reduction in total earnings to Kodiak residents was between $1 and 1.6 million that year, about 20% of that from fleet service businesses. The loss in crew positions (after two years) was 757 in king crab and 457 in snow crab--though many the same people. The Kodiak Borough mayor called it "a collapse, a wipeout." He was not talking just about Kodiak City or about fishing jobs, but all of the support businesses affected: providers of boat repairs, moorage, gear storage, fuel, food and drink, gear, and buyers/processors that had not been assigned boats.

The Gulf fishing villages and towns had now been hit three times--first their salmon fleets through Limited Entry, then their halibut and sablefish fleets and crews, now their crab fleets and crewmen, and in each case the support businesses that depended on them. The survey by Knapp and Lowe of effects on Aleutian East Borough, found that despite the Council's intent to protect smaller ports, the loss in jobs (processors and boats) was 900 for red kings, 450 for snow crab.

Despite continuing protests at Kodiak and elsewhere, the remaining crabbers voted to continue the program. The Council approved, as did NOAA. The Secretary of Commerce signed off, and Senator Stevens funded the start-up administration costs through another earmark. A year later crab ratz would be inserted in the reauthorized MSA 2006. I remember a newspaper photo from the winter after crab ratz began. It showed a large fleet of boats tied to the Kodiak floats that normally would have been out fishing.

Over the next few years, studies found that most transfers of crab quota had been from populations under 2500 in Native and mixed villages to larger towns and to non-local skippers and non-relatives. Not surprisingly, the greatest transfer of quota was from boats under thirty-six feet, clearly local boats. The price of quota, as with halibut, had quickly shot up beyond reach of most fishermen. Leasing of quota caused added heartburn for crewmen. When a skipper leased quota to fish, he took the cost of the lease off the top with other boat expenses before figuring the crew's share of a load.

The trend would continue and by 2012 only about 70 federal waters crab boats were still operating, half of them Alaska-based boats. Seventy percent of the red king crab quota was controlled by four companies, but could be leased out to others to fish and up to 75% percent of a boat's revenue could be going to pay the leasing costs. That could leave the crew with about half of what they had once earned. A former crewman at Homer gave me a cynical analysis: "Those guys with quota don't even have to fish anymore unless they want to. They can lease out their quota to a boat and just ride around on the deck and read magazines--or they can work aboard and earn that too." Since this is just what had taken place in the BC fleets earlier, the Council must have foreseen the likely outcome. On the other hand, some boat owners who had been allocated little quota wouldn't have been able to fish if they hadn't been able to lease more shares. But soon many fishermen found they couldn't afford even to lease, let alone buy quota in the market. Another survival mechanism evolved as it had in BC, a form of consolidation called "stacking". Several boats' quotas would be stacked onto one boat thus saving on crew numbers, gear, and fuel and saving on leasing costs, with the profits to be split among the stacked quota owners. It made sense except that it left crewmen without jobs.

There was still more for the non-skipper crewmen to be angry about. Even though with crab ratz they did win the right to buy quota shares, the needed loan funds didn't become available from the state until 2010, with the quota price by then incredible. It was left to the state of Alaska to figure out a system of verification of time-at-sea to determine a crewman's eligibility. If anyone with influence had cared about fairness, a solution to this paperwork issue would not have taken over five years.

Not all of the crabber comments for the Council's review were negative. Several sent positive suggestions to try to make the program fairer all around. And at least one organization of crabbers, the Alaska Crab Coalition, was pleased with crab ratz. Composed mainly of big Seattle-based crabbers, it commented that the consolidation that had taken place was part of remaining competitive and that the fleet was just facing "normal market forces" that occur in modern industry. I am puzzled just how the Coalition saw this government-engineered privatization as normal market forces. Others commented that reduced fleet quotas would have taken place regardless, due to downturns in crab stock. Probably, but a reduction in fleet quota wouldn't have caused such a radical tilt toward the loss of smaller operators and Alaska-based vessels.

Crab rationalization issues continued to enliven the pages of the *Kodiak Mirror* and other Alaska media regularly, but the Internet now provided new grounds for fishermen's communication, and was well used by the crabbers and their supporters to carry on about crab ratz. Today one can "Google" crab ratz and find a bookful of commentary.

The biggest outcry from the remaining fleet continued to be over processor quota and assignment. Skippers reported they had been forced to travel in storms, or because of ice, to distant locations to deliver their crab. This problem was corrected through mutual agreements that a boat during an emergency could unload at a different processor. But in 2008, a crab crewman told me his boat had to wait four financially precious days to unload at Dutch Harbor due to a pile-up of vessels at their processor. Apparently poor scheduling of unloading was not an emergency.

By 2010 the Council agreed to at least listen to complaints about crab ratz. But it wouldn't be easy to undo a privatized program even if the council chose to. Just as with the halibut IFQ, the preposterous prices of quota wouldn't self-correct with deep-pocketed entities cruising the ports. John explained the dilemma to me:

"Even if crab ratz was found to be illegal and they stopped it, how would they ever be fair to the people who've already paid half a million for quota?" He was close. And even though working crewmen are eligible to buy a total of 3% of the quota, few will, as very little of it is for sale. Crab owners realize the great advantage in holding onto it for leasing out.

A Pleasure Trip

On occasion I go for a short trip on one of my family's boats. I can't do much of the work except steer, and am one more body in the way, so I don't push for much time aboard, but enjoy my rides. As you leave the protection of the Nome boat harbor a little after midnight and head out for the pots, a quarterly slop is typical--for me a powerful *déjà vu* of swells forty years before, tossing that other boat, the *Nohusit,* as we pulled out of the shelter of an anchorage. On the *Nohusit* we would have soon been working our drag with the stabilizers out that made for an easier rolling ride. But these crabbers use no stabilizers, and we had fifty miles, six-seven hours, to run out to the pots. Crabbers can't use stabilizers, I soon realized, with the stop-start short runs they have to make between pots, and the likelihood of them tangling with pot gear. And for the long run to their grounds, with fuel close to $6 a gallon, who was going to add more drag?

The deckhands flopped down and immediately went to sleep; their arms and brains wouldn't be needed for hours. The boat was on autopilot, the GPS set for the first pot, the skipper needing only to spot logs or freak waves. I told them the truth when I said I never got seasick, and still didn't, but I dozed. On both trips the sun was just breaking over the horizon and the water calmed as we reached the first pot. The crewmen sprang up, pulled on their rain gear, and headed out on deck. They then became smooth human machines, not a motion wasted as the skipper kept sharp eyes on them, maneuvered up to each pot, and operated the hydraulic winch.

We rolled along as the sun rose, lighting our view and our search for pots. There was absolutely nothing to see outside the operation but sooty shearwaters--not a boat, not a log, just rolling ocean. The radio silence was spooky after the noisy Southeast trolling channels I remembered. We continued on the GPS courses, the crews hauled, emptied crabs into totes, and stacked pots. Both trips ending with fair loads.

I saw that crabbing had a monotony to it that was hard to escape. We were by regulation fifteen miles off shore, not much to see, but the shoreline wasn't very interesting anyway, no winding green inlets to explore as I remembered near Sitka, no dramatic new harbors to drop anchor in. You just hauled pots, emptied them, baited them, dropped them, and moved on. The thrill was a haul of lots of big crab in a pot.

The fishermen would be chasing halibut next, then a couple months break and they would be out on the ice crabbing again, and then if the ice cooperated and went out they could be herring fishing in June. So this fleet had one of the best attractions of small-boat fishing--variety, and with each new season a new chance to make it big. This time as I played tourist, the fall storms were already looming. The skippers had a couple days to sell, get their halibut gear, ice, and bait on board, spend a day with the family, and snag a crew for a new search.

Frank told me there was talk of IFQ among the little fleet. They had to think about their retirement years' security. But if they were to actually vote

IFQ in, the results were predictable. The originals would get their retirement nest eggs, and the younger, newer locals, our future fishermen, would be sunk. I told Frank, one who would surely benefit from IFQ, that they had better think about that more, about how long the quota would stay local. They might want to talk to my friend Sven from Old Harbor about how that town had to work to recoup from IFQ. But Norton Sound crab restructuring was not on the Council agenda. It had already moved on to its most complex challenge for privatizing yet: groundfish.

Chapter 30

GROUNDFISHERMEN'S TURN:
GULF OF ALASKA ROCKFISH AND COD

The passion to remake the fisheries kept growing, but no longer so easily. With the MSA 2006 re-enactment on the horizon, the regional councils had to deal with proposed amendments ordering stronger conservation, expanded enforcement, and more efforts to preserve of traditional fishing fleets and their communities. Once passed, the amended MSA impacted program planning even more than the 1996 amendments. Now national standards had to be met for all fleets. Of greatest long-range importance, the eco-scientists finally won their case, and ecosystem-based management (EBM) would replace management by species and maximum sustained yield. The eco-scientists had argued successfully that fish stocks were obviously influenced by far more impacts than their harvest levels. There were too many other factors in an ecosystem that could stop re-population, with Atlantic cod and salmon, king crab, and international waters "donut hole" pollock all famous examples.

EBM was a more complex and expensive way to manage yet in the long run would serve both fish and fishermen. But the cost of EBM was also one more argument for bigger boats, wealthier fleets that could cover a bigger share of the costs during a period when government each year was under more pressure to cut its budget. Congress wasn't awarding NOAA Fisheries what it needed, so Congress extended the 2008 deadline for ending overfishing and other goals to 2014.

In addition the MSA 2006 stated that all federal fleets were to have annual catch limits set at or below the "accepted biological catch" set by

348

scientists. This required that all commercial stocks, and the most important secondary stocks, have their levels assessed, a huge undertaking in staff time and funds. In the meantime, harvests would need to be set lower to allow for uncertainty. Much more attention to bycatch reduction and enforcement was also mandated, including that new "excluder" technology had to be developed. Better enforcement meant expanded monitoring, another expensive program.

Catch share, the offspring of IFQ, now received official endorsement in the MSA. Other choices of fleet structure were allowable, but the choice NOAA was willing to finance start-up costs for was some form of privatization, with more cost recovery later on being assumed. The government favored "sectors" similar to coop arrangements already in use, such as the Bering Sea pollock fleet's AFA. Vessels belonging to sectors would combine their allotted quotas through voluntary arrangements, and carry out much of their own management, paying for it from dockside fees.

The social protections in the MSA 2006 gave stronger direction. It asked councils to be more diligent in the fairness of original quota allocations, including mandated set-asides for community-based quota, new entrants' quota, and crew members' quota purchase, but it didn't mandate a percent. For this to be meaningful, state or federal government would need to provide low-cost loan programs for quota share purchases. To control too much concentration of quota by large entities, the MSA asked councils to include a 10-year sunset clause in any privatizing scheme, which would assure that no one was guaranteed their quota shares forever. But since no one had ever had quota rescinded except for illegal activity, it was unclear how a sunset clause would affect quota buying and selling. Once more the MSA had given a broad direction, leaving it to NOAA and the councils to provide the specifics within what the budget allowed. And so, again, much vacillation would prevail.

The Gulf Rockfish Program

The North Pacific Council already had a Gulf of Alaska Groundfish plan on its table as its next privatization effort. The groundfish fleet had experienced its own influx of displaced boats from salmon, halibut, crab, pollock, and sablefish privatization. Over a thousand boats with federal limited permits had descended on the open-access groundfish stocks like a flock of ravenous gulls. The fleet had an alarming vision of the future if this migration continued. The Council knew it needed to be innovative. Groundfish restructuring would be complicated because of the mixed stocks involved, the diversity in gear and sizes of vessels in use, and the vast area fished. About twenty groundfish species were fished off Alaska, each with its own state of health, and fished by trawl, longline, pots, and jigs. At the same time the Council was smarting from its wounds over crab ratz, with Kodiak still the most vocal port. But now objections to the Council's previous programs were more organized, especially from the smaller Gulf communities. The increasing protests against bycatch waste from both fishermen and ENGOs couldn't be ignored either.

One large groundfish issue was overfishing, a difficult problem as rockfish species grow slowly and will not reproduce for years. Stock assessments, so important for setting quotas, were often too old to be valid. Nationwide, of over 500 commercial stocks listed, less than 100 complete assessments had been done due to time and budget constraints, as well as other priorities the councils always had on their plates. Fishermen often objected to assessments in use, saying their input had not been taken. On both coasts they had been able to prove that they often knew things scientists didn't and had won quota increases. They argued that fleets needed to be included when stocks were assessed, not just ordered to "listen to the science". Groundfish planners would not have an easy ride. Yet once again the North Pacific Council had lessons available from British Columbia. DFO had already revamped its groundfish fleet with ITQ and had introduced many social protections lacking in its earlier halibut ITQ. But the social and economic damage already done to BC coastal villages was probably irreversible. An EcotrustCanada study found that rural flight of quota shares had resulted in 44% of all ITQ quota holders living in Vancouver or Victoria, with most fleet losses in largely Native communities like Kyuquot and Alert Bay. Though there were again organized fishermen's demonstrations over the social effects of ITQ, the government wasn't budging. It would increase social protections, but would continue its fleet restructuring, stating that its goal was self-supported fleets, that there was no choice. In fact, DFO's funding was being cut more each year and that would continue.

As the North Pacific Council's groundfish discussions began, the small Alaskan fishing communities complained that so far all of the fleet restructuring had been a disaster for them. Would the groundfish plan be any different? The radical changes had affected the larger towns too. Kodiak was the one most heard from, but at Petersburg, Icicle Seafoods (processor) representative, Kris Norosz spoke for the whole Alaska coast when he said, (*Pacific Fishing,* Sept. 2012), "… I can tell you that any time a vessel is sold out of our fleet, it has a huge negative impact on the city's economy…The basis of our economy is the boat harbor…." I doubt any small or medium fishing port would disagree with that. Local groundfishing trawlers, longliners, pot, and jig boats all worried how they would fare with IFQ, up against the corporate trawl and longline fleets from Seattle and south. History offered nothing to encourage them.

A North Pacific Council 2007 meeting held at Kodiak had the two-year review of the Bristol Bay crab program on the agenda. With so much anger seething over the extreme consolidation, effects of leasing, and neglect of crew issues, police were on duty for the meeting just in case. No physical violence took place, but dynamics had definitely changed in the eleven years since the halibut IFQ began. Complaints from the small ports and crews had risen to the top. Fishermen were communicating with each other more, having learned to use the new kind of web. Each month new fisherman and fisherman-advocate blogs appeared, and one could now read not only the upbeat governmental reports but the opposition views as well. The activists had even appealed to the state government to be their advocate and insist that the Council listen to their arguments of how unfair its programs had been so far.

The Council came up with a creative answer for the angry Kodiak community, a means to alleviate the economic distress there, a distress, as fishermen wryly pointed out, caused by its halibut, sablefish, and crab programs.

The "Central Gulf Rockfish Plan" (Rockfish) was a form of catch share--a voluntary, cooped-quota program similar to the AFA, with free quota to be distributed among catcher boats and catcher-processors that had sufficient history in the chosen years harvesting rockfish. The coops would allocate the quota and bycatch quota per member vessel, and trade it as they wished. The fish processing would be in Kodiak, now Alaska's second biggest fishing port. Like crab ratz, the boats were cooped with five of the larger Kodiak processors, ones they had used before, and were not allowed to sell to other processors.

Only a small amount of Rockfish quota was set aside for alternatives to trawling: the fixed-gear boats--longline, jig, and pots; "new entrants" otherwise not eligible; and independent-minded fishermen who were eligible but chose not to join a Rockfish coop. These were small set-asides that would comply with the intent of the MSA.

Rockfish did, however, include a big step forward for conservation, for many people the most important improvement. NOAA had changed the identity of many untargeted species from "bycatch" to sellable "secondary species". Hundreds of tons of previously wasted untargeted fish were now illegal to throw back, and instead were either to be processed aboard or brought to the docks and sold, even though they might not be worth as much as the targeted species and took up time and space aboard. This strong move for conservation was not only a regulation for Rockfish, it was for all the federal waters vessels large enough to manage the extra fish. Yet many of the fleet were not happy with the "secondary species" regulation.

When trolling in the 1960s, we were exhausted by all the "secondary species" we caught, cleaned, iced, and sold for .05/lb. It wasn't required, but we didn't want to waste them, yet the black rockfish and the hake did get thrown--there were just too many. Now trawlers and longliners would have to take care of their secondary species regardless of poor price. Some large boat owners sued over it but lost their case. Prohibited bycatch species like salmon, crab, halibut, and some rockfish would still not be sellable. To assure better enforcement, vessels in the Rockfish program would have to carry expensive fulltime on-board observers.

An advantage for those that joined Rockfish, and for the town of Kodiak was a longer season than normal, from May to November, putting not just fishermen but shore workers to work in the spring during what was a dry period in processing. The hope was that fresh fish could also bring a better price during more of the year. Thus the Council had come up with several innovations that it hoped would be positive for conservation and community issues and would produce a better quality product. It was privatization, but tailored more to some specific local problems.

After the one-year trial for Rockfish, my friend John at Kodiak said he thought the great reduction in bycatch waste was the most successful part of the program. The program also encouraged the use of off-the-bottom trawl and helped boats cut their unsellable halibut bycatch to a small fraction of that previous, while crab bycatch dropped to almost nothing. The overall harvest quota set also was barely exceeded. The Council's evaluators gave the program credit for several other positive elements that were actually well underway prior to Rockfish. Trawlers had already been increasing their use of off-the-floor gear, knowing that the MSA 2006 would require it. A number of boats fishing the Gulf had already switched from trawl to pots, a practical move as the trawl by-catch rate of juvenile halibut was subjecting the fleet to closures, and pot fishing cut halibut bycatch to almost nothing. The main social improvement was the drop in the unemployment rate at Kodiak as people worked at the plants a longer season.

Rockfish also had predictable disappointments, others not so predictable. No fleet consolidation took place that first year--a good or bad outcome depending on your point of view. The crewmen who had been left out of free quota allocation in previous programs were still left out, and still angry. Some processors complained that they hadn't been included, leading to a drop in profit for their season. Although almost all of the eligible trawlers had signed up for the program, only 26 actually participated and only one catcher-processor of the fifteen eligible joined in. Apparently there was more lucrative or less regulated non-program fishing available. Only one fixed-gear vessel joined, and only one boat owner chose to use the "new entrant" designation. Those quota allocations were so small that fishermen may have felt they weren't realistic for more than one boat each.

Rockfish members especially complained that the high cost of on-board monitoring by observers was going to be difficult for them to continue. This would become a major complaint for many fleets, and a serious problem for NOAA when monitoring became required for all fleets a few years later. The fishermen wanted the choice of cheaper electronic monitoring, but NOAAA denied that, saying it wasn't perfected enough. Rockfish members were also disappointed that rockfish prices didn't rise, something they had counted on through making more fresh fish available. They ended up making most of their revenue off sablefish, a secondary species for them that earned six times the value of rockfish. Despite these aspects, Rockfish was from many perspectives an improvement over crab ratz, and some sour attitudes regarding privatized quota programs possibly sweetened a little.

Alaskans Take a Stand

The North Pacific Council voted to extend the Rockfish pilot program to seven years, and in 2010 voted to extend it again through 2021. With leadership from its Alaskan members, the Council now improved Rockfish in more ways to help the small Gulf communities and their fleets. In the future, Rockfish members would be able choose their own processors for the year. (This led to a lawsuit in 2012 from Trident and other original processors, but it failed.) Allowing competitive processors brought a great improvement to the rockfish dock price--from .10-.12/lb. to .27/lb. However, coops still would not be allowed to negotiate prices and would turn over issues to arbitration. Still, there were real social improvements. The entry level would be for long-line only, and if entry-level boats caught 90% of their quota one year it could be increased for the next. Rockfish added several ways to protect the rockfish fleet from excesses in consolidation by limiting leasing, by placing limits on how much quota an entity could hold to only 4% of the overall fleet quota, and by limiting any processor to 30% of the landings. Though a quota owner could sell his rockfish quota, he would have to sell his Gulf federal permit at the same time and exit the fishery. The small-scale cod jig fishery remained open access, as in state waters. The Council also added a ten-year "sunset" clause, meaning no one had a right to quota forever.

These were all important corrections to two of the most objectionable parts of any quota program--the exclusion of groups and excessive consolidation. It showed that majority of the Council was taking the MSA more seriously and had come to understand the social ramifications of IFQ/catch share. Several of the changes were similar to the safeguards that the Council had fought through for the halibut IFQ in the early 1990s but had failed to include in crab ratz. Part of the progress was due to the MSA 2006, but part was that the Alaskans on the Council seemed to be listening better to persistent groups like the GOAC3. More robust advocacy should never have taken twenty years, but it was welcome.

The Alaskans on the Council had also successfully pressured it to move the "catch history years" that were used for original quota allocations to include more recent years, 2004-2006, thus opening up the program to newer, younger fishermen. Concern for the youth had been building in the fishing communities ever since the introduction of the state's limited permitting in 1975 and now was recognized as a problem nationwide. The qualifying years change was a start, but much more innovative work would need to be done to turn the "graying of the fleets" around. NOAA's offering of "Young Fisherman" workshops to draw them back in was not addressing the real issue, the economics of moving forward from deck hand.

A problem still remained with the sunset clause and the argument that there would be no absolute guarantee of future value, so why invest or speculate on quota that might disappear? One can see how it might work for Rockfish, where an inflated market had not yet occurred. Yet for older programs where people had already invested hundreds of thousands in traded quota, what council would be likely to announce, "Your program is now ended and your dearly bought quota is now valueless." But that problem was over the horizon.

Conservation had definitely gained with Rockfish amendments. The Council cut the Rockfish fleet halibut bycatch quota by 12.5%. Many of the regulations would soon apply to all fleets: the required keeping and selling of certain secondary species; and discards of other flatfish, the largest bycatch problem by volume, would be much more controlled; there would be no pollock roe stripping in the Gulf; the off-the-floor trawling rule would be expanded. The observer program wouldn't be universal but it would be increased to more fleets and boat lengths. Yet it would also arouse the most widespread complaint: the cost to the boat owner of the on-board observer monitoring. By 2012 this would become the biggest issue for many fleets.

All the improvements to Rockfish didn't mean that small fishermen and their supporters now accepted privatized fisheries. By now all of the US fleets were keeping up with events in other fleets. John e-mailed me, "Both the Kodiak City Council and the Kodiak Island Borough passed resolutions outlining their concern regarding the New England IFQ [catch share] program, based on the problems in Alaska's losses in the coastal communities since they [IFQ/crab ratz] were implemented." The message showed the growing understanding of the nationwide power behind privatization and its effect on all fishermen. The MSA each year gave more direction, but unless the Council radically changed its approach, MSA protections such as "increased Alaska Native participation and consultation" and contribution of "enhanced local and traditional knowledge from communities" would continue to be vague lip service and enforcement at best frustrating.

Cleaning Up the Cod Fleets

Another North Pacific fishery was also on the table for changes during this time--the Pacific cod fleet. As the Council became more cautious about the snares in IFQ, once again it was the small Gulf communities, and their organization, GOAC3, that helped steer the final result. Like salmon and halibut, cod has a long history in Alaska. The fishery began on Kodiak Island in the 1860s with dory fleets brought north by schooners. The fleet evolved to include locally-based dory fleets working for salteries in the Gulf that shipped to Seattle. Some codders decided to stay north, married local Native women, and became part of Native communities and little saltery settlements that later became cannery towns. Although cod didn't come close to salmon as a subsistence resource, Alaskans have always respected it. America's commercial Pacific cod harvest peaked in 1920 when it had already grown to pot, longline, and jig fisheries. Foreign fleets then dominated for decades, and the US didn't develop a strong market again for Pacific cod until Atlantic cod was almost belly-up.

In the 1990s the cod-fishing boats were included when the Council issued federal groundfish limited permits. Codding was still mainly a small-boat fishery by numbers, with sixty percent of the boats under sixty feet, and half the longliners under 32 feet. But a small fleet of big pot and longliner-freezer ships brought the majority of the fish into Seattle. According to the Council about half of the smaller boats had no cod landings from 2000-2006. For them it was an opportunistic fishery, a security fallback when other stocks or prices were weak. Because both state and federal managers have been careful to keep the total harvest below the "ABC", cod stocks off Alaska had stayed healthy.

Cod has its place in my own family history. One has to be young, strong, and fearless in these winter fisheries, yet there are always more people on the docks trying to join a crew than there are spots to fill. Frank's first commercial fishing experience as an adult was on a cod boat out of Kodiak. Called by a friend because a crewman had just quit a longliner and a place was available, he felt very lucky. He gives a picture of the winter fishing fleets about 1986 and how it can still be today:

"Once we left port, it was fishing conditions and schedules I never dreamed of. I was just a 22-year old winter crabber on the ice outside Nome until I got the call to Kodiak. I had no idea what I was headed for. On the *Van Elliott* we worked at least 18 hour shifts with only short breaks for food. You just grabbed a few minutes rest anytime you could. We had to race to catch as much as we could before the closure. Those conditions were normal

I found out. The crew accepted it as it was always this way, through the whole season, on every boat, all they knew.

"Everyone got so worn out they were constantly getting hurt, and it's a wonder no horrible injury happened on our boat. Just as one example of how things went, on our way out of town on one of our first trips gray codding, a new skipper was at the wheel, we hadn't even started fishing and he was so tired already that he ran the boat aground. Fortunately we weren't leaking, and it was low tide, so he could have waited for the tide to raise the boat, but instead he kept frantically calling every boat around to come get us off the beach. They came and kept tying onto us and couldn't pull us off. They kept ripping big chunks of steel out of our boat. Then he tied to the mast and when I saw that I ran for the bow; I thought this could be disaster. But the line broke.

"Finally a big dragger showed up, but they didn't feel like trying to come in where we were; they wanted us to row over and get a six inch hawser they had. Turned out our only life raft had a hole, and there was no paddle or oar. Jesus, that was great to find out! We fashioned a paddle out of a piece of plywood and my friend Greg, the one that got me this challenging job, somehow made it over to the dragger and got the hawser, brought it back, but then it snapped too. It was f--king ridiculous! The next high tide, a little boat came over and pulled us right off. But that's how crazy people got with lack of sleep.

"It took us a week to weld the repairs to the *Van Elliott* and off we went again. We never went aground again, but the rest of it never got any better. We never had any rest, because when we were in town we were getting ready to go out again. And then we had to change gear for other kinds of fishing. By the end of the season I was the only one who hadn't been injured somehow. I don't know how I was that lucky, maybe worried enough to be watching every minute. Afterwards I thought of that kind of fishing as hell boats. But I made $10,000 crewing that winter on the *Van Elliott*, for me very big money. I remember from that I bought everything I needed for the next winter's crabbing." (He smiles.) "I knew saner ways to fish than longlining...I thought then. Now here I am, chasing halibut!"

Frank, fishing halibut off Nome, had no serious race for fish with only six to eight boats to compete with, and no winter fishery, just stormy fall. Yet plenty of things could go wrong with his operation or with another boat in the fleet needing help. Privatizing would not end that. Technology could help--most of the Nome fleet installed satellite radios and stiffer Coast Guard regulations kept coming. But the nature of fishing still meant hard work, fatigue, and risk.

In the Gulf in 2005 there were over 1000 federal limited permit (LLP) groundfish vessels. Among the cod fleet there was again that fear of an over-crowded fleet if they should all choose to fish cod. The fleet wanted to shrink the likelihood but rather than go for catch share, they asked the Council to simply reduce the fleet to the regularly active boats. They asked for a new LLP with a specific cod endorsement that would be for active boats, based on their history of poundage landed during set years. The Council went for it, issuing a new LLP with 2002-2008 to be the qualifying years used, thus eliminating a lot of "latent" vessels--those not actively pursuing cod. But the Council left an open-entry breathing hole for boats under 26 feet, and after the objections regarding the Rockfish program, the Council didn't assign specific processors.

The cod endorsement had the potential to be more socially fair than catch share, but the Gulf's GOAC3 group, saw the proposal as negative one for the small ports. By using past catch history volume it would again be to the disadvantage of smaller boats and younger fishermen. By now the CQE eligible communities had gained more political savvy. They complained to the Alaska Legislature that the cod endorsement rules would exclude too many of their fishermen. Although there were many legislators that objected to being involved in a federal council's business, the House did pass a resolution in support of the CQE position. The Council listened. It decided to allow two cod endorsements (or 10 megaton harvest) for smaller non-trawl groundfish boats home-ported in each of 21 Gulf CQE communities, and regardless of harvest history.

The new cod endorsements were awarded and resulted in major con-solidation. Only about a third of the 1000 LLP applicants qualified for endorsement due to fishing history. But the boats disqualified for latency were spread across all sizes of vessels and homeports, most of them also active in crab, halibut or sablefish fisheries. NOAA figures show that the largest fleet reductions from 1998 to 2010 came in catcher-processors, large pot boats, and trawlers, while gains in endorsements came in hook and line vessels of all sizes. Thus the program was more democratic than the catch share programs. The decision to leave the small-boat jig fishery open access also had the potential to help many fishermen from the smallest communities. To discourage speculation, cod endorsements had to stay with their original LLP, not sold alone, and would require "owner on board."

Though Council had committed itself to a more socially fair program, in a later round of amendments, exemptions were granted to freezer-longliners that had not qualified because of latency. Thus the Council table was still the scene of much negotiation and compromise--and the big-boat fleets could still win the votes.

In 2015 the Council went even farther trying to salvage the small boat fleets of Alaska. It put the finishing touches on a CDQ allocation of Pacific cod for the local small-boat fleets like those at Nome. Prior to this only Frank and one other Nome skipper, because they had purchased cod quota with their boats, had been able to sell more than 200 lbs. of "incidental" cod from their halibut longlines. Soon other small halibut boats would also be part of a cod program and able to sell within a quota.

More Catch Shares Organized

The Council's next major re-structuring moves were for the Gulf pollock fleet and groundfishing vessels in the Bering Sea/Aleutian region. Huge freezer-longliners out of Seattle that targeted cod were so eager to be rationalized that they didn't wait for the Council, and had chosen to organize a coop themselves in 2006, to be formally recognized later. A main incentive was the growing pressure to reduce bycatch. It was easier for cooped vessels to avoid going over quota and being shut down. The roughly 35 Gulf pollock trawlers also began the cooping process themselves. Since neither of these fleets involved great diversity in size of vessels, nor many one could consider small, nor were they based in small Alaskan ports, it was much less likely to be a fleet consolidation that would hurt the traditional small-boat fleets.

In an important change from the earlier privatizations, in the new catch share programs a boat owner also had a right to remain independent. That was not a likely choice, however, since the quota set-aside for independents could not be very large or it would reduce too much the quota available and risk the profitability of the fleets fishing privatized stocks. Catch share, even with all the protections the Council might invent, would only go so far in being fair to independents, or to smaller vessels with less poundage landed, or to those that hadn't fished in the selected years. Whether there were boat owners that didn't wish to go to catch share, but felt forced to, I don't know. In any case, they would have probably been a small minority in the groundfish and pollock fleets.

Chapter 31

CDQ FORECAST: CLEAR, PATCHES OF FOG

With all of the re-structuring going on, the CDQ experiment continued unique for American fisheries. No consolidation threatened the small boats, and the CDQ revenues went not to mega-corporations, but to regional non-profits. But the shape of CDQ did change when NSEDC and others decided to not only take royalties but to buy into the industrial fisheries in the Bering Sea. That meant the CDQs formed profit arms in 2008 and would be obliged to pay the federal taxes on profits that were not part of the original royalty deal. The CDQs also paid millions in back taxes owed for those "secondary" profits. The new profit arms had their own boards of directors with any profits to go back into the CDQ non-profit activities.

Meanwhile the CDQs also continued to grow. As an example of what CDQs were typically active in, NSEDC's 2012 Annual Report to the regional public--not including the profit arm, SIU--listed its total unrestricted revenues as $52,987,565. The total investment the non-profit made for the year was $28 million. Equity in earnings of LLCs (the industrial fleets it had bought into) was $11,765,548. It passed on $13,305,811 in regional community services. These figures are not complete enough for analysis, but to show that it (and the other CDQs) were now big players in the Alaskan economy. SIU meanwhile partnered with the profit arm of another CDQ, Coastal Villages Region Fund, to purchase seven trawl vessels, a crab vessel, and almost three percent of the Bering Sea pollock quota. One vessel with partial NSEDC ownership, the 376 ft. *Alaska Ocean*, is the Bering Sea's largest pollock catcher-freezer ship.

Closer to the regional fishermen's hearts was the measure of their own small-boat fleets' success. Nine Norton Sound communities now had local commercial fisheries--herring, salmon, crab, or halibut--with a buying plant not far away. Efforts to do more with commercial fisheries for the northern villages of Teller and Brevig had so far not evolved. Worrisome stock shortages on the coast to the south, such as in crab, benefitted Norton Sound fishermen's prices, though they were still below those of ports to the south. Two new fishery support buildings were added in its member communities. Commercial fishing had expanded. The total number of commercial licenses active in Norton Sound in 2012 was 325, mainly in salmon at 129, but halibut was important for Nome and Savoonga, roe herring for Lower Sound, and there were over 30 active crabbers in the Sound. In 2012, $2.4 million was paid out to crabbers, $550,000 to halibuters, and $900,000 to salmon gillnetters, who were suffering the slump in Chinook.

In years when the ocean ice cooperated, NSEDC also subsidized herring roe prices and made arrangements with Icicle Seafoods to send up a processor and tenders to the lower Sound. Tomcod finally had their day too. The plant decided to buy them at .50/lb as bait-fish to freeze and sell back to the crabbers for .10/lb. more. On fall days ten or more families were out in the harbor jigging them through the ice for sale or personal use. Five interns had been funded for fisheries related positions, and 19 region residents worked in the pollock industry. NSEDC also hired local people for salmon research and stock development.

In non-fishing development, for 2012 NSEDC awarded 266 scholarships and grants for $292,000, increased the size of its grants to the region's cities to $300,000 each, gave several large infrastructure upgrades to villages, and additional community grants for substance abuse prevention. In 2014 it raised college scholarships to $2500 per semester per fulltime student. Energy efficiency projects continued at $1 million in grants available per community, along with savings through bulk fuel purchase. An energy costs subsidy of $500 went to each regional household that year. NSEDC also promoted local entrepreneurs through competitive "small business" grants, and special "Outside Entity" grant awards were distributed to regional cities and non-profit organizations.

The list of community and fisheries development is impressive, and Norton Sound was not unique. For example, Bristol Bay CDQ recently purchased 50% in a large shore-based processor including seven plants. It provided loans for salmon limited permits for those who couldn't qualify for a state loan and subsidized some of the interest costs. It began to provide ice barges for its gillnet fleets. Like NSEDC it financed fish-counting towers on the rivers and research. All of the CDQs provided a significant number of scholarships and training opportunities.

The CDQs had clearly made significant strides in community development, but the regions still had high unemployment rates; some of the fisheries still had serious problems; some of the villages still didn't have local resource development to speak of; and a few communities still didn't even have functional water and waste systems. The strategy of buying into the profitable Bering Sea fleets continued to bother some community members, especially the pollock fleets, as they had traded traditional reliance on salmon for royalties and investments in pollock, while pollock bycatch, they were sure, was hurting their salmon runs. Subsistence salmon fishing, the traditional base of their culture, to them wasn't tradable.

However, at least three of the CDQs had invested in support services that salmon fisheries needed, such as increased numbers of tenders and processing plants. Coastal Villages Regional Fund, for example, had built a $40 million salmon and halibut plant at Goodnews Bay. Such development was most unlikely to happen without the new CDQ revenues. At least one CDQ had members who declared that change meant they had to go with the best interests of their CDQ, not their cultural and economic history.

In 2012 the CDQs submitted five-year self-evaluations to the State. The four areas reviewed were: socioeconomic conditions in the region, financial performance, workforce development, and community development--meaning fisheries development. All six CDQs gave themselves "satisfactory" on community development and regional workforce development and financial performance. Even in the worrisome area of bycatch only one CDQ reported going over its chum salmon quota, and one of not having an adequate system of bycatch reporting. None of the CDQs that invested in trawlers went over their Chinook bycatch quota except in one year, 2007, when everyone did. But a growing portion of Alaskans said the bycatch rate was still too high. Improvement to socio-economic conditions--the most broad and long-range goal--was in every region left unmeasured because "reliable government statistics were not available". One of the most popular ways for CDQs to invest in the communities would pay back only in the future: scholarships. One of the reasons for high unemployment, though not the only one, was lack of training among rural residents in needed occupations.

Using NSEDC again as an example, self-evaluation included other interesting facts about its own workforce. Between 2006 and 2010 it had nearly doubled the number of its employees and paid out directly or indirectly $33.5 million in wages, an increase of 81% over the previous five-year period. Payments to regional fishermen delivering to its plants climbed by 98%. NSEDC listed its training grants at a total of $2.55 million for that period. Also of interest, though not from the report, the fishermen reported to me that of the six Nome men who got the first NSEDC boat loans (Frank was the youngest), only three were still in the fleet, but at least six men under 40 had joined the fleet as skippers--two of them passed down from their fathers' original loans. Savoonga's halibut skiff fleet is much a young man's fishery. CDQ local fleets were perhaps our only ones on the North Pacific not aging.

The Critics Speak

Not everyone is completely pleased with the CDQs' progress. Their critics believe they could do more for their local fleets. Local dock prices can be an issue. But you don't get far in the corporate world without attracting worry, envy, and even hostility. The CDQs will have their critics on one side or another for most moves they make. Some residents focus on all the huge gaps in development western Alaska still has. Others like what their CDQ has accomplished but worry about the risk involved in all the loans for large Bering Sea vessels and quota shares, and wonder if the boards understand the level of risk, or if the stocks could get overfished. How knowledgeable and reliable are the staffs, they wonder? Will the CDQ program just keep on sailing with fair winds?

Still other critics of CDQ like the concept and all of the regional development but oppose trawling, especially bottom trawling, for conservation reasons, including habitat protection, and question if CDQs should own and be profiting from that kind of vessel. Shouldn't the CDQs be investing in fleets using alternative gear? Or should they, as owners, demand that "excluder" gear be used to avoid untargeted species. As I write, much progress is being made with habitat protection and excluder gear.

So far I have listed reasonable concerns people have mentioned about CDQs. One less reasonable is the complaint that NSEDC by its existence discourages other private business in the region. In fact, Nome did have local private fish brokers in past years, and probably will again, but they have never sponsored new and improved regional infrastructure or other community development such as NSEDC has contributed. NSEDC does not forbid other fishery business to come in, and in fact has accommodated some. The real issue is, can competitive private business afford to operate in west coast Alaska? Several of the CDQs subsidize their local fisheries to keep them alive and rely on their royalties and other investment to pay for this.

There are other concerns, such as remaining gaps in Norton Sound services that fishermen need and that they say NSEDC is now in condition to provide. The local fleet is hurt in that there is no marine diesel mechanic in Nome and no place to do boat work, conditions one doesn't see in processing ports to the south. Frank grumbles, "I've brought it up ever year for three years. All I get is 'Thank you for your comments.' " The fishermen say it took too long for them to get a simple improvement like a safe way to haul their boats out for winter--a basic service for any commercial boat harbor. While such questions create healthy debate at NSEDC board meetings and assure that the board and managers will not get complacent. We would have more to worry about if there were no debate.

Local Entrepreneurs Arise

In an effort to get a better price for their crab, in the winter of 2012 Rob and Frank decided to try shipping to Anchorage wholesalers as they had years before. The NSEDC plant's cooked crab sections were not getting the price available for live whole crab.

Frank commented, "Three years ago I tried several times to get the Norton Sound plant to do what we finally decided to do for winter crabbing, to put a holding pen out under the ice to keep the crab fresh so it could be sent live. But we needed a tent over it. The plant always said no, so we did it ourselves. We held the crab fresh until we had a good-sized shipment ready to airfreight out. We got brokers' licenses, found a few other crabbers to sell to us, and passed along the extra dollars we got in Anchorage."

Rob says, "I didn't do it to make a lot of money but to show the plant that it could get a better price with some changes. It's a lot of extra toil. In winter 2013 I paid the crabbers that sold to me -- my son Chris for one -- from $6.50 to 7/lb and I got $8. I grossed a lot, but I only netted $13,000 after all expenses. But I made my point."

For summer crabbing Rob installed new holding tanks and pumps in the old Chukchi warehouse to keep crab lively, and as of 2015 the brothers were still shipping out at least part of their crab. Meanwhile the plant had ordered its own up-to-date tanks, and was able to ship live fresh crab. The local dock price also was considerably higher--the influence of competition possibly, but the price was better all over too. Korea, especially was happy for all the crab it could buy. Rob smiled, "The market can absorb both of us," but the crabbers had to recognize the tribulations of free enterprise once more when one week the Anchorage holding tank owner temporarily couldn't take more crab. To NSEDC's credit, its Nome plant manager accepted them, and continued to when there was a glitch in the air shipments. The plant also continued to sell bait to all the crabbers. But the realities of global competition became graphic for the small crabbers when illegal under-reported unlicensed Russian crab made their way to the Korean market. The Nome crab sellers counting on that market saw the Korean brokers' price drop from $7/lb to $4.75/lb and they were back to selling to the Nome plant.

My point in this story is that such small, risky enterprises would never have gotten their start under a catch share system, and indeed they are driven out. Under CDQ there is opportunity, though it may come slowly, and it doesn't go away at a whim.

CDQ's Wider Critics

Certain Bering Sea corporate fleets grumble over the advantages the CDQs have, such as their ability to have taxed profit-making subsidiaries that can pump funds back into their non-profit non-taxed activities. In 2011 various marine services in Seattle and the Port itself were furious when one CDQ announced it would move the berthing of several big vessels it had purchased, along with their support operations, from Seattle to Seward. Expected to be ready by 2016, the new facility means a big loss to the Seattle waterfront, yet was a normal business decision. As one Alaskan laughed, the Seattle waterfront was built on Alaska as its colony, and was bitter to see any part of that end.

Other critics believe that CDQs no longer have enough scrutiny and are too much like private corporations, yet with their first tier royalties still tax-free. Their management teams answer to their boards and the same state laws and regular audits like any corporation registered in Alaska, but they no longer receive the same oversight from government that an untaxed non-profit normally gets. They are unique hybrids. The CDQs' accountability is to their boards, representing the residents of the region, their stockholders. But their profit arms like Siu have the regional CDQs as their lone stockholder, and what a profit arm does is like any private for-profit business and doesn't have so much public scrutiny.

With corporate staff compensations a notorious issue in the US recently, that is one logical area to study regarding the CDQs' success in holding to their public mission. There are large differences. As examples, the NSEDC 2011 salary of the CEO was $160,102 (NSEDC Annual Report).

Meanwhile, the salary of another CDQ's CEO was recently $69,503 while another (the CDQ closest in size to NSEDC) was $832,367. Other administrative salaries in CDQs also reflect these wide differences between regions. This huge stretch could reflect a difference in duties, or regional philosophy, or board involvement in decision-making. One might argue, why shouldn't these CEOs make what the typical American CEO makes in salary and bonuses? But for much of the American public the broader question is, why should any CEO make the incredible amounts common now, four times or more what their average staff makes? The CDQs, as they grow in size and value, will inevitably have the pressure to be drawn into the "corporate mentality", like any large organization. They do need to be vigilant to avoid the tendency. A true conflict in philosophy obviously exists between those who believe growth and maximum profits, reinvested, is the path for survival, and those who believe that regional development is more important in the long run, even if not a money-maker for a time. A balance of the two is hopefully where the CDQs are headed.

Though all of the CDQs have benefitted their regions, each has developed its unique character. Complaints about the CDQs that I have heard and read are a mix of ethical and due diligence concerns, differences in goals, and in some cases simple envy. None are likely to get far with the North Pacific Council or NMFS, which see the CDQs as largely successful, legitimate business ventures. The state reported that it was quite satisfied with NSEDC's 2010 self-evaluation, and had little complaint about the others. Like all the rest of federal fisheries, the CDQ scorecard has many "howevers". Regardless, the CDQs have helped assure that the West Coast communities are not disappearing and that subsistence and commercial fishing can exist side by side. The North Pacific Council, because of the CDQ program, has so far done more than any other regional fisheries council for its small fishermen.

How much can the CDQs do about the serious socio-economic problems that still plague western Alaska? No one program can possibly solve deep long-term problems like unemployment, crowded housing, and youthful suicide. And the problems don't lessen. The CDQs each have to decide what is the right balance between building their bottom line and long-range social and economic development. How the CDQs shape themselves to answer the needs of their regions will be up to the leadership of their CDQ boards. If the boards are weak, the management team will run things; if they are strong, there will be conflicts, but in the end the residents will have done important work for the survival of small fishermen and their communities. The theme of this book is concern for the survival of small-scale fishing, and so far the CDQs, though they may move through troubled waters, are the best chance we have.

Chapter 32

BYCATCH TANGLES

Fish run interception used to be the issue that could infuriate fishermen. (Remember the fish traps? Remember the Canadian blockades?) Today, bycatch is, at least in Alaska, the winner for creating the most frustration, anger, media coverage, and conflicting solutions. It is always guaranteed to bring ENGOs to the front lines. It keeps the heat on the regional councils, and the MSA guarantees it will be high priority for fishery managers until every fleet has a bycatch reduction plan in place for non-targeted stock.

Though reducing bycatch is an essential commitment to conservation, there are many reasons why NMFS and the councils have procrastinated so long. Money and power rank high. The pollock fleet's bycatch of salmon, highlighted here, is just a sampling of the problem. Pollock mid-water trawlers actually have one of the lowest bycatch rates: two percent. The problem is that they are hauling in such a volume--pollock and everything caught with it. Bottom draggers globally are the worst with rates up to 50%. George says he worked on a petrale sole dragger once where they threw away 60% as too small to sell. Until 2013 halibut and sablefish fisheries had no by-catch quotas, so it is hard to say what their rate was. The UN figures the average across fisheries is eight percent.

Thirty-nine years have gone by since the passage of the MSA and we still don't have a firm bycatch plan for Bering Sea chum salmon, a situation unacceptable to the Council as well as the salmon dependent families. As of 2015, genetic identification now tells us that over 50% of the Chinook bycatch by Bering Sea trawlers comes from western Alaska.

In the most recent of many meetings over several years, the majority on the Council still conceded to the pollock fleet's choice: keep the hard cap high at 60,000 Chinook bycatch and instead work harder at pollock fleet self-policing, use of excluder gear, and avoidance of areas of congregated salmon. How could such a vote for 60,000 take place? An Alaskan, despite his protest, was recused, for conflict of interest. We are to accept that he was the only one with a possible conflict. But this was just the most recent meeting. If one reads the fishermen's blogs one finds that at every regional council it is very difficult to overcome the power of the big-boat fleets.

Every fishery is now by law (the MSA) expected to take action to reduce its by-catch. A groundfish trawl fleet, in particular, has every snag to complicate a bycatch program: fleet diversity, many mixed stocks, incidental stocks that may be of low economic value but by law must be kept and sold, protected stocks that can't be sold and end up dumped. The salmon bycatch of the pollock fleets has been the one getting heavy attention, but now the halibut fleet off Alaska is getting much coverage for its protests over halibut bycatch by the bottom trawl fleets. In spring of 2015 the Alaska Fish and Game Commissioner wrote the Council (of which he is a voting member) requesting a 33% cut in the trawlers' bycatch, stating that their present halibut bycatch quota was more than the halibut fleets' own catch quota.

The strategies in use for shrinking bycatch are many. In some cases it makes sense to make it a secondary species that must be kept and sold, but this won't work for a stock already at risk. Or managers can allow a small incidental amount that can be sold; this can help but not enough. Avoidance of bycatch through moving away or by setting seasons and areas is often used, but has not kept the bycatch numbers low enough. Special gear such as size of net mesh, barbless hooks, shape of net--all these can work in some cases. Excluder gear refers to adding some technology to the gear that causes the untargeted species to avoid it. Last priority, as far as the industrial fleets is concerned, should be hard caps, as there is no wiggle room. The industry prefers technical changes like excluder gear, like changing the shape of a trawl net, or avoidance. For example, draggers off Alaska have pretty much accepted a ruling that they must raise their nets off the ocean bottom. A NMFS article in July, 2012 reported that raising nets only two-four inches had reduced habitat damage 90% while still maintaining normal catch rate of flatfish.

In recent years many versions of excluder gear to keep untargeted fish (or birds) away from the gear have been developed and tested, some through NOAA pass-through ENGO grants to vessels and companies. The excluders have to be designed so that the vessel won't lose too much of its targeted stock--and profits. However, the Bering Sea trawler association says excluders, like other bycatch procedures, should be voluntary. The Bering Sea pollock trawler spokesman claims that only a very few identified vessels are guilty of high salmon bycatch, and that the association will deal with them itself. One wonders, should then the reduced, or eliminated catch of the salmon fishermen on the rivers also be voluntary?

In 2015 the value of a hard cap was brought home when a fleet of Gulf trawlers went over their bycatch limit and were closed down. Without it they would have kept right on fishing and tossing. Even the seemingly simple and productive solution of mandating that secondary catch be sold can run into resistance because the secondary species might be one of low value and yet take up space and crew energy better spent, in the skipper's opinion, on the targeted species. It can't be used in every case for a catch could conceivably not be worth catching, and it can't be used for prohibited species like salmon, king crab, or halibut. Yet it can work well with other stocks like ling cod.

None of these methods, so far, has been successful in reducing salmon bycatch to the level that west coast salmon fishermen feel they are treated fairly, and is why fish-ins and other protests over Chinook bycatch will continue as long as salmon fishermen can't catch one Chinook while the allowed bycatch remains at 60,000--even if the actual bycatch is only 5,000 in a certain year. What about the next year? Salmon move around. But there will always be someone on the Council who will say: "Who has proved that the bycatch is to blame for the disappearing Chinook?" And that's part of the reason the arguments have gone on for an incredible 39 years. Yet the North Pacific Council has the reputation of being the most conservation-minded of the eight councils. To hold on to this reputation, it must overcome the salmon bycatch problem in a way that stands the test of fairness for the small as well as big operators.

In 2013 Alaska Fish and Game tried its own technical strategy and outlawed chum salmon gillnetting on the Yukon. It ordered subsistence fishermen to use dipnets only, which would allow their Chinook bycatch to be released to go on upriver. According to the agency the scheme worked and would be used again on the Kuskokwim. That year some fishermen found it a workable solution; others on the Kuskokwim said it was a failure at their location, and they wanted to work closer with Fish and Game to improve the method, or some other one. In spring of 2015 the Federal Fish and Wildlife was forced to rule that there would be no commercial fishing on the Kuskokwim again to protect the projected low run of Chinook. However, very limited amounts of Chinook subsistence catch would be available to selected villages to share as community catch, monitored by those villages.

The technical problems of reducing bycatch don't explain the many years of delay. The Council has had many brilliant people to work out the technical barriers. The delay is due to costs and political power more than technical puzzles. The reason NMFS took so long to rule in favor of the sale of incidental (secondary) species was clearly because of industry resistance and its power on the Council. The new rule turned out well in cases where a secondary species was a valuable one, like sablefish. But related to stronger restrictions is always the problem of adequate enforcement or there is little gain. Enforcement using on-board observers is always going to be very expensive and impossible for the smallest boats. A more difficult problem is required releasing of illegal or undersized fish in good condition. This can be done, always with some loss, by care in handling the fish, but it will slow down the operation. It won't work with groundfish with air bladders, as few of them can survive being tossed back. The "crucifiers" still used on some halibut boats tear the mouths of undersized halibut destined to be dropped back in the ocean. Why isn't snap-on gear, which can prevent this, the only legal gear?

A popular solution for controlling bycatch, usually tied to privatizing or cooping programs, is to allow a fleet to coop its vessels' bycatch quotas. This is one of the biggest selling points for catch share sectors--that sectors make it easier to manage bycatch quota. But bycatch quota trading is like carbon trading; it doesn't in itself actually reduce bycatch; it helps control it within the assigned fleet quota.

Even with the stronger guidance from the MSA 2006, the composition of the regional councils guarantees that few bycatch plans of any kind will be passed unless the industrial fleets agree with them--a clear example of how politics can dominate the councils. The west coast salmon fishing families and tribes' request to drop a proposed chum hard cap from 60,000 to 30,000 didn't get far. The trawlers argued that the amount of revenue that would be lost to them at that level would put the Council at odds with the MSA's rule that "optimum yield fisheries must be maintained within conservation guidelines". To them a cap of 30,000 wouldn't allow "optimum" catch. But what level is optimum? The MSA is frequently vague enough to encourage this sort of debate.

The huge related problem is in monitoring and enforcement. It took fifteen years from the start of the EEZ for the US to have 100% monitoring in place for just its largest vessels, those over 125 feet, with only about 30% coverage for boats 60 ft. to 125 ft. The partial coverage was so easy to manipulate that skippers made a joke of it. Today, stalling no longer works so well. Following the guidance of the MSA 2006, if the council process doesn't produce what NOAA needs and wants, NOAA can mandate. But though the problem can be kicked upstairs, it is still unsolved when NOAA can't get the congressional appropriation it needs for such an expensive program as on-board monitoring. It's possible that the trawl industry's lobby has something to do with that, but we now face a composition in Congress that has budget cutting a top priority.

Nonetheless NMFS did institute increased monitoring that began in January 2013. The new rules leveled the fishing grounds, though not as much as originally planned. An important improvement was that instead of skippers deciding, for their own advantage, when and where a monitor would observe the gear hauls, a government-supervised contractor would choose the schedule. Now halibut and sablefish boats for the first time must also have observers. But due to a shortage of funds, NOAA decided that all boats 40 to 57 1/2 feet would carry an observer only 1/3 of the time, randomly selected, while boats under 40 ft. would not carry observers at all in 2013 and might not for some time. Jig boats also were not included. To cover the program, all boats paid an added 1.5% dockside landing fee. The fees charged that year would pay for most of the following year's costs and so on, year by year. The percent charged would gradually rise.

Many boat owners protested the plan, saying they couldn't afford the fees for expensive on-board observer/monitors and requested cheaper electronic monitoring (EM) that is accepted in BC for all but the largest trawlers. Smaller boats also argued that they couldn't physically manage having a non-crew person staying on board. NOAA rejected on-board electronic monitoring until it could met the needs of research, but the fleets argued that they had to have priority consideration. It becomes ever clearer why NOAA favors consolidated fleets of larger vessels that can more easily pick up increased administrative costs. The good news was that both the west coast councils moved ahead with testing of EM and said that they would have it approved and put to work by at least 2017.

That will be very good news to small boats. The cumbersomeness of the on-board observer program hit home when Frank got a notice that his *Mithril* would have to carry an observer part of the time when he fished halibut in fall 2013. The fleet had worse than usual weather that fall, and he took advantage of a brief calm spell to rush out and get his fish aboard rather than to call and then wait for an observer to fly into Nome and climb aboard. Sure enough, NMFS hit him with a fine. Then, in 2014, somehow he was "randomly" picked again for an observer. One group of Alaskans has sued NOAA over the observer issue.

Another bycatch success story is not about reduction but about use of the prohibited fish like salmon that can't be sold by the trawl vessels but must be brought to the dock for research. Once the researchers are through with the salmon, a company called "Seashare" is authorized to process and distribute it to food banks and humanitarian aid organizations. This year the company was authorized to expand its distributions. Fishermen, big and small, could feel better when they at least knew their unsellable fish was going to feed people in need.

In 2015 the battle in the Council moved from salmon to halibut, and was at least as fierce. The bottom trawlers for flatfish, fishing off Alaska but based in Seattle, were taking and dumping too many halibut, said the halibut fleet, already suffering poor harvests for years. In the last decade, they claimed, it was "82 million pounds of dead and dying halibut". How could this be right? But the trawlers again prevailed. The Council's cut in bycatch rate that passed was 21%, far too small all of the Alaskans but one agreed. They voted against that and lost. As on earlier votes our representative from Norton Sound was recused for conflict of interest, this time as NSEDC has an interest in halibut boats.

Rather than leaving such battles to politically stalemated councils, NOAA itself could have mandated stronger bycatch controls. It could have ordered this for the federal fleets starting in the 1990s when the North Pacific Council began privatizing, a reasonable trade-off for the windfalls those originals received. Surely the commercial fish stocks would be in better condition today if the great time and resources spent on IFQ/catch share schemes had instead gone into more buy-backs, stricter bycatch limits, and new stock assessments, management by ecosystems, and the expanded monitoring and enforcement we finally see. Still, the Council's 2013 monitoring program is a gain for conservation, covering at least 1200 vessels previously unmonitored.

Long term, all fishermen have everything to gain from improved bycatch programs. But meanwhile other changes were also in progress in the council meetings. The catch share spotlight had swung its focus beyond Alaska's federal waters to the south and east. All of the maneuvering over bycatch was one act in the bigger drama.

Chapter 33

New England's Bitter Stew

Ew England is a useful region to compare with all the management experiments on the North Pacific and how they have affected the small fishermen. I didn't personally know a groundfisherman to interview, but there are several books that look closely at the terrible problems that arose in the New England fisheries, many fishermen's blogs, and a new growing collection of community-based pro small-fishing websites in the region and just to the north in Canada.

The region's groundfish fleets were, since industrialization hit, especially diverse. Like Alaska, still holding on are numbers of small trawlers operated by extended families out of small ports, many mid-sized, and a growing number of huge trawlers that are only ten percent of the fleet but catch 90% of the fish. The same diversity, and the conflicts between small and large, is true of the herring fleet. In the 1990s the east coast saw the consolidation of clam dredges from many down to a few owned by less than a handful of companies, so when New England groundfishermen heard that a form of privatization was proposed for them for 2010, they already knew what it could mean.

By the 1990s the New England Council and NMFS had already tried any number of strategies to create sustainable groundfish fisheries: seasonal fleet quotas, area and time openers and closures, buy-backs, gear restrictions, limited permitting (1995), and most recently trip limits and days at sea were all tried with little success at rebuilding or maintaining sustainable stocks. Buy-backs and other forms of retiring had cut the fleet to half what it had been in the 1970s, but the stocks, with a couple exceptions, were not rebuilding and harvests kept going down. Cod, the icon fish for New England, in particular continued to be a concern, though it seemed to be slowly rebuilding from the crash in the 1990s. NOAA wanted the Council to prepare a plan for a form of privatized system that seemed to be working off Alaska for several fleets. It would be similar to IFQ, but organized into sectors, similar to the pollock fleet's AFA.

Two regional councils embarked on catch share sector plans about the same time: New England and the Pacific Council that covered waters of Washington, Oregon, and California. They both had groundfish stocks and trawl fleets in trouble, but their histories were quite different, New England's dating back to the 1500s. (The outcome for the Pacific Council's fleet follows this chapter.)

When in 2009 the new NOAA Administrator Dr. Jane Lubchenco, a PhD in Marine Ecology, greeted our commercial fleets, she found many of them in high stress over conservation issues and ever-heavier restrictions, with New England the worst. Fisheries management was just one of Lubchenco's many responsibilities, but one she had special interest in because of the broader environmental issues in the oceans. Coming from the Northwest, she found herself in much wilder political waters in New England. Support for NOAA's new director from the fishing fleets was mixed in that she had recently held a position on the board of the Environmental Defense Council, the organization that many east coast fishermen believed dipped all too much into industry affairs. This added to a fairly popular suspicion among commercial fishermen that anything the federal government proposed for their management would work against them.

Dr. Lubchenco was launching the biggest US catch share project yet, and in New England, home to some of the feistiest fleets in the country. How she would juggle all this would be a drama to watch as she said, "The scientific evidence is compelling the catch share can also help restore the health of eco-systems and get fisheries on a path to profitability and sustainability." Possibly one reason she was willing to give catch share a try in such a polarized environment was that coming from the Northwest she had already seen so much failure in traditionally managed state fisheries. She was free of vested interest in keeping such failures going. But of course, no system was going to work if science wasn't listened to, if fishermen's knowledge wasn't given a place at the table, or if short-range goals or agency protectionism superseded long-range health of both fish and fishing communities. These issues would exist in any system, but New England certainly had its share.

The MSA 2006 re-enactment, meanwhile, had included a victory for eco-scientists in the long battle for ecosystem-based management (EBM). All fleets were now to begin moving to this form of management. EBM makes sense scientifically but is a more complex and expensive way to manage, requiring a shift from management by an individual stock's "maximum sustained yield" to management dealing with the interactions of all the elements in that stock's ocean ecosystem from coral and plankton on up the food chain. New England was where the most overfished stocks had not yet recovered, though many had. Lubchenco believed that moving trawl fleets to privatized catch share sectors had the best chance for meeting the MSA 2016 deadlines on overfishing and getting EBM underway. But the startup administrative costs for catch share had to come from NOAA, which didn't have all that was needed. Added funds for fisheries were always hard to squeeze from Congress, now under even more stress from an economic downturn.

By now, no New England small trawler who kept up with industry news should have missed the warning lights flashing in catch share. Though organizing small boats into coop-type sectors might help more of them survive the conversion to a catch share fishery, many probably would not survive the change. In Maine, especially, the fleets were already greatly shrunk. In 1994 the New England groundfish fleet had been 3033; of that Maine's was 19%. By 2007 the N.E. groundfish fleet was a little more than a fifth of that at 574, with Maine's portion at 7%. Ever larger sections of the coast were completely closed to commercial fishing. The Council had come to the end of its rope trying to pull its voting factions together. The issues included the old ones: overfishing, stock assessments' accuracy, ever-increasing harvest restrictions, management of mixed stocks, and what fishermen saw as harassment by NMFS. Now global warming was probably affecting stocks like cod. So the situation was like that off Alaska in the 1990s but even more polarized.

Of the twenty-seven stocks historically trawled, NMFS listed about half as still overfished. But with assessments years behind their quotas were often set with a large "uncertainty factor" meaning harvest limits were lowered. To the fishermen, overfishing, always used as an argument for reducing harvest quotas, was now being used as a justification for completely restructuring fleets. Many fishermen, however, argued that the stocks were in better shape than the scientists said. They complained that they were not being actively involved in the assessment work, but involving them undoubtedly would mean more argument, and more time and expense for NMFS and the Council.

Lubchenco set summer season opening of 2010 as the startup date for the New England Council's region-wide catch share sectors for groundfish trawlers to go into effect. For startup costs she shifted funds, some of them from research, and went shopping for more funds from the ENGOS where she had influence. Thus, two major efforts would be going on at the same time--rebuilding stocks using EBM and start up of catch share sectors--neither of them projects simple or cheap, and facing fleets full of anxiety or outright hostility.

The government's "overfished/ overcapitalized" label was not just for the New England fisheries. Similar problems had been building over a decade in the mid and south Atlantic regions where traditional management had imposed restrictions that had reduced the number of active vessels and affected coastal towns utterly dependent on their small-boat shore fleets. Now catch share planning was in process there too, for grouper. As the MSA deadline for overfishing crept closer, the battles in New England would eclipse those to its south or in Alaska over crab rationalization. There was no proof as yet that catch share was the cure for overfished stocks. What was for sure is that the previously privatized fleets had been much reduced and their profile much changed.

The New England fleets were not only much larger and more diverse, they were more accustomed to venting publically. No one on the east coast liked the management they had, such as "days at sea" but could they be sure catch share would be better? Catch share still implied consolidation, and consolidation meant winners and losers. Who was behind the movement? However, the North Pacific Council had, by now, happy privatized halibut and sablefish fleets, and the Pacific Council's plans for trawler catch share, expected to be approved for a 2011 start, were coasting along with little turmoil.

Though New England fishermen often saw big ENGOs as working against them, Lubchenco had influence with several ENGOs and could expect some financial help with pilot projects. The Pew Foundation (Pew) already sponsored two fixed-gear catch share sectors on the Georges Bank that starting in 2004 had used hook-and-line and gillnets to demonstrate alternatives to trawling, and were functioning well as role models. It cost fishermen between $5,000-$10,000 per vessel to join, not counting the new gear and vessel refurbishing. But it was possible. Still, the hook-and-line coop had only 24 members and the gillnet coop was similar. The Environmental Defense Council (EDC) was another group that Lubchenco counted on. But even some scientists viewed EDC and others with skepticism because of their commitment to a market-based means of managing conservation of resources rather one based entirely on scientific principles.

On-line eastern newspapers like the *Gloucester Times* and eastern fishermen's blogs were so full of the issues that it was no problem to follow events in the groundfish trawl fleet, still four times the size of the Pacific fleet. There were plenty of reasons for discontent with traditional management. By 2010, the year set for conversion to catch share, there was an average of 35 "days at sea" per trawler, hardly enough to pay boat expenses unless one was especially lucky. In some areas the overloaded near-shore fisheries had closed entirely, meaning small day boats were then forced to fish offshore on a coast worse than the Pacific for vessel losses. Many years the trip limits on "choke stocks" meant that the quotas on important healthier stocks were not caught. The problems may have been largely a product of weak management but demographics also played a big role on a coast much more populated with dependent fishing towns than Alaska. The scene was more like that in eastern Canada, where the major cod collapse of the early 1990s had sent an entire region into economic crisis. Both Canada's DFO and the New England Council had instituted limited permitting too late.

Due to growing restrictions on groundfish many fishermen in the Gulf of Maine were shifting to snow crab and lobster. Farther south they shifted to dogfish and skates, once considered scrap fish, and now even those were being restricted. But not every fisherman was financially able to switch to another fishery, nor was there room for all of them. As smaller trawlers took a buy-back or simply gave up, they were replaced by larger fishing ventures better able to take the financial risks as well as the physical risks of offshore fishing as inshore grounds were closed. All this had created a situation so economically distressing that most fishermen were willing to at least look at catch share. How could it be any worse?

In the opinion of many fishery scientists, not just Lubchenco, the New England trawl fleet, as in other regions, needed to be retooled for a more sustainable kind of fishery. But that required funds that few trawler owners could undertake, or wished to. Trawling, under good conditions, was where the money was, and a huge political battle was guaranteed if it were threatened with complete closure. Those willing or forced to change found their own way to do it. I can understand resistance to switching to an entirely different gear, unless as desperate last resort to stay fishing. When Jack and I bought the troller *Nohusit* there was a halibut reel that went with the sale, but though it was an open fishery at the time we never did try it; we didn't know a thing about it and didn't have the confidence. Those Maine trawlers that did switch to lobster pots were that desperate. The Pew sponsored pilot projects had been a good idea, but the second stage, expanding the projects significantly if the first stage was successful, would require government support. Instead, the Council went the direction strongly encouraged by NOAA, a program for the entire groundfish trawl fleet that would still be trawling. Thus a good idea from Pew couldn't rise to its potential.

The NOAA boss knew it was to her advantage to move rapidly on catch share sectors before more resistance could build, and that tide was coming in fast. To expedite the move, NOAA announced a new arrangement that changed fishery politics hugely. From now on, in considering a move to IFQ or catch share, a Council would not have to require a fleet referendum to accept or reject the program. A boat owner could either join the new privatized fleet, or could stay independent. Independents would get their own fleet quota and all the customary restrictions. Eliminating the referendum was a smart move in a fleet so diverse and so fraught with old and new dissension. Many New Englanders protested, arguing that the MSA required a fleet referendum's vote for yes for a new program to go through. But NOAA stayed by the new arrangement, and a court case to stop it failed.

Before the Council could even finish its catch share plan, old issues had re-ignited. Several east coast fishermen's associations and seafood marketers entered lawsuits accusing NMFS of an inaccurate and harsh, to the point of vindictive, system of enforcement. Excessive fines were applied for "over the line", "fishing during closure", and other violations that skippers argued were hard to avoid with the rapidly changing rules. Fines collected were being misused. NMFS admitted that its data collection system was faulty, and that it would take some time to correct it. The Dept. of Commerce Inspector General investigated and recommended that the agency update its stock assessments, improve communication efforts, and in effect (my paraphrase) act more like a public servant, less like a police force.

The trawl fleet split in its view of the proposed catch share sectors. Once again quota shares would be allocated through history of catch and a free market in shares would operate, so the usual risks of excessive fleet consolidation and rapidly inflating share prices were assured. The boats would need to pay to join cooperative sectors (similar to the Bering Sea AFA) that would do much of their own management. The public joined in the arguments, much more than in Alaska, complete with candlelight vigils in the big fish ports like New Bedford and Gloucester. Widely read *Gloucester Times* fisheries reporter, Richard Gaines, took the fishermen's side and roused public sympathy for them. A few skippers did have fines cancelled. Lubchenco announced that the New England Council had better have its catch share plan completed for the 2010 season, or fleet restrictions would have to be increased.

The New England debates drew in fishermen from across the country. A group of Kodiak fishermen, funded by a sympathetic ENGO, flew to New England to share the negative effects that privatization had brought to their fleets and communities. But fervent promoters such as the EDC insisted there could be great advantages for conservation. The debate grew hotter. An independent investigation, the "Lenfest Ocean Program Study" about the same time evaluated fifteen fisheries in US and Canada and came to the conclusion that privatized programs like IFQ could stabilize fisheries to avoid erratic swings in harvest, but generally did not lead to more fish to catch. Ecotrust Canada in 2009 released "A Cautionary Tale About ITQ Fisheries", which looked at the results of privatized systems so far, and didn't reject ITQ entirely, but pointed out its many common pitfalls, and offered specific recommendations for corrections to the west coast programs. Ecotrust (US) and other critics believed that the earlier absence of national standards for privatized programs had caused much of the trouble.

Other critics stated that catch share programs needed to be crafted more carefully to meet the needs of specific fish stocks and specific fleets. But in fact, both the halibut and crab plans from the North Pacific Council had been carefully constructed to meet specific stock and fleet needs as the Council majority saw them. The problem was that the "fleet needs", as perceived by the powerful factions in a fleet, were not seen the same by the less powerful.

New England Catch Shares Launch

Keeping afloat despite the uproar, the New England Council produced a groundfish catch share plan on time in 2010. Like the Gulf of Alaska Rockfish program, it minimally followed the social guidelines from the MSA in Standard 8 in providing set-asides to protect small fishing and small communities. Like Rockfish it allowed fishermen the choice to stay independent. But for many the elimination of a fleet referendum was dead wrong. A fractious crowd greeted the Council's presentation of the plan. Fishermen raised practical questions such as cost to join a sector.

They worried that the quota share allocations could be too small for any but the biggest producers. Some declared that 50% or more of the small boats wouldn't survive. Since that did occur with the Bering Sea crab fleet, it wasn't a wild prediction. Nonetheless the catch share plan passed the Council with a single dissenting vote, from a fisherman, and the Secretary of Commerce approved it as expected. The anti-catch share faction protested, stating that if the usual fleet referendum had been carried out that catch share would have failed.

Private business saw a useful niche for itself. The Northeast Seafood Coalition brokerage bought up 34 required limited permits it could lease out to prospective fishermen lacking them. The Council then allocated the Coalition a pool of leasable quota for those permits. The Coalition also helped sponsor start-up costs for most of the new sectors; very likely they wouldn't have started without that support. About 75% of the fleet found a way to join a sector. The fleet quota remaining for independents wasn't encouraging: a little under 1/4 of the yellowtail sole, but only 5% of any other stock. The independents would continue with "days at sea", but with the days halved.

Belatedly, NMFS expressed concern that excessive fleet consolidation could be a risk under the Council's plan and asked it to work on ways to prevent this. Yet NMFS was where the push for the plan began and it could have raised this concern early on; how to make sense the delay? Work on Amendment 18 began, and the complexities of how to make catch share, or any "dedicated access program" more socially fair after the fact were much exposed. As I write, Amendment 18, five years later, is still a work in progress.

The trawl fleets went fishing on schedule. From a fleet of 600, 432 had joined one of the 17 sectors. All American fleets should have been watching, as the program involved such a large area and such a large, diverse fleet. How many fishermen would find the program worked for them? In smaller ports many trawlers stayed tied to the docks, their allotted shares not realistic to fish. Some who had said they wouldn't fish in the end did so. Either they decided they couldn't afford not to, or couldn't bear sitting on the beach.

Some had leased more quota shares so they could fish; others had leased out their shares and switched to lobsters and pots--one fishery that was doing well for the time.

Fleet consolidation had already been going on for years. Now it was a clear winner. Figures from a NOAA review showed the number of active trawlers, already much reduced through buy-backs and discouragement, dropped from 600 at the start of catch share to 450 two years later. Sixty-five percent of the revenue was taken in by 20% of the fleet. The number of boat owners owning three or more vessels increased by almost a third.

When smaller trawlers sent up a chorus of complaints, Lubchenco laid the responsibility for program design on the Council, stating, "Many fishermen are concerned with the problem of consolidation where a few people buy up all the shares. You can structure the rules of fisheries to limit that happening." Apparently if regional councils' plans had problems, it was their own fault, as they were free to choose other models. But it had always been difficult for small fishermen to have influence on regional councils. How could they be heard now, when the rush to install the sectors had created a scramble for survival across the whole fleet? No one was in the mood for thoughtful collaboration. Many New Englanders, not just fishermen, continued to cry that catch share would never have passed if the usual MSA rule for a fleet referendum had been required. That can't be known. Such program referendums always had passed, though New Englanders could have been different.

NOAA urgently needed additional help for the sectors' startup costs, so Lubchenco turned again to private foundations and was able to draw in millions more from groups like Pew, Moore, Walton (Wal-Mart), and World Fish and Wildlife Foundation. She would continue to use some of these funds for alternative fisheries experiments, but they could only be small pilot projects. The anti-catch share factions continued through street demonstrations and protests in meetings, and through their politicians as reported in the *Gloucester Times* and other regional media. They created the term "catch share-cropping", a repeat of Iceland's popular labeling, "feudalism".

More Trouble for Catch Share

In the spring of 2011 the South Atlantic Council, responding to rising complaints over the red snapper, grouper, and tilefish catch shares it had initiated, voted to stop any further catch share programs in its region. This was a first for regional councils--to stop a NOAA promoted program. This council's reasons reflected complaints from other regions but with important added protests from the numerous charter fleets that feared catch share programs would take quota from them to allocate to the commercials. Required by the MSA 2006 to be included in council planning, the charter fleets used NOAA's own figures to argue their importance to the national economy--$50 billion annually. Alliances shifted: that formidable opponent of commercial fleets that targeted fish that sportsmen liked, the Coast Conservation Association, now joined the anti-catch share fight.

The New England cities of Gloucester and New Bedford filed a lawsuit against NMFS, later joined by amicus briefs from Massachusetts Congressmen Barney Frank and John Tierney, and Food and Water Watch, a consumer protection group. Charges of unfairness in the quota allocations, and the lack of a referendum were major points. Barney Frank stated that it was a divergence from the original intent of the MSA, and so was the transfer of funds to catch share from research needed for accurate stock allocations. The trawlers claimed there was no authenticity in the underfunded, incomplete stock counts and unscientific to have a single deadline for all stock restoration. Frank insisted that the MSA was not a bad law; it was how it was being interpreted and manipulated that angered so many. The anti-catch share forces lost this lawsuit and appealed and lost the appeal. More lawsuits against NMFS surfaced, some regarding catch share, some regarding other management issues.

The catch share opposition expanded to include more mayors, governors, and congressmen. The NMFS general response to these complaints was to plead, "listen to the science". But the science was too politicized to be defensible in some cases. "Who paid for the study?" was a common question. Or people might agree the science was right on, but the solutions weren't available or affordable. Often fishermen were too upset by management chaos or too cynical to listen to the science. The scientists themselves were split on the issue of catch share. Politicians were roused by their constituents to insist the local economies had to be taken into account.

Fisheries scientist Dr. Brian Rothschild demanded an investigation by the Secretary of Commerce "…as we translate dry statistics into lost livelihoods and collapses of small businesses." On the west coast, Dr. Daniel Pauli, at University of BC, wrote of BC catch shares, "…The transferability of annual quota [makes it] a commodity like anything else--like pork bellies…In BC where I live, 80% of the fishermen are employees of firms that own shares [that are] out of the fishing sector and now in the hands of the banking sector."

In March 2013, NOAA published the findings of a study that collected 40 oral histories from New England trawlers by Pat Clay and others: "Understanding the Impact on Fishermen and Their Families from Southern New England Catch Share". They interviewed 36 trawlers and four former trawlers. Twenty-six said they were doing worse, five were doing better, seven were neutral. Their main complaint was that the program was good for a very few people. It must have been the five.

The eco-scientists meanwhile, were becoming a stronger voice in New England fishery management and were frustrated with the continuous debate over catch share, as it diverted councils and NMFS away from whole ecosystem-based management required by the MSA and toward market solutions for fleets and stated that any radical management change needed to take place after such studies, not before. They pointed out cases where IFQ/catch share was in effect, but stock decline had not stopped even after many years. Thirty-two stocks had been rebuilt in New England since 2000, but was it realistic to think in 2012 that had been much influenced by the 2010 catch share start?

Being slammed from so many directions, NMFS ordered an independent review of the Council's work by biologist Dr. Preston Pate. He found that the region's fisheries management was "void of leadership, lacking direction, guilty of poor collaboration, poor data management," and that "the Council overstepped its authority… that its deliberations were unfathomable." The Council responded that it would take the lead on improving communication and collaboration with stakeholders. This would include re-designing meetings, creating a strategic plan, and designing a cost-effective management system--all things one would expect from an organization managing regional fisheries.

New England's dramas continued to get attention from big ENGOs--the EDC, in particular. Years before, EDC had contributed much to the catch share design and would fight to keep it afloat, even though bottom trawling and dredging were the gears that conservationists hated most for the way they tore up the ocean floor and left it a desert. Others, like the Ocean Foundation, were becoming much more skeptical. The New England chaos was unbelievable unless you accepted my cousin George's popular cliché to describe all fisheries management, "Just follow the money". Eric Osborne, Nome fisherman, agreed that was the case, and shared his more detailed analysis:

"These national environmental organizations like EDC have grown huge and are dependent on their big sponsors like Wal-Mart and oil companies for support. They have to attract wealthy foundations that are supported by large industries. Like Pew--you know it's dependent on big oil. They all have to please Wall Street to keep themselves afloat, so they leave the big fishing companies alone and go after the small fishermen. And the smaller environmental groups have to go along with the big outfits' agenda if they want grants from them."

Many of the pro-nature public, however, took big ENGOS as their sources of gospel for saving the fisheries, and believed that the programs they funded did use the "best science available". Scientists complained that the ENGO members flooded the Internet with polls demanding this or that change to fishery regulations instead of demanding funding for better research.

I asked a local fish biologist, a former New Englander, why the ENGOs didn't go after industrial trawling full throttle. He was easier on the ENGOs than Eric. "Those court fights are very expensive to engage in. It makes sense to them to use their resources to choose smaller fights they can afford and have a chance of winning." The New England catch share battle was bound to be a big one.

Another scientist, Nils Stolpe, writing for the website "American Institute of Fishery Research Biologists" in March 2014 presented NOAA statistics showing that the predicted increase in New England harvest value through catch share did not take place, dropping from 82 million in 2009 to 70 million in 2012, even though prices had gone up. By the next year it cost more to lease quota shares per lb. than in many cases what one could get per lb. for fish--similar to what had emerged in BC. No wonder so many skippers had turned to lobster. Belatedly again, the MSA mandated duty to protect traditional fisheries caused both state and federal governments to work with several ENGOs on making permit banks and quota leases available for "boats under 45 feet and communities under 35,000".

Another Cod Collapse

Just when it seemed that the tensions in the New England offices of NMFS and the Council couldn't stand any more strain, in 2012 came shocking news of another cod collapse. While NMFS, the Council, and the fishermen had battled over the accuracy of stock assessments and catch share, the cod had practically disappeared again and no one noticed. In 2008 scientists had reported the Georges Bank cod stocks to be rebuilding well, and a third of the fishery was reopened. Now it turned out that the information was wrong. The consensus of government scientists now was that cod was not recovering and was still seriously overfished. Earlier Canada's industrial lobbyists had been able keep the cod processors running until almost the last of the cod went through a filleting knife. We may have done a little better as the Council convened an emergency session. The message was clear that stock assessment was not getting the attention it needed, or the managers would have known cod was still in trouble.

The Council saw it as economically and socially impossible to suddenly stop all mixed-stock trawling or even make immediate major quota cuts. It decided on a one-year cut of about 20%, and then, for 2013, an incredible

61% cut for Georges Bank cod and 75% for Gulf of Maine. It wasn't enough; cod grow slowly. In January 2015, NMFS decided to close the Gulf of Maine cod entirely for six months. The economists' admired free market in quota shares responded. Value of groundfish catch shares on the market plummeted and the once hot commodity was soon hard to sell except to speculators who would take the gamble that the cod could recover and that they would then make a killing. Places like Gloucester and New Bedford, the oldest fishing ports in the country, were outraged. New England's coast economy, especially the small ports of the Gulf of Maine, was soon suffering in the same way that Alaska's had but multiplied many times for numbers of small fishermen and communities affected.

Dr. Lubchenco's public statements showed her mastery of turning bad news into good. The bad news that many boats didn't fish because they hadn't received enough quota to make it profitable she turned into good news, stating that those boats had received quota they could lease out while they waited for the stocks to recover so that they could fish again. And how long should they wait? The Massachusetts Governor requested federal disaster aid for the groundfish fleets in 2012 due to the number of fishermen on the beach or with low catch. It was the third aid request since the 1990s. The Commerce Secretary approved it; Congress needed to fund it.

Now anti-catch share activists pulled in more powerful fighters. The House of Representatives, led by east coast Congressmen, passed a surprising resolution that no new catch share programs would be funded for the entire east and south coasts, and confirmed it later by formal vote. Further fleet privatizing on that coast was dead in the water at least for a time.

One fisherman on the New England Council commented that the cod would have declined anyway due to warming waters, but that catch share sector organization had occurred at the worst possible time. But was there ever any good time for catch share in New England or other areas with such diversity in stocks and vessels, and with so many small communities utterly dependent on their small-boat fisheries?

As the Council debated over how to patch up the problems catch share had created--all of the problems that Alaskans had earlier experienced, and even traveled to New England Council meetings to warn them of, many questions were still hanging: Would there enough be fleet quota over time for the sectors to be profitable? Would they be able to pay for expensive monitoring when the government phased out its support as planned? Would Amendment 18, now in its fifth year of discussion, accomplish what was needed regarding excessive consolidation and losses to small communities, with so much of it already history?

Chapter 34

THE PACIFIC COUNCIL CLIMBS ABOARD

While the lovers and haters of catch share battled it out on the east coast, in the west catch share planning for Washington, Oregon and California trawlers was also underway. So far, the Pacific Council's catch share plan had received little attention outside its region. But the western fisheries definitely had problems. The salmon and the fleets chasing them, once the pride of the west coast, had shrunk to a shadow. Years of overfishing of rockfish had caused some to receive ESA listing. Trawl fleets had gone through a large buy-back process and were finding it hard to meet their loan payment schedule. Ports like Newport with large processors, having lost their salmon support, were now dependent on hake and albacore. By 2000 the smallest coastal economies, so tied to salmon fishing, both commercial and recreational, were barely surviving, and the federal government had declared several of them, and even larger Newport, eligible for disaster relief more than once.

The Pacific Council's trawler associations had taken their own initiative to request a catch share program. Several of the hake companies that mid-level trawled offshore--some of which also fished pollock in the Bering Sea-- had already voluntarily set up coops to help manage their bycatch problems and share the costs of observer monitoring. These seemed to be functioning fairly well, though the voluntary nature created internal fleet issues, meaning the next step could be to formalize their coops. The groundfish fleet had similar problems to New England's with its mixed-stock fishery. The nets frequently brought up a rockfish species that was a choke stock, and if caught still fishing, the skipper would be hit with a fine and a shutdown.

Or a trawler could be landing too many of a prohibited stock like halibut. The trawlers thought the Pacific Council's proposal to shift into groundfish catch share sectors could be a good one as sectors could include cooping of both targeted and non-targeted quotas, allowing more flexibility with quotas and with movement to avoid unwanted stocks.

This regional council also had a special problem. Already by 2003 it had reduced its groundfish trawl fleet from about 400 to less than 155 active vessels, much of it through voluntary buy-backs in the 1990s costing $46 million through a fleet loan from the government. The fleet was concerned that due to interest owed it was not making headway on the payoff. Perhaps catch share would help.

NOAA Throws a Bone

Responding to the MSA 2006 guidelines, the Council proposed to keep a few small fisheries available as open entry in its catch share plan. Visiting in Newport again--I had a hard time staying away from fishing ports--I was surprised to find that George had reached that stage of desperation that the Maine trawlers felt. He had finally given up on salmon and was taking advantage of a new open entry in black cod the Pacific Council offered the starving salmon fishermen. I saw the *Helen McColl's* deck fitted out with strange new gear: pots! George laughed, "Yeah, are you amazed? I'm black cod fishing! It's just a part-time fishery, all we're allowed, but I like it." He pointed out to me the handsome vessel tied up next to him, its gleaming bow looming over the ever-more salty *McColl.*

"Take a look! Pretty nice, eh? That's a real black cod boat, what they call sablefish now, one of those IFQ fisheries like you have up in Alaska. Well, NOAA threw us trollers a bone and saved out some quota for a little open-access black cod pot fishery! And that's what I'm doing. I think about a half dozen of us trollers here at Newport are taking advantage of it, and more at Coos Bay. I can't make any real money at it but it helps. I hear it's the Magnuson Act that forced them do something to help us out. To really make a living with it, to buy into an IFQ black cod tier, you'd have to pay $80,000 to $100,000. Then you can catch a personal quota of over 150,000 pounds per boat. This boat here can gross $500,000 or more a season. Some even have three tiers stacked on one boat. Think of it!

"But what ordinary fisherman can afford to get into the tier fishery? Anyway, since they allow this part-time fishery for us, I'll do it. It takes me just three days to get my little weekly quota--a day to get out to the grounds, a day to put out my pots, then go back in a day and pick them up and go home. Right now the limit is 700 pounds a week, just enough to pay for my gas and groceries and a few bills, not enough to keep up my boat. I wouldn't mind at all doing this for a living, but they won't let us--the real quota is all doled out."

We stood on the float, eyeing the black cod boat, then the *Helen McColl*. George said he would retire from fishing soon. I stared at him. No, he said he really meant it. It was difficult to talk about. What would become of the *McColl*? I thought maybe it could become a salty tour boat on Puget Sound.

George continued to fish black cod for a couple years and liked it well enough. He kept up the *McColl* with some essential re-planking, satisfied Coast Guard requirements, and passed by cosmetic painting. Then the tide ran out on the *Helen McColl*.

An Obituary

One fall morning in 2009 when George phoned, I thought I'd enjoy the usual update on the troubles of Oregon fishermen to exchange with my report on our subsistence season. But he wasn't fishing anymore; he wasn't even black codding. The *Helen McColl* was no more; she'd sunk at the dock one October night.

That was hard to hear--any boat sinking is, but George's legendary old boat? After a time I asked him to write about the end of his historical wooden schooner-troller, one of the last of its kind. He had always refused to write about fishing, but he agreed now to a bit of an obituary, and got in a few more jabs on how salmon and other small-boat fishing fits into our larger political economy. His words:

"A really good story requires a meaningful ending. The *Helen McColl*'s long-storied existence ended when it was crushed at its moorage and sank on Halloween night, 2009. It was built in East Booth Bay in 1911 and functioned as a salmon troller for its last forty years, sunk overnight after nearly a century.

"Like so many other salmon trollers--about 12,000 licensed in 1970 in the lower-48 alone--the *Helen McColl* contributed two or three million dollars to the coastal communities during those forty years. It has always been understood that every dollar sold from a salmon troller created six more dollars in the community before the fish reached the consumer's table. This doesn't include the supporting industries: the shipyard haul-out every year, busy fuel and ice docks, gear stores, crew shares, licenses, and state and federal taxes. Neither does it include the significant poundage fees we paid at the dock (6 cents plus) that should have left the hatcheries a tidy profit. The outgoing hatchery costs for a returning adult hatchery salmon ran about 18 cents a fish for years. For the returning fish we caught, trollers paid average of 50 cents in dock fees. The troll fishery was a wonderful enterprise all around--respectable profits for the state hatcheries, supportive businesses throughout the communities, taxes collected at every transaction, intrigued tourist industries, young peoples' daydreams, and so on--all the stuff of healthy societies.

"So here is the ending. The sinking of the *Helen McColl* represents the last of an era. It is doubtful that privately-owned offshore salmon rearing pens or private fish ranches in other countries will promote prosperity for any of our coastal communities. It is nearly certain that the new business models for salmon production will not create the happy communities of the 12,000 small salmon boats that operated not so long ago out of every western coastal port."

I thought, not for the first time, George should have written this book. The Internet and local press were full of commentaries of appreciation for his old boat, with lots of photos and history from "wooden boat lovers" and other people that cared. He said to just Google *Helen McColl* and I could read it all. I did, and saw all kinds of nostalgia for an era about gone in most regions. I'll add that the *McColl* wasn't insured, and that he won a long, hard battle with the Port over salvage costs. A few of the old wooden halibut schooners the age of the *McColl* are still fishing, but you just about have to go to Alaska to see working wooden trollers slipping into an unloading dock. What next? I thought, well, George could put his creative energies into the cheese house, but it didn't turn out that way. Four years later he was commercial fishing again, not in Oregon, in Alaska. How could I have been surprised?

Pacific Council Gets the Green Light

As expected, the Pacific Council got the green light on its groundfish catch share plan to begin in 2011. It included the formal coop arrangement for its offshore hake fleet and processors, but left the shore-based hake boats independent and free to set up coops as they wished. Its groundfish trawlers became IFQ, and included shore-based processors. To assure compliance, all vessels would carry one 12-hour-day observer with the costs to be borne by the vessel owners after a phase-in period. The Council followed the MSA direction, the program including protections against a rushed, excessive consolidation, as the quota wasn't transferable for the first two years. To protect traditional fishing communities the program allowed a few small open-access fixed-gear such as the black cod pot fishery, and an amount of quota was also set aside for community non-profits and for crew. The Council also allowed the fleet to fish other species such as shrimp when their groundfishing quota was filled.

But these social considerations were offset by the worst of free-market driven catch share strategies--that anyone could buy up quota shares, and the usual fleet of speculators was assuredly going to develop. It wasn't too hard to picture the consequences, since we'd seen them repeatedly off Alaska. In 2010, knowing what was coming, three fishermen's organizations had filed suit against the Dept. of Commerce and NOAA/NMFS to halt the program, adding to the growing list of lawsuits coast-to-coast against NMFS. Several congressmen from California supported the suit, protesting that catch share was destructive to small fishermen and small ports. The lawsuit was a strategic move, but it failed. The judge stated that it had not proved catch share would harm the fleets, that indeed it appeared it would help them to survive. Like the government, the judge discounted the fishermen who would not survive.

At the one-year program review the Council found that the offshore hake fleet produced one of the best results for catch share. Its cooped by-catch program had saved individual boats from being shut down over a haul of protected rockfish, as their coop allowed them shift quota among them as needed. In contrast, the shore-based hake boats, which had stayed in a traditional independent system, faced the threat of shutdowns over bycatch. Both fleets had made a stronger effort to avoid this, but the cooped ones had s much better outcome with the bycatch rate dropping from as much as 35% of the take for some species to less than 2% overall. The active hake vessels also reported a large revenue gain, with the fleet average of several years' dockside value of $273,000 rising to $775,000 per vessel for 2011. It was not clear from the report whether the gain per vessel was because of catch share consolidations, or an improved run of hake, or both. These are obviously not small boats, but an industrial fleet like pollock boats, fishing a low-value fish that must be caught in quantity.

Catch share could probably work well for such a fleet with little diversity if the quota allocation was large enough per vessel and sector and if there were protections against excessive consolidation and against the quota leaving the active fleet. But these were not all included. Yet the trawlers understood the cautions needed with privatized programs. Steve Bonder, a Coos Bay trawler, as quoted in the *National Fisherman* June, 2013, said, "We have a very hard cap on ownership of quota so as not to allow big boats, the owners with deep pockets, to buy up all the quota." Still, by 2012 the fleet included only about 100 active vessels from a recent 150.

Also, there was little incentive to trawlers to switch to more sustainable gear. The original windfall quota allocations gave 70% to trawlers, 10% to fixed-gear boats, all based on selling history in set years, and 20% to the processors. There was probably much economic pressure for the processor quota, as Oregon's coast is dotted with small fishing towns dependent on local processor contribution to the economy. Some had already closed down when the salmon fisheries collapsed.

With the outcomes from the north available for study, I'd hoped to see more innovations for protecting the small-boat fleet communities. Had the others felt forced to drop out? How had catch share affected the small ports? No report was offered on that. After two years the quota shares would go on the open market. The price for purchase or leasing would doubtlessly rise fast, and for most fishermen and small communities, would be a pipe dream. Thus, the Pacific Council's groundfish catch share was in several ways a retreat from the early IFQ plans for halibut and sablefish. It had made a weak effort for social fairness, but with all the experience of other fisheries to apply, it could have done better. It could have set aside a set percent to be allotted to small trawlers with smaller landings history. It could have required "owner on board". It could have set aside quota for long-term crewmen with a subsidized fixed price. It could have guaranteed some business for the smallest ports and their processors? Before long, even survivors of this catch share had new worries and complaints.

On conservation issues the Council also could have been bolder. Here was a chance to provide financial and other incentives to move trawlers toward use of more ecologically adapted gear like pots and longlines. When the Pacific Coast Federation of Fishing Associations (PCFFA) asked the Council why only 10% of the quota was allotted for more ecologically friendly fixed-gear, the Council responded that it was too difficult to carry out enforcement on fixed-gear like pots. If so, the plan was giving up long-range stewardship for short-range expediency, and often the smaller fixed-gear fishermen weren't awarded enough quota to continue. In all the debates the fact gets lost that fleets do not have to reorganize as catch share in order to form coops on their own if they follow MSA guidelines, and they can lobby their regional councils or NMFS to cut their coop a major slice of the fleet quota. However, they may not get what they ask for, and they may not get any startup funding.

I could see only two improvements for stock conservation through this catch share--the same as for the earlier IFQs--but they were important ones for NOAA. The bycatch reduction program was successful, and the more lucrative seasons expected for the consolidated big vessel owners meant they could be expected to pay more for an expensive enforcement program.

West coast small fisherman David Helliwell, as quoted in *Pacific Fishing*, November, 2011, echoed what we heard many times from Alaska:

"Due to groundfish quota being given to big draggers, small boats up and down the coast have lost access to the public resource and can no longer serve their communities. Through money, political influence, and the multi-million dollar horsepower of corporate environmental groups, fisheries management brought forth the catch share solution. Somehow, however, the shares all ended up with the big operations that caused the problem in the first place."

Soon complaints began from the trawlers still active. Were the free original quota allotments large enough for fishing to be profitable, or did the fleet need to be smaller yet? Would the fleet ever get the loan paid off from the earlier buyback? So far, they were sailing backward; the originally $28.5 million was now at $30,000. To get anywhere on the loan, they could hardly afford to reduce the fleet more through another buy-back. They had more worries. They said that costs of the on-board observer program had gone up beyond what \they had expected and asked for cheaper electronic monitoring. But NMFS estimated it would be four more years before electronic monitoring would have the features needed to replace on-board observers. The fleet also complained that the rockfish protection areas that the Council/NMFS had created had cut the fleet out of too much area, and some needed to be re-opened. Otherwise, the trawlers said, they might not be able to catch enough to afford the catch share program.

The Environmental Defense Fund, so diligent in promoting catch share, saw its favorite fisheries project in trouble and officially supported the trawlers in all their demands for corrections. Soon Congress did pass a bill refinancing their old buy-back loan with a lower interest rate going forward, but as of 2015 the funds had not been transferred. Some rockfish preserve areas also got reopened. But the fleets would have to wait a few years on electronic monitoring.

Without an increase in federal funding for startups it is likely that fleet restructuring will slow down. Dorothy Lowman, an Oregon representative on the Pacific Council, commented unofficially in the *Pacific Fishing* August 2013 issue that NMFS didn't have the resources to cover all the requests for changes flooding in. "NMFS isn't moving forward….A lot of work is just sitting on the shelf."

Meanwhile, in the East, retired congressman Barney Frank, Dr. Bryan Rothschild, and others, convinced that neither NMFS nor the Council could give the leadership needed to solve commercial fisheries' problems, formed a new group, the Center for Sustainable Fisheries, with the goal of rebuilding sustainable fisheries and protecting fishing communities through better science.

Chapter 35

Fish Management and its Discontents

I was in Juneau, a trolling haunt of forty years before, headed for the airport. It was inevitable that I brought up commercial fishing to the middle-aged cabby.

"Yeah, I crewed out of Dutch for fifteen years," he said. "Cost me a marriage, and more."

It was a common enough outcome for fishermen, but still, personal. I sighed my sympathy and changed the subject.

"So what do you think of IFQ?"

"IFQ?" He snorted, "It's just a way for the rich to get richer and the poor to get poorer."

He said in one sentence what has taken me half this book to relate.

Talking with people like him about fishery problems left me with more questions. Why did our federal government persist in promoting a restructuring of fleets that eliminated one small-boat fisherman after another? Why did the regional fishery councils agree to develop the proposals for such changes? Why did the Secretary of Commerce sign off on them and why didn't Congress at least appropriate adequate funds to allow the programs some social fairness? Why didn't the big ENGOs see the fallacy, that privatizing fisheries had little to do with their goal of saving nature, that it was an economic strategy? Why didn't the general public. as they rallied around conservation slogans like "Save the Fish", consider saving the fishermen too?

The growing disconnect between what the government saw as a model for modern fleets and what the majority of fishermen needed to survive came to a head with the New England Council's plan. After three years of turmoil NOAA Administrator Lubchenco gave up on the project and headed back west to lead a state university program focused on ecosystem sustainability, leaving NMFS and the Council to work out a truce with the fishermen. As a marine ecologist, Lubchenco had overseen a program to shrink and restructure the fleets, only to find out that it was not the silver bullet to protect fish stocks, despite what some big ENGOs believed, and that scientific management still was the answer. Off Alaska and in other countries where forms of IFQ/catch share had been in effect longer, the fish stocks were worse off than our own, and the human communities that depended on them were in social disarray or outright despair. When the New England cod stocks collapsed a second time, scientists in numbers finally spoke up. All that energy going into restructuring fleets had drawn managers away from taking care of fish.

The cultural and economic value of our small-boat fleets, and the adversaries they face, have been my topic in these pages. But today's battle to save our small fishermen obviously has to be tied to the protection and rebuilding of the fish stocks under siege, not only from the promoters of industrial fleets, but from far more complex problems. The fish face warming oceans, acidification, drought, human population growth, and other gigantic trends that are beyond what fishery managers alone can influence, or even Congress. In July, 2015 as I send off this manuscript at last, the news gets grimmer: returning Chinook spawners dying by the score in tributaries of the Columbia, the Willamette and Deschutes rivers, apparently due to excessively warm water. All the hard work to rebuild other Columbia-Snake runs seemed doomed when by late July both federal and state scientists estimated that at least 500,000, half the returning sockeye, were dying due to warm water. This condition is caused by a combination of drought (climate change) and channel damming. Only major social movements can reverse these trends. Other adversaries our elected leaders should be more able to control, like our own toxic dumping, ruined river habitat, and trade agreements that help big corporations, not ordinary people. But the fish, like the fishermen, are doomed by public passivity and the brand of politics that generates. With so many competing issues only iconic fish like the Chinook salmon grab the public imagination, and even that species is in trouble today on Alaskan rivers, the waters that, until now, had escaped biological and social disaster.

While the picture I've painted shows the small fishermen as likely losers, we need to remember that Rachel Carson was widely ridiculed and attacked when *Silent Spring* debuted. How dare this isolated, unrecognized (and female) scientist stand up to huge corporations? And yet she initiated a movement that still grows. In the last decade more and more people have come to recognize that endless growth via the great god technology has gotten us in a trap that Luddites recognized centuries ago. The industrialized nations will have to make radical changes, and not just in commercial fisheries, or will see their coastal cities underwater, their rivers dried mud. The decline of Alaskan rivers' Chinook salmon and the desperation of western Alaskan fishermen today helps to remind us that everything in our fisheries, and our societies, is interconnected, that industrialization always has a social price, and that a lone Chinook spawner heading for the Yukon has a world of eyes on it.

Fish stocks in general have been moving north since 1970. Scientists say it is due to warming waters. Acidification, another growing problem, will be the worst in the northern waters. It primarily affects shellfish, but that affects the whole ocean food chain, from tiny shelled copepods to pink salmon, cod, pollock, and other species that feed on them. Pacific halibut keep getting smaller in size; breeding females that used to average 50 lbs. now average 20. Baltic salmon is so full of dioxin that several EU countries have banned its sale. Due to years of spreading drought and dried rivers, California resorts to trucking its salmon smolt to the estuaries. But how will it truck the spawners? Most US fishermen and all fishery scientists are well aware of these huge problems--their websites are full of them, and also the coastal public media. One would think even the most blasé coastal person would be aware of them, if not concerned.

The first fishing story I remember hearing--I was four or five--was about my Norwegian grandfather I never met, capsized in a storm off Lofoten Islands, hanging by his knife blade stuck in the hull of his dory in bitter cold water, waiting for rescue, but to die of pneumonia. I tell that story--of one boat, one fisherman, one grandfather--as a metaphor for whole fleets. My hope is that the lives of today's small fishermen won't turn out, in their own way, so lost. But the waters I see out front are ever rougher.

The fish-in by the Marshall gillnetters, where this story began, turned out to be a harbinger of more to come. The poor Chinook runs spread and so did the anger, including to sports fishermen who couldn't get their annual two fish. Replayed television documentaries ten years old extolling the great health of Alaskan salmon were laughable as the Chinook problem appeared to be spreading from the Yukon and the Kuskokwim to the other big Alaskan rivers. And where were the larger halibut? Where were the herring schools? For that matter, where was the California/Washington/Oregon/British Columbia's mountain snowfall?

We have an advantage in Alaska in that here it is easy to see how essential fishing is to family survival, and realize that it is still that way over much of the world. I recall George's comment that if we look at the global picture, subsistence fishing is by far the most important to save. Perry's cousin Enoch is writing of the traditional subsistence life he remembers growing up on the Noatak River north of Kotzebue. "The next generation may need to have this information." Yes, they may. I think of a group of crabbers I watched in 1991 on the ice in a Magadan Bay, Russian Far East. In ten degrees and wind, clad in thin jackets and leather boots, they were hand-lining with bare, cracked hands for a few small crabs we would have thrown back. The men had walked several miles in the snow without snowshoes to get to the bay, then punched their holes in the ice by hand. I visited the cabin they lived in, very much third world, though lots of firewood. Is this what is in store for us if we don't take care of our oceans?

In 2010 NOAA reported that the arctic fisheries off Alaska are unexploited, apparently not counting subsistence fishing by people living along the arctic coast. In northern Canada, arctic fish stocks are already dominated by industrial fishing. In Greenland the largest commercial contributor to the economy is "Royal Greenland", a fishing company wholly owned by Inuit. Industrialization has no one language or color, and I would guess that in every northern indigenous coastal community a debate goes on right now, I know it does here, about long-range costs versus short-range benefits of trawling, especially bottom trawling. Every indigenous community will put subsistence needs and protection of the environment first, but also want some form of fair equity in the commercial economy. The unexploited northern seas off Alaska offer a perfect time and place for our federal management to stand firm and with the other nations bordering on the Arctic Ocean, agree to keep it closed to commercial fishing until more stock research can be done, the indigenous residents heard from, and policies for truly sustainable fisheries developed.

Two Different Ways to Manage Fisheries

This book's two parts describe the effects of two different strategies of fish management: the traditional one still most used by the states, and catch share--the one the federal government has succeeded in installing in about 30 fleets. The question of which is the best system for the fish isn't the right question. Neither can work without policies based on science, not politics. The question of whether science could still win came to a head with the catch share plan proposed by the New England Council. At the time the condition of the fisheries in New England was disastrous. The message, not just from the federal government but from the big ENGOs and the industry itself, was to try something new. The replacement chosen was wrong, as history clearly showed, but satisfied the goals of our federal government. The US, as the world leader, had to have fleets to compete: bigger vessels with more technology, more efficiency; fishing run more like a business; add to that the free market and we would solve the fisheries' problems, including the problems of the fish stocks. This package of beliefs grossly oversimplifies the problems of the fisheries and has been repudiated by a growing number of coastal communities, fish biologists, fishermen, politicians, ENGOS, and recreational fishing associations, but still has the government and too many fishery leaders committed to it.

The more I read about what has taken place since the birth of the MSA, the more I see that we Americans need to train our eyes to see the big picture, the whole world's seas of fish and fishing people, and observe the patterns evolving. The worldwide commercial fisheries have grown hugely: from 19 million tons harvested in 1950 to 90 million in 2005. Think what it must be now. But at the same time, according to World Watch Institute, that figure hides realities like West Africa where ocean fisheries operated by local people have declined 50% in last thirty years as industrial fleets moved in. The UN calculates that the livelihoods of 540 million people depend on small fisheries, commercial and subsistence, and are being lost to mega-corporations. This year the UN's Food and Agriculture Organization passed the first global guidelines for bycatch and discards and emphasized that care must be taken that these guidelines would not "place an undue burden on poor and artisanal fisheries and developing states." I can't help but wonder if these protections, as interpreted by countries, will be real, or will they be like those in the MSA, "to the extent practicable", interpreted by managers and regional councils to be almost anything.

Our northern fisheries, in their mix of small-boat commercial with subsistence fishing within a family, have much similarity with those from Africa to northern Russia, and face the same threats. The difference between the US and most of the world is that we do have the MSA, regional councils, and fish boards, all charged to look after our fish, our fisheries, and our fishing communities. But only in the North Pacific Council's CDQ program is there a fair guarantee that small fishermen will still be afloat twenty years from now. The success of this experiment must have frightened corporate interests to the point where the model could never be tried again. It still could be if small fishing communities could rally a broad support for it.

The Commons and Overfishing

Without doubt overfishing begins with fishermen and their managers. However, fishermen are simply exhibiting universal human greed when it is uncontrolled. You take all the fish you can, now, because they may be gone tomorrow, or someone else will catch them. I've been there. But history and studies have shown that in isolated fisheries a traditional, informal group of people can control greed and manage a commons long-term in the interest of themselves and their descendants for sustainability. When individuals get out of control with their methods or in the amounts they seize, the group brings them back to reality. But when the population becomes much larger and/or influxes of new people move in, we do have to have formalized fishery management. This is especially true when new forms of technology come along with the increase in harvesters. I talked to a biologist from Mongolia recently who described this very process going on now in their grazing areas, where the incomers don't know or pay attention to the old understandings the commons operate with, and she believes a new form of management will be necessary. Though the MSA says sustainable fisheries must have priority, I hope the rest of our laws, and Mongolia's, are not as easy to maneuver around as the MSA has been.

The states haven't dumped the ocean commons, though limited permitting clearly does modify it. They have stayed way from privatization of fleets for reasons probably both economic and political. It is not that they are above the influence of factions and lobbies, but that they feel less pressure from industrial fleets--that is, until fish farms captured their imagination, and so far most states have said no to the farms. The other related reason the states have stayed with traditional management is probably that state politicians are more influenced by local and regional constituents and their small-fishery interests, both commercial and recreational, and less by our federal and international politics. The states could have done better, though, by keeping a limited bank of permits available to new entries, especially to aid young entry-level fishermen, and indeed some New England states are now doing just that to try to mitigate the destructive social effects of catch share and limited licensing on their coastal communities.

The Puzzling Regional Councils

It is easy to understand why industrial fishing corporations favor privatization, more puzzling to understand the role of the regional fishery councils in promoting it. Why isn't sustainability for small fishermen along with fish a priority? The regional council members have not been uniformly anti-small fishermen, nor lazy with the tangled web handed them by NOAA, so why do the smalls always lose? I go back to the cabdriver's words at the beginning of this chapter. He was saying what I heard over and over from the small fishermen, their favorite cliché being: "Follow the money!" It's hard to hear that about a system of regional councils that were supposedly created and structured to juggle all the interests in the fisheries and promote a democratic process for management changes. The council structure as laid out in the MSA is intended to provide a balance between big and small-boat fleets, and between districts--in the case of the North Pacific Council, usually between Alaska and Seattle--and between the industry and other interests, such as ENGOs. But it takes only one vote to swing the majority from Alaska-dominated to outside-dominated, and it happens frequently. The North Pacific Council recognized the progressive concepts used in its unique CDQ program, but it was prey to the same conflicts as the others, and it is as if it had burned its own charts.

When the small fishermen say money speaks, they aren't pointing at anything as crass as bribes but at more subtle forms of pressure, including from NOAA. As a comparison, the states, too, have their boards or commissions with voting representation from bodies frequently embattled. Yet there one doesn't see a particular group always the winner. In the North Pacific and Pacific Councils' voting, catch share and similar socially destructive plans almost always win when it comes to a vote. In New England a broader support to the small fishermen is evident, not especially from NOAA, or even their council, but from political leaders and communities. If all else fails, the defenders of New England's small fishermen will take an issue to the court. And they insist that if that council had used a fleet referendum that catch share would never have been approved. We can't know.

The North Pacific Council has much to its credit when it comes to sustainable fish stocks. It makes a respectable effort to combine privatization with scientific management, but even so it hasn't succeeded, as seen in the 20 years of battles over bycatch, the related battles over enforcement, and the decline of several of important stocks that the Council states are not overfished. In every region, money and energy that should have gone to research, with NOAA's encouragement was shifted to introducing and administering privatized fleets. The councils have gotten important stocks off the overfished list. But what can one say of council (and NMFS) efforts, overall, when you look at the state of Atlantic cod, Pacific halibut, Alaskan Chinook?

As one goes up the management ladder, the same puzzle and answer seem to apply to NOAA. The amended MSA of 1996 and 2006 mandated that councils provide serious protections for fish stocks and traditional fleets and their communities, but in almost every situation the implementation has been watered down by lack of zeal usually combined with lack of federal funding. NOAA's broad goals are not the problem: "Prepare for a changing environment; Support healthy ecosystems, communities and economies; Support marine fish, habitat, and biodiversity in a healthy ecosystem; Support productivity and sustainability of fish and fishing communities through science-based decision making." Who would argue with those? The problem has been in converting them to action.

Why did NOAA promote privatization so insistently to the councils, to the point where by 2006 it would fund startup administrative costs only for catch share programs? Why did it remove the fleet referendum? Why didn't its advisory scientists fight catch share as a solution? I can understand how its scientists might have been mislead by the first years of privatization, as it would take several seasons for it to become clear that IFQ was not, in itself, going to rebuild or protect stocks. But I don't understand how the scientists continued to be misled, with the cynical small fishermen again shrugging and saying, whose payroll are they on? Fishery scientist Jim Lichatowich, in his second book, *Salmon, People, and Place*, speaking of his colleagues, says many of the salmon problems in the Northwest are their own fault, that they have been far too passive in fighting for what they know to be true, too willing to not rock the boat and jeopardize their program funding, state or federal.

The more kind explanation for council scientists who conceded to catch share plans is that they saw for decades the failures of traditional management and so were willing to support something new. But though IFQ was new for the US, the record was already there to study in Europe and Canada. Specialization in the sciences may also take part of the blame, with social scientists strangely having little to no involvement in the council's catch share planning process. Typically in Alaska, and possibly elsewhere, they were brought in to do community surveys or impact studies, but only after a privatized program had already been operating at least a year and back-treading was difficult. Most fish biologists on the scene stuck to fish issues; it's always simpler to stay within your own area, but often taking risk is what's badly needed.

The ENGO Role

Certain large ENGOs, often via NOAA, have had considerable influence on the councils. At the beginning of the privatizing movement, organizations like the Environmental Defense Council who promoted IFQ as a way to save fish didn't spend enough energy looking at the record. Their lobbies, like industrial fishing's, are powerful. Yet some ENGOs such as Ecotrust, and Friends of the Earth, and I'm sure others, never did buy into the catch share silver bullet. Nor did Food and Water Watch. A few big foundations like Pew offered positive voluntary projects in alternative gear as a substitute for trawling and dredging. Today fewer ENGOs are hypnotized by government hype about the promises for sustainability through catch share. Many now recognize there is no built-in conservation ethic in privatization schemes, that scientific management is just as important as ever. More ENGOs see economies like small commercial fishing as not the enemy but part of the ecosystem to be saved. They see that fish stocks and their ecosystems don't get the attention they must have as resources are diverted to catch share. Mark Spalding, president of the Ocean Foundation, on its website, March 13, 2013, summarizes the problem:

"ITQs/Catch Shares, as a tool by themselves, lack the means to address issues like conservation, community preservation, monopoly prevention, and multispecies dependencies....an unintended consequence in many places has been the increasing monopolization of the fishing business in the hands of a few politically powerful companies and families."

Alas that it took so long to get public statements like this from the big ENGOs.

Since 2010, several, like Ocean Conservancy and EarthJustice have used federal and state grant money, to help small fishermen get back on the water. ENGOs have also joined with fishermen's groups in lawsuits to stop irrigators' excessive water draw downs, destruction of stocks by industrial fleets, and huge mining projects like "Pebble" that would affect spawning areas. Most ENGOs, however, continue to stay away from the conflicts over privatization.

The Public's Foggy View

Small fishermen need a larger body of supporters if they are to win their battle for survival. They don't have a powerful national organization to fight for their interest. They have never been very good at organizing themselves for action, clustered as they are with five or fewer crew on a boat, many with only two. They need a broad community of supporters. But the general pubic, who can get so excited over threats to a spotted owl, a whale, and a Chinook salmon, doesn't make the connection between admirable Nature and traditional fishing communities. Fishermen, of all the players, are the most visible and easiest to blame for threatened fish stocks. Usually the public hasn't made a distinction between industrial fishing that destroys habitat and small fishermen that barely touch it. The public, except for that of coastal towns, gets its information about fishermen from television, which has generally painted the picture of them all as enemies of the natural world. Lately it might offer popular reality shows, but so far the subjects have not been what we would call typical small fishermen. In contrast, one small-scale trawling group at Port Clyde, Maine has through a grant from an ENGO produced an informative video explaining what ruined their traditional on-shore fishery and what they are doing to save their livelihood through "community supported fisheries". Why haven't more such videos been done for PBS and similar channels? But there has to be public demand for government or ENGO funding for such serious documentaries. Privatized fisheries aren't going to provide it.

For small fishermen to salvage their way of life the public can't be left at a level of information no greater than what primary school children are taught: "We must take care of nature". They need to recruit broader, sophisticated public support, but small fisheries are rather obscure causes to rally around. When the coastal public does see before its eyes how the waterfronts have become gentrified and that its fishermen-neighbors are no longer fishing, it is late in the day to begin a protest movement. A small public has started to pay attention to the "sustainable fisheries" label they see on a fish fillet, but for non-coastal people to attach that to the lives of small fishermen takes more education.

Small fishermen, too survive, need to draw in the level of the public energy we saw three decades back when Willie Nelson took "Farm Aid" on the road across the country to publicize the plight of the small farmers being squeezed out. Nelson's Farm Aid is still going. Fishermen have nothing like that kind of public awareness and support, and no national organization like the old Grange to speak for them. But especially, in today's media-saturated world, they don't have a celebrity like George Clooney taking on their cause. We may laugh at this sort of politics but that's what is required to get attention today. Though I don't know how many small farmers Nelson's "save the small farm" movement actually saved, it pointed the way that much public action is formed today. Fishermen need to swallow their proud independence and analyze how the public becomes engaged in popular causes, and where the financial backing could be for engaging the public.

The public's lack of concern for commercial fishermen, however, goes deeper than missing information. I believe it is partly due to our fisheries never having been part of a glorified myth of the frontier taught every child, the players being adventurous pioneers, fur trappers, cowboys, Plains Indians, soldiers, gold miners, lumberjacks--everybody gets their glory, even gamblers, even small farmers/homesteaders, but I never heard anything about commercial fishermen in school, neither a glorified nor a realistic version, no fisheries romance embedded in us to make us rally to its cause as Willie Nelson was able to organize for small farmers. *"The Old Man and the Sea* was as close as we came for high schoolers; *Moby* Dick for the minority in college literature classes. For the public to see the full import of losing our small fishermen it needs to recall a tragic social loss people are more familiar with that took place a few decades back when the "efficiencies" of big agriculture drove out our small farmers' coops and individual farmers. The revenues from today's mega-farms don't stay in the county where the crops grow, and when the owners of fish quota are not local, the profits from the fish unloaded at the dock will go elsewhere too. Distant stockholders of any commodity typically don't have much concern over "preserving traditional communities" that are invisible to them.

The "public commons" itself is a foggy concept to the public today. People take commons like our federal park system for granted, assuming that a commons is just there--until it is taken away. Public outrage did erupt when it first viewed clear-cutting in public forests. The media made the difference. For our small fishermen and their communities there is no vivid image like a clear-cut foothill to shock them into action. Publics need to feel physically hurt or morally insulted to take on a battle.

Our Elected Leaders' Hypocrisy

The loss of our federal small-boat fisheries is also due to the sort of people we elect to Congress. To win an election today requires that a candidate be a millionaire or have deep-pocketed donors. We have few politicians left who prove with their sponsored bills and their votes that they truly represent people in work jeans and rubber boots--small farmers, miners, laborers, and fishermen. Government staffers' lack of knowledge is also part of the problem. I suspect few of the economists attached to regional councils or NOAA have ever been part of a rural economy where families' lives were, and perhaps still are, tied to small commodity production. Engaging our political leaders in the cause of small fishermen, or small anything, gets more difficult with each decade as our elections become more dominated by big lobbies and the money behind them. Overcoming this can still happen, but only with a groundswell of public support. And, apparently, celebrity promotion.

There is a flip side to this used by governments: blame the voters. Leaders in both the US and Canada insist that fishery management must be self-supporting, i.e. its costs covered by industrial-sized vessels, because "the taxpayers" are tired of supporting fisheries, and at an unsustainable level. But commercial fishermen today are literate and are aware of government subsidies to many sectors. Why, they ask, should they be chosen for the goal of self-support when "the taxpayers" don't stop our government from subsidizing oil companies; auto companies; corporate miners, grazers, and agriculture; most noticeably banks and investment firms--all at levels apparently unsustainable without government help. All the rationalizations one hears for eliminating small-boat fishing are simply in the service of industrial fleets.

The councils and federal government, following the MSA 2006 amendments, have recently put in place important improvements, most of them in bycatch waste reduction and protection of the sea bottom. Bottom trawlers in much of federal water off Alaska now must use off-the-bottom trawl gear. The flatfish trawler coop, Alaska Seafood Cooperative, reports that 94% of their catch is now retained and sold, and that its habitat destruction has been reduced 90%. Harvest quotas, reduced waste, and better enforcement are in effect for all regions. The state of our commercial fish stocks does slowly improve. I wish we could say the same for more of our fishing families and communities.

The MSA 2016 Amendments

The House has passed its amended MSA for 2016, the Senate must submit its version, the combined version will follow, and then we should have passage of an updated, improved MSA. However the advisors to the president on fisheries say he will veto the bill. The title warns of the battle resurfacing: "Strengthening Fishing Communities and Increasing Flexibility in the Fishery Management Act." An introduction by Rep. Don Young from Alaska says:"…will allow the councils to consider a community's economic need when setting annual catch limits." The Pew Charitable Trust said such an amendment will "undo all the gains of the original MSA", a feeling no doubt many scientists and ENGOs agree with, believing it takes back some of the power of science included in the 2006 version, giving too much back to the regional councils. Alaska's Don Young, who authored the amendments, followed the recommendations of the Center for Sustainable Fisheries that it was scientifically poor to have one deadline for all stock rebuilding, since stocks grow at different rates, and similar inflexibility, and would put unnecessary hardship on fishing families. Much of the proposed amendment follows suit--flexibility, not deadlines, and also calls for more transparency and better communication with fishermen on the part of the councils and managers, including more involvement of fishermen in research. Opposition to this flexibility is based on the fear that councils will succumb to pressure from the industry and make timelines too liberal for the particular stock rebuilding. The North Pacific Council has followed the policy of using its scientists' recommendations, such as ABC limits for setting quotas and other biological issues, but not every council does. On the other hand, a focused attention on stock rebuilding, while downplaying Standard 8 and community protections had had terrible social consequences. So we seem to be back to the same debate, with the ENGOs and some scientists on one side and industry, with some scientists on the other; most Republicans on one side (for flexibility), and most Democrats on the other. One must hope that a reasonable compromise will be reached.

I had hoped to see more direct protection of fishing communities, not just as a byproduct of stock rebuilding positions, but that seems to be expecting too much on an obscure problem when it is so much simpler to square off on fish management. Other concerns from the councils were over the costs and availability of some other changes being put forward, such as increased communication and vessel monitoring.

The amendment proposals from the senators showed the interesting influence of Alaska in requesting that wherever "state" or "federal" is mentioned that "tribal" be added, and where "commercial " or "recreational" fishing is mentioned that "subsistence" be added. Furthermore that if an EEZ zone is created in the US arctic waters, that ten percent be set aside for a CDQ program. Could we be so bold a second time? I hope so.

Unfortunately the MSA congressional debate will once more pit sustainability of stocks against sustainability of communities. Much of that, not all, could have been avoided by including the fishermen more--small as well as large--in councils' planning and research. That has to take place more in the future.

The States' Kettle of Fish

Though both federal and state management sail a deep sea of troubles, the state fishery managers get a double dose of complexity to deal with. Like all commercial fisheries they must answer to federal laws like the MSA and ESA, listen to NOAA, and also follow their own state laws and regulations. In addition they have to deal with the biological/ecological/political problems peculiar to anadromous fish like salmon that have been growing since the late 1800s: urbanization, changing culture, hydropower, ruined habitat, foreign salmon farms, and the great number of stakeholders with competing agendas. The salmon fleets, especially the gillnetters, also face the building political strength of sport fishing associations determined to have an ever-increasing share of the harvest.

This gillnet versus recreational fishing, like so many fishery problems, is beyond what the state fisheries managers can solve alone. Until about fifteen years ago the clear message from the states was that they wanted to protect their fishing commons, their fish, and their fishermen, both commercial and sport. But with each decade, as a reflection of demographics and growth of affluent middle-class outdoor interests, the states embrace more the values of recreational fishing. The arguments the Coastal Conservation Association puts forth, a mix of truth and propaganda, have often been successful with voters and political leaders. Even when initiatives fail, they have great value as a way to gradually change beliefs. While state managers have rejected fleet restructure or outlawing as a solution for poor runs, sport interests have taken it up. Commercial fleets are now the Northwest states and BC's last priority in fisheries. Without federal trust protections, Indian fisheries would no doubt be at the bottom too.

Our salmon fishermen have adversaries with great staying power. The big western federal dams are not likely ever to be breached, nor is the ESA likely to be rescinded. We have little power to stop the effects of drought on the western rivers. The states also see federal agencies like NOAA/ NMFS and NEPA taking on an increasing role in the salmon management as required by the ESA. The sport-fishing sector grows in power. State-managed salmon and shellfish commercial fleets, though not privatized, have paid heavily in other ways: reduced fleet quotas, extreme restrictions, and buy-backs that in the end hurt their communities. Still, though many fishermen have given up, fishing permits keep selling.

The survivors in state-managed fleets have this consolation: they remain a fleet of independents fishing in a commons, a restricted commons it's true, but all of them know that they have a chance, poor as it may be, to make a season. That, for small fishermen, is the difference between federal and state systems of management. State regulations in a commons can lighten up as fish stocks rebuild. Fishermen who have held on will see brighter days. That is the way the states have managed over many decades,, and as scientists learn more about salmon and what is needed to restore the runs, the problems will be the traditional bewildering haul: changes in the climate and ocean, industrialization/hydropower demands, interception, trawler bycatch, and of course politics--including the special politics of the recreation fishing industry, but even more the engagement of the voting public to demand that fishery managers get the funds they need to do their job right. Certainly it's rough waters, but state-managed fishermen, as always, will hang onto hope.

Chapter 36

ROUGH WATERS: FISHERMEN AND CHOICES

O f all the questions in my mind about the small fishermen and their battles, I have left the toughest question of all to the last. Why have the federal waters small fishermen voted against their own interest in IFQ fleet referendums and even when there is no referendum, as in New England, chosen to join catch share sectors? The BC troll fishermen's survey that I mentioned points to the answer: they believe themselves under serious attack and have seen IFQ/catch share as the only way to protect themselves. People are vulnerable when in crisis, and one crisis after another had hit the fleets as the government planned its move into catch share: the grounds too crowded, the fish stocks shrinking, and weak, underfunded management forever introducing increasing restrictions. The fishermen could view from their cabin windows industrial waste from mining and logging, and easily imagine the inevitable destruction of sea habitat. Global markets and foreign farmed fish were eroding their prices and on the competitive big-boat fleets they saw continuously installed new technology they felt forced to keep up with.

Then, if they needed more reasons to worry, they saw or read of evidence of ocean warming, dried up rivers, and creeping acidification with jellyfish invasions. They heard of the widely circulated reports that the world's fisheries will collapse shortly from overfishing, that framed fishing was expanding rapidly, and that Congress and DFO (Canada) would be handing down budget cuts for fisheries funding.

Finally, they saw the traditional management of fisheries as too often a failure. As an example, the members of our Deep Sea Fisherman's Union, representing the west coast halibut crewmen, opposed IFQ generally, but when it came down to the vote they said yes to IFQ, and set the table for more such plans. They explained that they felt they had no choice. So the strategy of divide and conquer worked as well as ever. Some would find catch share worked out well for them, others would abandon their livelihood, lease out what quota they had, and sit on the beach. Whether that was fine with them was an individual matter. Others found a way to get into another fleet, even part-time was better than giving up their identity as fishermen. I don't think that was a romantic choice; it was as basic as you can get.

Even the referendums themselves were not clear decisions. We don't know which way the entire fleet would have voted in the North Pacific referendums, as they took place after the end of the pilot years and included votes only from boat owners still actively in the fleet. The ones who had decided to sell their quota shares, almost all of them with small, unviable allocations, no longer counted. Yet, even later, when small fishermen knew about the halibut and sablefish IFQ outcomes in BC and Alaska, they still would vote to abandon their commons and vote for IFQ. They explained that they had no retirement fund beyond social security, and for most of them even that in modest amount. Their retirement support, as they saw it, would be in those marketable quota shares--to be sold and invested, or better yet, leased out. They would at least have something for their old age. And that is what has happened, so far.

There are variations on the theme. In BC, salmon fishermen also believed that the First Nations and sport fishermen had gained so much political power that they would soon leave little fleet quota for the commercials. In Oregon the trawl fleet owed so much for buy-backs that they were sliding backwards on their debt. In New England even though fishermen had the choice to stay independent, the restrictions they would endure were onerous, with more promised if they didn't join catch share sectors. Roughly 75% there chose to join. People believing they are in crisis can often be manipulated to make choices against their own interest.

That doesn't mean the fishermen are happy now. It has been twenty years since the first big IFQs were signed off. Observers might think that residents of healthier Alaskan fishing ports like Kodiak would by now have seen a positive impact of IFQ/catch share and that their earlier anxieties were just that--anxiety over change. Yet in 2012 a new survey by C. Carothers, the University of Alaska researcher, reported that still only 13% of Kodiak respondents (residents, not just IFQ holders) found that fisheries privatization there had been positive.

When my cousin George started commercial fishing in 1954, it was still possible to start out with a very small boat and motor, a few supplies, and a couple trolling poles or gillnet, and survive, moving up to better outfits as you were able. That's hardly possible now. However, in a commons I believe the stubborn will survive, just as small fishermen always have, maybe by working at something in the winter, or you borrow money and buy permits and various types of gear, or lease quota for different fisheries at incredibly inflated prices and chain yourself to loans. You are now a so-called part-time fisherman: a gillnetter/longliner, or a Dungie crabber/winter troller, or even a summer troller/ trawler winter crew. At least you are out there fishing.

Recognizing the Universal Rights of Small Fishermen

The problems for small fishermen and their management stretch across the world. Of course every group and region wants to do a better job of protecting its fish. But this will only come about when everyone faces the reality that better scientific management, adequately funded and uncluttered by political pressures, and not economic schemes, is what rebuilds fish stocks. It seems obvious that must have priority over catch share or any other management structure, but to be done right is expensive. Rebuilding river habitat, for example, is not just a management problem but a great cost problem. Traditional management, that used by our states, can be continually corrected and improved to work for everyone when everyone "listens to the science", managers don't allow themselves to be intimidated by factions and lobbies, and when active fishermen are a serious part of co-management teams. Funding will take serious political pressure.

Along with the fish stocks at risk, so are small fishermen wherever industrialization is chosen or forced. But there is growing recognition that deliberately drying up the small-boat fleets was not, after all, the best economic move. In the United Kingdom belated efforts now go on to halt the consolidation of quota and dismantling of small-boat fleets. These still comprise the large majority of boats, almost 80%, and 2/3rds of fishing employment, while 95% of the fleet quota has migrated to large vessels. (This is similar to the ratio in New England groundfish and herring.) In a turnaround in 2013 the British high court ruled that unused ITQ could be redistributed to smaller boats. It added that "…[fish stocks are not] a private commodity but a public resource…" I'm glad a court reminded us of that and of the history behind our English common law heritage. On the international level, the Food and Agriculture Organization, a branch of the UN, recognizes the right of small fishing to survive. The UN's Convention of the Law of the Sea also asks states to protect the rights of these groups.

The big question is, can our federal system work for more than a select few if it insists on catch share structure? Or, more to the point, what can be done to make it work better if we are stuck with that? Is it inevitable that when you are out, you're out? Is it possible that you could climb back aboard, but only through an unlikely dismantling of the privatized fleet, or some other major change in NOAA policy such as repurchase of quota by the government, requiring funding in the millions? In 2012, when Congress ruled that no new catch share programs would be funded for the east coast, it didn't reflect a change in philosophy at the Dept. of Commerce, but a response to heavy political pressure. Even if no more catch shares were funded, here we are, stuck with about 30 of such programs, regardless of the never-used "sunset clause".

Catch share can work, apparently, where a large majority of the fleet want it, and where there are fish stocks stable and healthy enough that fleet members will be awarded enough quota to be profitable. The fleet must be uniform enough that members will get reasonably similar allocations of quota, and there must be little interest in or need for future generations' opportunity to enter the fishery. But even where every member of a fleet wants catch share there are still social protections that need to be built in, or the communities depending on that fleet will soon suffer as the quota market takes over. If new fishermen can't afford to enter the fleet, the fleet will age and the quota eventually be owned by outsiders.

It doesn't have to happen. The government can buy back quota from retiring fishermen and keep it as a bank for communities and new entries; this is being done in some regions. There has to be control on purchases and sales of quota and on keeping the quota in the hands of active fishermen, not speculators. There are many ways to do this, already described by fishermen and managers, for both new and continuing catch share programs: keeping a limit on the amount of quota an entity can own or lease, keeping it within its original geographic area, freezing the market price, setting aside a certain percent for fixed gear, community-sponsored quota, and new entries, and so on. If the fleet referendum isn't used, those fishermen staying independent need to have equity in terms of fleet quota and government assistance. Modifications of existing programs can be made through the councils to include these social protections; some are expensive, others not so.

Such actions can be demanding of government funding assistance, but much money went into creating socially destructive catch share programs and fishermen can insist that a decent amount go to fixing such programs to be fair. The cost would be no more than the level of assistance government has given in the past to other small business caught in radical technological or economic changes. Today, the amount of federal subsidy given to business, large and small, is huge.

These proposals given above are meaningful because our small fishermen are not yet obsolete. NOAA says that in 2010 there were still 210 commercial boats under 30 feet operating in federal waters off Alaska. The majority of vessels that unload in Alaska are still less than 60 feet, though many of the smaller ones are salmon boats that are by state regulation limited to 32 and 58 feet. In the east, such as in the Gulf of Maine, there are also many small commercial boats, using many gears, still fishing in both state and federal waters.

To survive, small fishermen need to seize the initiative themselves and study the strategies of groups who have developed political action skills, like recreational fishermen, tribes, and ENGOs. Our west coast small fishermen could take lessons from New England fishermen. Closer to home, an example of an earlier battle won by small fishermen was the culmination of all the fish-ins and court cases that led to the Boldt Decision in the Northwest in the 1970s. Another was the battle for co-management of sea mammals that the Inuit of Alaska won. Today those whale stocks are in sustainable condition. The Gulf of Alaska (GOAC3) communities' years of protest did force more social fairness in the crab, rockfish, and cod programs. Today it is probably harder to win anything the establishment opposes, but it can still be done with "significant mass".

With the old Boldt Decision stand-offs fading in the mist, Native and non-Native are seeing that working together they have a better chance of winning on the rebuilding of the Columbia runs. In both the Northwest and Alaska, facing another year of weak Chinook runs, Alaska Native groups have demanded and will win more say in management of the rivers' salmon fisheries. Today, because of the combined protests of commercial, recreational, and Native subsistence fisheries, millions of dollars, state and federal, are being poured into needed Alaskan Chinook research. On the east coast the federal government has made small positive moves that need encouragement to grow. NOAA obviously anticipated problems for New England's small fishermen facing privatization, as shown by the $5 million fund it had already established in 2010 for grants to three states and to large ENGOs to be used to set up permit and quota lease banks for communities. It amounted to tokenism, but showed what could be done. For small fishermen everywhere, the same strategy applies: competing factions--commercials, sport fishers, ENGOs, treaty Indians--all need to put aside their differences and combine forces.

Though it's very possible that NOAA will not be awarded the additional funding truly needed to make corrections to the federal programs, quite a few improvements are possible without great financial commitment beyond normal staff time. They do require take determination from the fishermen and their communities.

Five years ago I had little hope for our small fishermen. I knew of niche fisheries and marketing schemes that were working for a few. But direct sales at the dock, even though they can work for a fishing family, can't be a solution for whole fleets. Yet lately the direct fish-to-consumer movement, like the direct farm-to-consumer movement, has spread so rapidly that I see hope in it for many more small fishermen. An early direct sales movement through roadside fruit stands invented by young people during in the Depression of the 1930s saved many western fruit growers. We take those profitable fruit stands for granted now. Through marketing collaborations like "Community Supported Fisheries" and also joint lawsuits it is clear that small fishermen have begun to see they can hold onto their fisheries only if they join forces with other groups with political and financial clout. The "sustainable fishery" labeling some fleets have invested in is also paying off. In 2012 when foreign farmed salmon flooded the US market again, the prices for domestic "caught fish" held up, including that which we exported. Certain wholesalers had chosen to buy the caught fish rather than only cheaper farmed fish.

Without doubt catch share systems and other new fishery management schemes will be the focus of worldwide debates for the next few years, along with the traditional fishery battles over regulations. Will closed-down fishermen get government help to make a transition to other gear or stocks, or will they, like their boats, simply be added to the growing scrap pile? Or will fish farming take over entirely, as some economists predict, and the catch fisheries become, what economists showed John--an empty place on the chart? Fortunately in this country we haven't reached the stage yet where pounds of protein produced per pounds of feed, along with the profit earned, takes precedence over all other concerns. Eric, the widely-read member of the Nome fleet, in speaking of industrial fishing and other modern impacts, gave me this prediction: "Before long environmental criminals will be tried along with war criminals at The Hague. Then we will see things get better." I would like to be here to see that.

But fishermen can't wait for that. They do have to reach the general public, and soon, with their information, and the public needs to reach its political leaders. Clearly, if small fishermen don't win more support, most of them--like my codfishing grandfather--will hang by a knife blade.

In writing this book I wanted first to pass on stories from small-scale fishing lives, both subsistence and commercial, that still survive. I wanted to show what is lost if small fishing should disappear, especially the social cost if young people can no longer find a livelihood in the fisheries. My other purpose was to show how our government, starting with narrowly focused, poorly funded fishery management, bought into the magic cures of limited permitting, then of privatization and changed forever small fishing in ways that hurt fishing families and their communities. In this chapter I've described changes more hopeful for small fishermen through a stronger commitment by government and councils to the intent of the MSA, and through niches such as community supported fisheries.

But the rebuilding of small fishermen's livelihoods means to oppose the free-marketers and other big-industry interests. They have been overcome in the past only when a public voice was strong enough to overcome powerful lobbies. Think of child labor laws, anti-trust laws, miner safety laws, food safety laws, water rights. All of these and so many other social improvements put aside industrial efficiency in favor of quality of life for families and communities.

I don't want my story simply to come across as an argument that more people need to be able to make a living fishing, or as a Luddite-like hatred of industry or technical progress. I know that the erosion of the small-scale fishing has not been more tragic than the sorrows of the small farmer, the small logger, the craftsman, or small local market owner. Except for the rigorous outdoor activity, small fishing has much the same virtues and risks as any small family business. Obsession with growth, technological advances, and false efficiencies, and of course profit, has undermined all of those livelihoods. When society and nature eventually rebel, it soon becomes everyone's problem, everyone's risk of sinking. The lure of independent work and new challenges are what draw people to small ventures everywhere, and we Americans traditionally honor that. Add to that the smell of fresh sea wind, the promise of fish flashing silver before your eyes, and the chance that this very trip may be the best one of your life. If our small boat fleets go under, we lose one more story of the United States as the special land of opportunity for everyone.

Much credit should go to all those who believe small fishermen's survival is important and have the courage and persistence to use our political processes to salvage it, crying out against, in John's words, "the extraordinary loss of so many who inherently owned that resource too." I hope I can speak for more than myself when I say thank you to all those who follow in the spirit of Rachel Carson, taking on powerful interests in the cause of both nature and human societies.

To non-fishermen readers, I hope you will find out more of the problems small fishermen face. Visit on the docks, read what they say in their blogs and on-line periodicals. You may feel concerned enough to send a message to your Congressman that you would like to see the MSA more diligently enforced, and that we should try to save not only fish but fishing communities. While you hope for action you might go down to the nearest fishing port and buy a fresh fish--any kind but farmed--to enjoy for dinner.

I don't fool myself; I know there are fishermen who will fight to fish, but who will never go to a meeting even if their livelihood depends on it; it's not in their nature. My cousin George may be one of them. He called me again in summer of 2013, but not from Newport. He was at Sitka, working as a boat-puller on an old friend's troller. I burst out laughing. But what should I have expected? He had worked ashore for three years, separated from the cheese house venture that his wife and grown daughter run successfully on their own. He took a break from the ocean, went on long hiking trips alone in the mountains in the summers.

"But I couldn't just keep hiking in the mountains forever. And I found out I didn't have any identity, just being a retired guy ashore. So here I am, fishing again. I love Sitka--I think I want to live here. There are lots of boats for sale, even a few I can afford."

A month later he bought a 40-foot gas-powered wooden troller, modest, but with all the modern gadgets he had said he'd never need. The old owner said, "Well, you can take them all off if you want!" George's oldest son loaned him the funds for an Alaska limited permit, and he headed north, to fish out of Sitka in 2014 with another at-loose-ends aging cousin with him as boatpuller. The next spring he was ready to go fishing again, and with no more worries about lost identity. He already knew many of the boats in the Sitka harbor, and their histories, if not their present owners. Trolling was looking up in Southeast Alaska.

It was an incredible regeneration story. So I had to ask George again for his explanation of the powerful attraction commercial fishing has for so many, most of whom will no longer make a full living at it, or never did. What makes them untie and head out once more, and once more? His answer was not, "Oh, the thrill of landing a big lunker," or "The chance to get rich." It was, after a long moment: "It's that you have real choices. You're not just carrying out someone's orders; you're out there making your own real choices. That's what draws us." I would add to that my own answer: fishermen's incurable hope.

Part II

Our Federal Small-Boat Fleets Photos

Winter crabbing near Nome, Dan Thomas and Frank McFarland, 1990

Winter crabbing outfit, Nome, Rob Thomas and Frank McFarland, 2012

Rob Thomas and Audrey Aningayou on *Island Girl*, Nome, 1993

Frank McFarland crabbing with *Island Girl*, near Nome, 1996

Crewman Fred Topkok, Frank's skiff, about 1996

Nome's small-boat crab fleet, 2010

Frank's *Mithril* at sea, Nome, 2013

Frank unloading *Mithril*, 2010, before extension

Crewman Conrad Klemzak, left, Frank in door, *Mithril*, 2010

Johnnie Noyakuk's Bristol II, crewmen Robert Noyakuk, Garrett Adsuna, 2013

Mithril crewman Frank Kavairlook, subsistence halibut, 2012

Rob's *Stephanie Sue* fishing halibut off Sledge Island near Nome, 2011

Stephanie Sue, Nome, 2012

Stephanie Sue crew, Scott Aningayou and Dan Apok, 2011

Rob and Scott; this makes it worth the storms, 2007

Stephanie Sue herring roe fishing, lower Norton Sound, 2013

Icicle Seafoods' herring roe processor; a contrast in vessels, 2013

Acronyms Used

ABC Acceptable Biological Catch
ADFG Alaska Dept. of Fish and Game
AFA American Fisheries Act
AFS Aboriginal Fisheries Strategy (Canada)
ANILCA Alaska National Interest Lands Conservation Act
BC British Columbia
BIA Bureau of Indian Affairs
BLM Bureau of Land Management
BOR Bureau of Reclamation
BSFAAK Bering Sea Fishermen's Association of Alaska
BS/AI Bering Sea/Aleutian Islands
BPA Bonneville Power Authority
CBD Center for Biological Diversity
CDQ Community Development Quota
CFEC Commercial Fisheries Entry Commission (AK)
CQE Community Quota Entity
DFO Dept. of Fisheries and Oceans (Canada)
EDC Environmental Defense Council
ENGO Environmental non-governmental organization
EEZ Exclusive Economic Zone
ESA Endangered Species Act
IFQ Individual Fishing Quota (see ITQ)
IPHC International Pacific Halibut Commission
ITQ Individual Transfer Quota
LEP Limited Entry Permit (State)/ LLP, FLP Limited Permit (Federal)
MMPA Marine Mammal Protection Act
MSA Magnuson-Stevens Act (earlier the Magnuson Act)

MSY Maximum Sustained Yield

NEPA National Environmental Protection Act

NMFS National Marine Fisheries Service (branch of NOAA)

NOAA National Ocean and Atmospheric Administration

NSEDC Norton Sound Economic Development Corp

NPFMC North Pacific Fisheries Management Council (federal waters off Alaska)

ODFW Oregon Dept. of Fish and Wildlife

PSC Prohibited Species Catch

PFMC Pacific Fisheries Management Council (federal waters off WA, OR, CA)

PST Pacific Salmon Treaty

SARA Species at Risk Act (Canada)

TAC Total Allowable Catch

WDFW Washington Dept. of Fish and Wildlife

Appendices

A. A Bare Chronology of Fishery Management Events (incomplete, mainly west coast)

1858 Treaty awards several Washington tribes fishing rights at their traditional sites.

1879 First salmon cannery opens in Southeast Alaska

1914 Northwest canned salmon pack starts to shrink.

1920s Commercial herring fishery begins in Gulf of Alaska

1920s Federal support to hatchery system in Pacific Northwest begins.

1930s Bonneville Power Authority begins federal dams on Columbia-Snake system.

1933 Pacific halibut fleet begins a voluntary conservation plan using "Lay-ups".

1934 Washington State outlaws fish traps via citizen initiative, Oregon follows.

1950 Diesel-powered foreign stern trawlers arrive in the Gulf of Maine.

1958 New State of Alaska outlaws fish traps; protects sustainable resources for residents.

1960s-1970s Northwest Indians protest non-enforcement of treaty fishing rights.

1964 Major earthquake decimates much of fishing industry in Gulf of Alaska.

1967 Stratton Commission recommends a shrinking and modernizing of US fishing fleets.

1969 Moratoriums and limited permitting starts in BC.

1973 The Endangered Species Act passes (first Pacific salmon listed in the 1990s.)

1974 Judge Boldt decision gives Washington treaty tribes 50% of salmon quota.

1975 Limited Entry Program begins for Pacific salmon: AK, WA, OR, CA.

1975 to present: Government/fleet buy-backs of licenses and boats begin.

1976 Magnuson Act (MSA) passes. The 200-mile limit Exclusive Economic Zone passes.

1976 Regional Fishery Councils created under MSA.

1978 Alaska legislature passes "subsistence priority" for fish and game resources.

1980 Federal ANILCA law creates a "rural subsistence priority" on AK federal lands.

1980-1996 Feds and Alaska battle over the conflict between two versions of subsistence policy.

1981-82 The Kodiak red king crab fishery and others crash; many closures.

1984 Iceland introduces IFQ to its fleets.

1985 The Pacific Salmon Treaty signed, divides fish 50-50 between Canada and the US.

1986 New Zealand introduces ITQ.

1986 Canada begins BC ITQs with scallops.

1987 Fall run Sacramento Chinook ESA listed.

1989 *Exxon Valdez* giant oil spill in Prince William Sound ruins many fisheries.

1990s Dungeness crab fleets of WA and OR become Limited Entry.

1990-91 The first US IFQs begin with Mid-Atlantic Council: surf clams, quahogs, wrackfish.

1991 Snake River sockeye is first Northwest salmon to be ESA listed. 13 stocks soon added.

1992 Canada's Atlantic cod crash.

1991-1992 Canada begins BC ITQ for sablefish (black cod), then halibut.

1992 The Community Development Quota (CDQ) program begins in Western AK

1992 Norton Sound Economic Development Corp, the CDQ for Norton Sound, begins.

1994 The Oregon Coast Coho harvest closes.

1994 Federal limited permitting (LLP) of cod fleets begins in Bering Sea.

1995 "Days at Sea" regulations begin on east coast.

1995 North Pacific Council begins IFQ for Alaska halibut and sablefish fleets.

1995-present: Numerous lawsuits entered against NOAA/ NMFS by ENGOs and fishermen.

1996 Feds take over subsistence management of Alaska's federal lands and waterways.

1996 Ninth Court of Appeals finds IFQ programs legal.

1996-2002 A moratorium by Congress on new IFQs. More research ordered.

1996 MSA reauthorization adds three new standards; a 2014 deadline for all stocks' recovery.

1997 Canada rationalizes BC trawl fleet.

1997 BC gillnetters hold AK ferry *Matanuska* hostage; AK governor sues Canada.

1998 The American Fisheries Act passes, rationalizes Bering Sea pollock trawlers.

2002 US moratorium on IFQ ends.

2002 Pacific Council Sablefish IFQ begins for WA, OR.

2002 Klamath River low water Chinook disaster; restrictions/closures on trollers.

2002 Gulf of Alaska Coastal Community Coalition forms to represent small communities that lost fleets through IFQ.

2003-2012 ESA Columbia/Snake salmon cases before Judge Redden.

2003 Dam breaching movement active on west coast.

2004 Pew Environmental Trust funds two pilot non-trawl fisheries at Georges Bank.

2004 Canada's "Species At Risk" Act installed; provincial acts also in effect.

2004 North Pacific Council's Community Quota Entity program approved for Gulf of Alaska.

2005 N.P. Council rationalizes Bering Sea/Aleutian crab fleets and processors.

2006 N.P. Council review of "crab ratz" turns into major protest at Kodiak.

2006 Magnuson-Stevens Act is reauthorized with increased emphasis on "catch share" option, ecosystem-based management, bycatch reduction, and protection of fishing communities.

2006 Old Harbor begins first Community Quota Entity halibut program

2007 Central Gulf of Alaska Rockfish coop program begins.

2007 BC hook and line groundfish catch share begins.

2008 Bering Sea non-pollock groundfish catcher-processors cooped.

2008-2011 BC pilots a salmon catch share program.

2008 Sacramento River fall Chinook crash; drought and ocean conditions blamed.

2009 Ecotrust Canada, Food and Water Watch protest effects of catch share.

2009 Frazer River sockeye run crashes. Court review follows.

2009 N. P. Council passes first salmon bycatch hard cap: Bering Sea pollock fleet's Chinook bycatch.

2010 NOAA proposes leasing US federal waters to private fish farms.

2010 Aquabounty announces first genetically modified farmed salmon.

2010 NOAA releases National Policy on Catch Share.

2010 AK crabber crews get federal loan program for quota purchase.

2010-15 Western Alaska Chinook runs in crisis, cause unknown.

2010 More N.P. Council catch share planning in progress.

2010 New England Council groundfish catch share sectors begin.

2010 NE mayors, Congressmen Frank and Tierney, sue NOAA/NMFS over "misinterpretation of MSA". Suit lost and appealed.

2010 West Coast Fed. of Fishermens' Assns. joins lawsuit to halt Pacific Council catch share; suit lost.

2011 Chile salmon viruses lose strength and farm fish again flood market.

2011 Pacific Council groundfish catch share program begins.

2005-2013 Direct sales of fish through "consumer supported fisheries" grows in popularity.

2012 U.S. House of Rep. votes no new catch share funding for east and south coasts.

2012 President signs new National Ocean Policy.

2012 US imports 17 billion lbs. of fish, mainly farmed fish.

2012 NMFS postpones exploratory trawling opening in the Northern Bering Sea after west coast Alaska Natives protest.

2012 N.P. Council orders AK halibut charter boats to limited permitting, and catch share with commercial boats.

2012 AK Native organizations demand more share in river fisheries management.

2012 N.P. Council catch share programs improve options for small fishing/ communities.

2012 Canada's official report on missing Frazer sockeye: "multiple causes… in the ocean".

2012 BC fish farm groups begin "enclosed pens" away from inlets.

2012 New England Council orders huge cut in Atlantic cod quota.

2012 Major dam removal on Elwha River, WA; important Chinook stream rebuilding.

2013 WA and OR governors' phase-out of Columbia gillnetting begins.

2013 Initiative to close AK gillnetting near urban areas is declared illegal.

2013 Major studies released on the growing acidification of the oceans.

2013 Pacific Council's trawlers demand more NMFS help for their catch share program.

2013 and 2014 Record salmon catch in Alaska; Chinook runs still very weak.

2013 Canada's DFO proposes catch share for all BC salmon fisheries.

2015 Congress debates new MSA amendments; large issue: sustainability v. social protection.

2015 N.P. Council cuts trawlers' halibut bycatch by 25%. Too little, say protesters.

2015 BC Indians refuse $1 billion for land trade/construction of LNG plant.

B. MSA 2006, A Summary of Revised National Standards

1. Prevent overfishing while achieving optimum yield, based upon the best scientific information available.
3. Manage individual stocks as a unit to the extent practical; interrelated stocks as a unit.
4. No discrimination between residents of different states.
5. Promote economic efficiency, except that no such measure shall have economic efficiency as its sole purpose.
6. Minimize costs and avoid duplications where practicable.
7. Take into account and allow for variations and contingencies in fisheries.
8. Take into account the importance of fishery resources to fishing communities to provide for the sustained participation of, and minimize adverse impacts on, such communities (consistent with conservation requirements.) *Note the revision from earlier versions; this new version omits phrase "to the extent practicable."
9. Minimize bycatch or mortality from bycatch.
10. Promote safety of human life at sea.
11. Sec. of Commerce will: establish guidelines for management plans based on national standards, a balanced apportionment on councils from the fisheries, technical assistance to councils as needed.

Sources Used

Books and Video Tapes:

Allison, Charlene J., Sue-Ellen Jacobs and Mary A. Porter. *Winds of Change: Women in Northwest Commercial Fishing,* Seattle: U. of Wash. Press, 1989.

Anderson, James. *Decadal Climate Changes and Declining Columbia River Salmon,* Seattle: U. of Wash. Press, 1997.

Arnold, David F. *The Fishermen's Frontier: People and Salmon in Southeast Alaska,* Seattle: U. of Wash. Press, 2008.

Bailey, Kevin. *Billion-Dollar Fish: The Untold Story of Alaska Pollock,* Chicago: Univ. of Chicago Press, 2013.

Bell, E. Heward. *The Pacific Halibut: the Resource and the Fishery,* Anchorage: Alaska Northwest Pub. 1981.

Brown, Bruce. *Mountain in the Clouds,* Seattle: U. of Wash. Press, 1995

Brown, Dennis. *Salmon Wars: The Battle for the West Coast Salmon Fishery,* Madeira Park, BC: Harbour Publishing, 2005.

Caldwell, Francis. *Pacific Troller: Life on the Northwest Fishing Grounds,* Anchorage: Northwest Pub. 1978.

Carson, Rachel. *The Edge of the Sea,* NY: Houghton-Mifflin, 1955.

Carter, Bill. *Red Summer: The Danger, Madness, and Exaltation of Salmon Fishing in a Remote Alaskan Village,* NY: Scribner, 2008

Carey, Richard Adams. *Against the Tide,* NY: Houghton Mifflin, 1999.

Crutchfield, James and Guilo Pontecorvo. *The Pacific Salmon Fisheries: A Study of Irrational Conservation,* NY, RFF Press, 1969.

Cullenberg, Paula, ed. *Alaska's Fishing Communities: Harvesting the Future,* (Essays by Linda Behnken, Courtney Carothers, Fred Christiansen and

Gale Vick, Steve Landon and Emilie Springer, Marie Lowe) Fairbanks: Univ. of Alaska Seagrant, 2006.

Fields, Leslie Leyland. *The Entangling Net: Alaska's Commercial Fishing Women*, Urbana: U. Illinois Press, 1997.

_____. *Surviving the Island of Grace*, Kenmore, WA: Epicenter Press, 2008.

_____. *Hooked!* Kenmore, WA: Epicenter Press, 2011.

Haig-Brown, Alan. *Still Fishin': The BC Fishing Industry Revisited*, Madeira Park, BC: Harbor Publishing, 2010.

Hawley, Steve. *Recovering a Lost River: Removing Dams, Rewilding Salmon, Revitalizing Communities*, Boston: Beacon Press, 2011.

Haycox, Stephen. *Frigid Embrace: Politics, Economics and Environment in Alaska*, Corvallis: Oregon State Univ. Press, 2002

Hylen, Walter. *The Fish Belong to the People,* (video documentary), Port Clyde, ME, 2008.

Klein, Naomi. *Shock Doctrine: The Rise of Disaster Capitalism*, New York: Picador, 2007.

Krueger, Charles C. and Christian E. Zimmerman, eds. *Pacific Salmon: Ecology and Management of Western Alaska's Populations,* Bethesda MD: American Fisheries Society Symposium 70, 2009.

Kurlansky, Mark. *Cod: A Biography of the Fish That Changed the World*. NY: Penguin, 1997.

_____. *The Last Fish Tale*, NY: Penguin Books, 2008

Lichatowich, Jim. *Salmon Without Rivers: A History of the Pacific Salmon Crisis,* Wash. D.C.: Island Press, 1999.

_____. *Salmon, People, and Place: A Biologist's Search for Salmon Recovery*, Corvallis: Univ. of Oregon Press, 2013

Lord, Nancy. *Fish Camp: Life on an Alaskan Shore*, Wash. DC: Counterpoint Press, 1997.

McCloskey, William. *Highliners, Breakers; Raiders* (fiction triad), Guilford, Conn.: Lyon Press, 1979-2004.

Molyneaux, Paul. *The Doryman's Reflection: A Fisherman's Life,* Boston: Perseus Book Group, 2005.

_____. *Fishing in Circles: Aquaculture and the End of Wild Oceans*, NY: Thunder's Mouth Press, 2009

Montgomery, David R. *King of Fish: The Thousand Year Run of Salmon*, Cambridge: Perseus Books Group, 2003.

National Research Council. *Upstream: Salmon and Society in Pacific North Northwest*, Wash. DC: National Academic Press, 1996.

Playfair, Susan R., *Vanishing Species: Saving the Fish, Sacrificing the Fisherman*, Lebanon, NH: Univ. Press of New England, 2003.

Roberts, Callum. *The Unnatural History of the Sea*, Wash. D.C: Island Press, 2007.

Rose, Alex. *Who Killed the Grand Banks?* Vancouver: John Wiley $ Sons, 2008.

Safarik, Norman and Alan Safarik. *Bluebacks and Siver Brights: A Lifetime in the BC Fisheries*, Toronto: ECW Press, 2012.

Ulrich, Roberta. *Empty Nets: Indians, Dams, and the Columbia River*, Corvallis: Oregon State Univ. Press, 1999.

University of Alaska Marine Advisory Program, UAF (Film) "Community Development Program", 1999.

Wickham, Eric. *Deadfish and Fat Cats: A No Nonsense Journey Through Our Dysfunctional Fishing Industry*, Vancouver. BC: Granville Isl. Pub. 2002.

White, Richard. *The Organic Machine: The Remaking of the Columbia River*, NY: Hill and Wang, 1995.

Williams, Richard N. *Return to the River: Restoring Salmon to the Columbia River*. Burlington, Maine, Elsevier Press, 1991.

Articles, Reports, Websites, and Blogs

Alaska Commercial Fisheries Entry Commission. "Changes Under Alaska's Halibut IFQ program, 1995-1998, 96-10N," 1998.

_____. "Limited Entry Permits Issued, 1976-2009"

Alaska Dept. of Fish and Game website. "Commercial Fishing Effort 2000-2004; "Hatchery program, history, goals, successes, and statistics, 2005."

_____ "History of Alaska Salmon Fisheries." No date.

Alaska Marine Conservation Council. "Standards for IFQ or other Dedicated Access Programs", report, June 2005.

Alaska-Yukon-Kuskokwim Sustainable Salmon Initiative/ "Protecting the Future of Salmon: Annual Report", Anchorage: Bering Sea Fishermen's Assn., 2006.

Anderson, J. "Decadal Climate Cycles and Declining Columbia River Salmon", *Sustainable Fisheries Conference Proceedings, Victoria BC, 1998*, Bethesda MD: American Fisheries Society, 2000.

Arundel, Barbara. *Emagazine.com*. "How Factory Fishing Decimated Cod", March-April, 2001.

Agar-Kurtz, Breen. "Slipping Through Their Hands", *Pacific Fishing*, May, 2013.

Bardot, Bruce. "A Novel Approach to Fisheries", *YaleEnvironment360*, Jan. 2011.

_____. "Hatch-22: The Problem of Pacific Salmon Resurgence", *YaleEnvironment360*, Nov. 1, 2010.

_____. "Fish Council to Review Controversial Crab Rationalization Program", *Alaska Journal of Commerce*, June, 2009.

Bernton, Hal. "Alaska, Washington in Fishing Fleet Tug of War", *Seattle Times*, Oct. 1, 2011.

_____. "Dividing the Catch", *Seattle Times*, July 1, 2012.

Blackford, Mansel. "A Tale of Two Fisheries: Fishing and Overfishing American Waters", *Origins*, Sept. 2008.

Blume, Johanna. "Offshore Profits, Onshore Communities: Historical Perspective on the Effects of Federal law on Subsistence Fishing in Bristol Bay", *Fishing People of the North*, Fairbanks: UA Sea Grant, 2012.

"Boldt Decision": see: US Federal Court V. Washington

Brewer, Jennifer F. "Paper Fish and Policy Conflict: Catch Shares and Ecosystem-Based Management in Maine's Ground Fishery", *Ecology and Society*, Vol. 16, #1, 2011.

Butler, Caroline. "Paper Fish: The Transformation of the Salmon Industries", American Fisheries Society, 2008

Carlisle, John. "The American Fisheries Act: Special Interests Politics at Its Worst", National Center for Public Policy Research #209, Wash. D.C., 1998.

Chrisman, Gabriel. "The Fish-in Protests at Frank's Landing", Seattle: Univ. of WA, The Seattle Civil Right and Labor History Project website, 1998.

Clark, Michael. "Experiences with Individual Fishing Quotas and Their Potential in Japan", NOAA Website/Fisheries Agencies, Jan. 29, 2010.

Davidsuzuki.org (website). "An Opportunity for BC Salmon Fisheries to Catch Up with the Pack", August 21, 2013.

Dew, C. Braxton and Robt. McConnaughy. "Did Trawling on the Brood Stock Contribute to the Collapse of Alaska's King Crab?" (tech. paper), Alaska Fisheries Science Center, 2005.

Dinneford, E. and, K. Iverson. "Changes Under Alaska's IFQ, 1995-1998", Alaska Comm. Fish. Entry Commission, 1999.

Dory Associates. "Access Restriction in Alaska Commercial Fisheries: Trends and Considerations", (Report), Kodiak, Jan. 2009.

Doughton, Sandi. "Bristol Bay Sockeye Thrives on Diversity UW Study Says", *Seattle Times*, June 3, 2010.

Ecotrust (website), "National Panel on Community Dimensions of Fisheries Catch Share" (16 recommendations), March, 2011.

EcotrustCanada (website), "Catch-22: Conservation, Communities, and Privatization of BC Fisheries", 2004, 2009.

Enge, John. *Alaskacafe* (website) fishery commentary, 2004-2010.

Environmental Defense Fund (website). "Catch Share Key to Reviving Fisheries", Oct. 2009.

_____. "Catch Shares Benefit Fishermen and the Environment: A Scientific Compendium, Nov. 11, 2012.

Eythorsson, Einer. "A Decade of IFQ Management in Icelandic Fisheries", Alta, Norway: Finnmark Research Center, 2006.

_____. "Coastal Communities and ITQ Management: The Case of Iceland", Alta: Finnmark Research Center, 2005

Elfring, Chris. "The CDQ Program: Summary of Review", Report, Wash. D.C: National Research Council/Nat. Academy of Sciences, July 19, 2001.

Fisheries and Oceans, Dept of. (Canada) Official website. (Misc. reports, program evaluations).

Fish News EU. (website). "Arctic Fisheries Catches 75 Times Higher than Previous Reports," Feb. 2011.

_____. "Community Dimensions of Fisheries Catch Share Programs: Integrating Economy, Equity, and Environment", March 15, 2011.

Flournoy, Peter. "Mail Buoy" letter, *National Fisherman*, July 2012.

Food and Agriculture Organization (UN) Media Centre. "Fishing Experts Agree on First Guide- lines on Reduced Discards", Jan. 14, 2011.

_____. "A Global Assessment of Fisheries Bycatch and Discards", 2011.

Food and Agriculture Organization/Fisheries (UN) (website): FAo.org/fishery/ssf. (Misc. articles on importance of small-scale fisheries).

Food and Water Watch (website). Misc. articles on catch share issues, 2009-2012.

GAO Highlights (US GAO Office) ""Protecting Economic Viability of Fishing Communities and New Entrants", 2004.

Frula, David and Shaun M. Gehan. "New Stream of Litigation Hits NMFS", *National Fisherman,* Sept. 2013.

_____. "Turning the Tide", *National Fisherman*, May, 2014.

Gaines, Richard. (misc. fishery news and commentary) *Gloucester Times,* 2008-2012.

Gharret, Jessica and Phil Smith. "Community Quota Equities: Technical Support Workshop" NMFS, Feb. 2009.

Gilden, Jennifer and Courtland Smith. "Survey of Gillnetters in Oregon and Washington," OSU Seagrant, 1996.

Glavin, Terry. "Transferable Shares in BC Commercial. Salmon Fishery", Vancouver, Watershed Watch Salmon Society, Sept. 2007.

Grader, Zeke. "Choices: Groundfish, Which Direction", *Fishermen's News,* Sept. 2010.

_____. "Standards for IFQ: Open Letter to NMFS", *Fishermen's News,* Oct. 7, 2002

Grafton, Quentin and Harry Nelson. "Effects of Buy-back Programs in BC Salmon Fisheries", San Diego: Univ. So. Cal., 2004

Grimm, Dietmar, and Judd Boomhower, et al. "Can Catch Shares Reduce the US Federal Deficit?" *Journal of Sustainable Development,* V. 3-4, Dec. 2010.

Goldfarb, Ben. "The Catch-22 of New England Fisheries' Catch Share Scheme", *Earth Island Journal,* April, 2013.

Grossman, Zoltan. "The Cowboy-Indian Alliance Rises to Protect Our Common Land and Water", *Common Dreams,* April 23, 2014.

Gulf of Alaska Coastal Communities Coalition (website). "Community Purchase of Halibut and Sablefish IFQs," May 30, 2000.

Haig-Brown, Alan. "BC Salmon Catch Increasingly Controlled by a Few Corporations", *The Tyee,* April 24, 2011.

Haines, Terry. *Alaska Report* website, (commentary), April 15, 2006-2010.

Halvorsen, Robert. "Implications of Seafood Processor Quotas", Senate hearing testimony, Feb. 2004.

Hanna, S.S. "User Participation and Fishery Management Performance in NPFMC," *Ocean and Coastal Management 28 (2-3) 23-94,* 1998.

Harsila, David. "Threats to Bristol Bay Fishery Continue to Cloud Future," *Pacific Fishing,* Jan. 2012.

Healy, Michael, "Resilient Salmon, Resilient Fisheries for British Columbia, Canada," Vancouver: UBC, 2009.

Heard, William. "Alaska Salmon Enhancement: A Successful Program for Hatchery and Wild Stocks", UJNR Technical Report No. 30, 2002.

Helliwell, David. "Catch Shares Move Money to the Already-Rich", *Pacific Fishing,* Oct, 2011.

Hume, Mark. Commission Presses Ottawa to Look at Fish Farm Impact on Salmon," *Globe and Mail,* Vanc. BC, Oct. 31, 2012.

Jensen, Andrew. "Rockfish Program Signals Shift in State Fisheries Policy", *Alaska Journal of Commerce.* June 23, 2010.

_____. "Washington/Oregon Crab Interests Take Aim at CDQs", *Alaska Journal. of Commerce.* Sept., 2011.

_____. "Coastal Villages (CDQ) Exploits Salmon Shortage to help 'Pollock Provide'," *Alaska Journal of Commerce,* July 19, 2012.

Johnson, Terry. "Climate Change Will Change You," *Pacific Fishing,* Aug. 2012.

KNOM Radio, Nome. Profiles "Interview with Nick Andrews re: Protest fish-in at Marshall", July 2, 2009.

Knapp, Gunnar. "Attitude of Alaska Halibut Fishermen Toward IFQ Management", Anchorage: UAA/ISER, 2000.

_____. "Comparison of Recent Sport and Commercial Fisheries Economic Studies", Anchorage: UAA/ISER, 2009.

_____. "Economic and Social Impact of BSAI Crab Rationalization on Aleutian East Borough", Anchorage: UAA/ISER, 2006.

_____. "Economic Impacts of BSAI Crab Rationalization on Kodiak Fishing Employment and Earnings", Anchorage: UAA, ISER, 2006.

_____. "Local Permit Ownership: Alaska Salmon Fishery", Anchorage: UAA/ISER, 2009.

Knapp, Gunnar and Dan Hull. "The First Year of the Halibut Quota Share Holders", Anchorage: UAA/ISER, 1996.

Knapp, Gunnar and Cathy Roheim, James Anderson. *The Great Salmon Run: Competition Between Wild and Farmed Salmon,* Anchorage, UAA/ISER 2007

Kresta, Joey. "Codfish Stock Report Sparks Fear in Fishery", Seacoastonline (website), Dec. 9, 2011.

Kawerak, Inc. "Norton Sound Fisheries Enhancement Summit" Final Report of Conference, Nov. 4-5, 2004.

Langdon, Steve. "The Community Quota Program in the Gulf of Alaska: A Vehicle for Native Village Sustainability." *Enclosing the Fisheries:People, Places, and Power,* Bethesda, MD: American Fisheries Society, 2008.

Lewis and Clark Law School, (no author given) "Alliance Against IFQ v. Brown", 9th Court. *Environmental Law,* 2006.

Lowe, Marie. "Crab Rationalization and Potential Community Impacts", *Enclosing the Fisheries: People, Places, and Power,* Bethesda, MD: American Fisheries Society, 2008.

Loy, Wesley. "Highliner" columns, *Anchorage Daily News,* 2008-2009.

_____."Deckboss" columns, *Pacific Fishing,* 2009-2015.

Madsen, Stephanie. "Alaska's Groundfish Fisheries: A National Model for Sustainability for More Than Thirty Years", *Alaska Business Monthly,* Dec. 2006.

Magdanz, Jim and Annie Olanna."Controls on Fishing Behavior on the Nome River", Technical Paper, AK Dept. Fish and Game, Nome, 1984.

Mansfield, Becky. "Neo-liberalism in the Oceans: Rationalization, Property Rights and the Commons Question", (paper), Columbus: OSU, rev. 2003.

Martin, Jim. "A Perspective on Coho Salmon Management in Oregon", *Pacific Salmon: Ecology and Management,* Bethesda, MD, American Fisheries Society Symposium, v. 70, 2009.

Matulich, Scott. "Implications of the Halibut and Sablefish Study for Crab Rationalization", paper presented to NPFMC/SSC, April, 2002.

Mayo, Ronald D. "An Assessment of Private Salmon Ranching in Oregon", (Report, no sponsor listed), Nov. 30, 1988.

_____. "Did Processing Quota Damage Alaska Red King Crab Harvesters?" *Marine Resource Economics,* Vol. 23, 253-271, 2008.

McDowell Group. "State of Alaska Economic Seafood Strategies", (report for State of Alaska), 2006.

McGee, Steven. "Salmon Hatcheries in Alaska--Plans, Permits, and Policies Designed to Protect Wild Stocks", Bethesda, MD: American Fisheries Society, 2004

McNicholas, Laurie. *Nome Nugget,* fisheries related articles, Nome, AK, 2004-2013.

Menard, Jim, and Chas. Krueger, John Hilsinger. "Norton Sound Salmon Fisheries", *Pacific Salmon: Ecology and Management,* American Fisheries Society, 2009.

Meridian Institute. "Catch Share in New England: Key Questions and Lessons Learned from Existing Programs," Feb. 2010.

Miller, Scott A. "Economic Transition in Western Alaska Communities", *Fishing People of the North: Traditional Salmon Fishery Dependence and Emerging Groundfish Fishery Dependence,* Fairbanks: UAF Seagrant, 2012.

Mittal, Anu E. "Economic Effects of IFQ on Processors and Methods Available to Protect Communities", U.S. Government Accounting Office, Feb. 25, 2004.

Muse, Ben and Kurt Schelle, et al. "Changes Under Alaska's Halibut IFQ Program, 1995-1998," AK Commercial Fish. Entry Commission Report 96-10 N, 1999.

National Fisherman, 2010-2012, misc. articles.

National Marine Fisheries Service/NOAA (website), news, minutes, reports, rulings, 2000-2014.

National Resources Defense Council (website). *"Bring Back the Fish: An Evaluation of U.S. Fisheries Rebuilding Under the Magnuson-Stevens Fishery Conservation and Management Act",* March, 2013.

North Pacific Fisheries Management Council. *"Bering Sea Chinook Salmon Bycatch Management,"* Vol. I and II, Dec. 2009.

_____. "Fishing Fleet Profiles, 2012".

_____. "Gulf of Alaska Groundfishery Management Plan, June 2012".

_____. "Review of the CQE Program of the Halibut/ Sablefish IFQ Program; Final Report", March, 2010.

_____(website). (minutes, reports, program evaluations), 1995-2014.

Northwest Atlantic Marine Alliance (NAMA.org) (website). A Community Supported Fishery network with links to the west coast.

Northwest Fishletter (website). # 291, July 15, 2011.

Norton Sound Economic Development Corp. (newsletters, annual reports), 2004-2014.

Oliver, Chris. "Summary Report to Council Coordinating Committee" (NPFMC report to regional fishery councils), May, 2009.

O'Neill, Catherine. "Fishable Waters", *American Indian Law Journal*, Spring, 2013.

Oregon Dept. of Fish and Wildlife (website): misc. reports on fisheries issues, 2004-2012.

Pacific Fishing, misc. articles 2006-2015.

Pauli, Daniel. "Aquacalypse Now", *The New Republic*, Sept. 28, 2009.

_____. (as quoted by Cassandra Profita in "Catch shares: How to Split a $40 Million Slurpee", *Oregon Public Broadcasting*, Feb. 19, 2013.

Pew Environmental Group. "Design Matters: Making Catch Share Work", Wash. D.C.: Pew Charitable Trusts, 2009.

Pinkerton, Evelyn and Danielle Edwards. "The Elephant in the Room: Hidden Costs of Leasing ITQs," *Marine Policy Journal*, 2008.

_____ "Alternatives to ITQs in Equity-efficiency-effectiveness Tradeoffs", Simon Frazer Univ., 2013.

Platt, Ben. "A Fisherman's Perspective: Catch Share", *Institute for Fisheries Resources*, 2008.

Reedy-Maschner, K. "Eastern Aleut Society and Three Decades of Limited Entry", *Enclosing the Fisheries: People, Places, and Power*, Bethesda, MD: Amer. Fisheries Soc., 2008.

Rust, Suzanne. "Catch Share Leaves Fishermen Reeling", *The Bay Citizen*, March 12, 2013.

Sackton, John. "Battle to Save the Small Boat Fleet in New England", *Seafoodnews.com*, May 13, 2010.

Save Our Wild Salmon Coalition (website), misc. commentary.

Schwagerl, Christian. "Will Reform Finally End the Plunder of Europe's Fisheries?" *Yale Environment 360*, Feb. 28, 2013.

Schwindt, Richard and Aidan Vining, David Weiner. "A Policy Analysis of the BC Salmon Fishery," *Canadian Public Policy*, Vol. XXIX No.1, 2003.

Seattle Times, S.T. Editorial, "An Overdue Apology for Native Fishing Activists", Jan. 21, 2014.

Sewell, Brad, et al. "Bringing Back the Fish: An Evaluation of US Fisheries Rebuilding", National Resource Defense Council, Feb., 2013.

SmallScales.ca (website), a Canadian Community Supported Fisheries network.

Sousa, Michael (website). "Why Some Say Catch Shares Work/Some Say Do Not Work", 2011.

Spalding, Mark. "Catch Share: Not the Silver Bullet They Hoped For", The Ocean Foundation (website), March 13, 2013.

Stolpe, Nils. "So How's That Catch Share Revenue Working Out for Groundfish?" American Institute of Fishery Research Biologists Website, March, 2014.

Suder, Rob. "It's Time to Fight for Your Right to Fish", *Pacific Fishing*, Jan. 2012.

Tamm, Eric. "Catch 22: A Cautionary Tale about IQ Fisheries", EcotrustCanada (website), July, 2009.

Taufen, Stephen. *Groundswell* (website), commentary 2009-2012).

Tizon, Alex. "The Boldt Decision/ 25 Years: The Fish Tale that Changed History", *Seattle Times*, Feb. 7, 1999.

Tolley, Brett. "Reject Distractions; Fix Catch Shares Now", Small Scales Website, Dec. 10, 2014

Univ. of Alaska Marine Advisory Program. "Charting New Courses for Alaska Salmon Fisheries: The Legal Waters", *Alaska's Marine Resources,* November 2003.

U.S. Dept. of Justice, / Bruce McDonald, "The Implications of Seafood Processor Quotas", Feb 25, 2004.

U.S. Federal Court V. Washington: ("Boldt Decision"), 384 Supp 312 W.D. Wash. 1974.

_____ V. Pacific Coast Federation of Fishermen's Assns. etc. v. Sec. of Commerce Locke, NMFS and NOAA, Oct. 22, 2010 (9h Dist. Court).

_____. Sohappy v. Smith (U.S. v. Oregon) 302 F supp. 899 D. OR 1969

_____. David Sohappy, Sr. v. Washington, 384 F supp. 312 W.D. WA. 1974 _____v. civil # 9213 phase II, W.D. Wash., Jan. 16, 1981, 594,464 b.

Washington Dept. of Fish and Wildlife (website), Misc. reports regarding salmon, 1995-2012.

Welch, Craig. "Sea Change: The Pacific's Perilous Turn", _Seattle Times_, Sept 11, 2013.

_____. "Sea Change: Vital Part of the Food Web Dissolving", _Seattle Times_, April 14, 2014

Welch, Laine. _Anchorage Daily News_, (weekly fishery columns), 2004-2015.

_____. "Permit Bank Encourages Fishermen and Coastal Communities" _Sitnews_ (website), Aug. 9, 2010.

Wingard, John. "Community Transfer Quotas", BNET (website), Spring, 2000.

Wood, Melissa. "Share America, Part 2: West Coast and Alaska." _National Fisherman_, May, 2012.

_____. "U.K. Ruling Gives Unused Quota to Smaller Boats", _National Fisherman_, Oct. 2013.

Worm, Boris, "Impacts of Biodiversity Loss on Ocean Ecosystems", _Science_, Vol. 314, 2006

Interviews

Haakanson, Sven
Ilutsik, Esther
"John", a businessman at Kodiak
Kavairlook, Frank
Kline, Linda
Lean, Charlie
McFarland, Frank
Mendenhall, Perry
Morford, George
Osborne, Eric
Shiedt, Enoch
Sloan, Mike
Smith, Phil
Thomas, Lesley
Thomas, Robin
Thomas, Daniel
Murphy, Travis
Waltz, Ken
Wise, Nate and Virginia

Photo Credits

Aningayou, Audrey
Crow, Rich
Danielson, Torvald
Fagerstrom, Peggy
Ilutsik, Esther
Kavairlook, Cherilyn
Kline, Linda
McFarland, Frank
Mendenhall, Perry
Mendenhall, Nancy
Morford, George
Noyakuk, John
Thomas, Daniel
Thomas, Lesley
Thomas, Marilyn
Thomas, Robin
Thompson, T.

Index

About the Author

Nancy Danielson Mendenhall came to Alaska from the Puget Sound area in 1961 and stayed an Alaskan. She trolled for salmon commercially in Southeast Alaska for ten years - her children along - among the first women in the fleets. Since the mid-1970s she has spent summers subsistence salmon setnet fishing with her family near Nome. Until retirement she worked as a teacher and administrator for the University of Alaska.

Mendenhall comes from many generations of commercial fishermen on her Norwegian father's side; three of her sons carry on the fishing tradition in Western Alaska. She wrote *Rough Waters* as a reflection of the tough realities facing the small-boat salmon, crab, halibut, groundfish and herring fisheries. She enjoys cross-country skiing and excursions with grandchildren and in the winter she writes, mainly social history her families have been involved in.

Her books:

Storytellers at the Columbia River (a novel) (Far Eastern Press, 2020)

Beachlines: a Pocket History of Nome (George Sabo Press, 1997)

Orchards of Eden: White Bluffs on the Columbia, 1907-1943 (Far Eastern Press, 2006)

She coordinated an oral history project for Nome, part of a statewide project: "Communities of Memory", 1996-98

Her website: **www.nancydanielsonmendenhall.strikingly.com**

Made in the USA
Middletown, DE
28 April 2022